Review and Evaluation of Alternative Technologies for Demilitarization of Assembled Chemical Weapons

---

Committee on Review and Evaluation of Alternative Technologies
for Demilitarization of Assembled Chemical Weapons

Board on Army Science and Technology

Commission on Engineering and Technical Systems

National Research Council

NATIONAL ACADEMY PRESS
Washington, D.C. 1999

NATIONAL ACADEMY PRESS • 2101 Constitution Avenue, N.W. • Washington, DC 20418

This is a report of work supported by Contract DAAG55-97-C-0044 between the U.S. Army and the National Academy of Sciences. Any opinions, findings, conclusions, or recommendations expressed in this publication are those of the author(s) and do not necessarily reflect the view of the organizations or agencies that provided support for the project.

International Standard Book Number 0-309-06639-5

Limited copies are available from:

Board on Army Science and Technology
National Research Council
2101 Constitution Avenue, N.W.
Washington, D.C. 20418

Additional copies are available for sale from:

National Academy Press
Box 285
2101 Constitution Ave., N.W.
Washington, D.C. 20055
800-624-6242 or
202-334-3313 (in the Washington Metropolitan Area)
http://www.nap.edu

Printed in the United States of America

COMMITTEE ON REVIEW AND EVALUATION OF ALTERNATIVE TECHNOLOGIES FOR DEMILITARIZATION OF ASSEMBLED CHEMICAL WEAPONS

ROBERT A. BEAUDET, *chair*, University of Southern California, Los Angeles
RICHARD J. AYEN, Waste Management, Inc.(retired), Geneva, Illinois
JOAN B. BERKOWITZ, Farkas Berkowitz & Company, Washington, D.C.
NOSA O. EGIEBOR, Tuskegee University, Tuskegee, Alabama
WILLARD C. GEKLER, PLG, Inc., Newport Beach, California
HANK C. JENKINS-SMITH, University of New Mexico, Albuquerque
JOHN L. MARGRAVE, Rice University, Houston, Texas
WALTER G. MAY, University of Illinois (retired), Urbana
KIRK E. NEWMAN, Naval Surface Warfare Center, Indian Head Division, Yorktown, Virginia
JIMMIE C. OXLEY, University of Rhode Island, Kingston
WILLIAM R. RHYNE, H&R Technical Associates, Inc., Oak Ridge, Tennessee
STANLEY I. SANDLER, University of Delaware, Newark
WILLIAM R. SEEKER, Energy and Environmental Research Corporation, Irvine, California
LEO WEITZMAN, LVW Associates, Inc., West Lafayette, Indiana

Board on Army Science and Technology Liaison

LAWRENCE J. DELANEY, Delaney Group, Potomac, Maryland

Staff

ROBERT T. BAILEY, Study Director
HARRISON T. PANNELLA, Research Associate
JENIFER M. AUSTIN, Senior Project Assistant

BOARD ON ARMY SCIENCE AND TECHNOLOGY

Staff

Preface

The United States has been in the process of destroying its chemical munitions for more than a decade. In keeping with recommendations from the National Research Council (NRC), the U.S. Army selected incineration as the destruction method at all storage sites. However, some citizens near those sites are opposed to incineration because they believe that the exact nature of the effluents escaping from the stacks is unknown. Because of public opposition and a report by the NRC on alternatives to incineration for destroying bulk agent, the Army has selected chemical hydrolysis for destroying the VX and mustard stored in one-ton containers at Newport, Indiana, and Aberdeen, Maryland.

In 1996, persuaded by public opposition to incineration in Richmond, Kentucky, and Pueblo, Colorado, the Congress enacted Public Law 104-201, which instructed the Department of Defense (DOD) to "conduct an assessment of the chemical demilitarization program for destruction of assembled chemical munitions and of the alternative demilitarization technologies and processes (other than incineration) that could be used for the destruction of the lethal chemical agents that are associated with these munitions." DOD established a program manager for Assembled Chemical Weapons Assessment (ACWA) to respond to the Congress. Public Law 104-208 required that the new program manager "identify and demonstrate not less than two alternatives to the baseline incineration process for the demilitarization of assembled chemical munitions."

The mandate from Congress included a provision that DOD "coordinate" with the NRC throughout the program. In response to this mandate, the NRC established the Committee on Review and Evaluation of Alternative Technologies for Demilitarization of Assembled Chemical Weapons to oversee the ACWA program. I was requested by the NRC to chair this committee. Being aware of the urgent need to destroy these weapons as soon as possible, I enthusiastically accepted. I have frequently been asked if I believe that assembled chemical weapons should be destroyed by incineration, but the capabilities of incinerators are not at issue in the ACWA program. Many citizens of the states where the weapons are stored are strongly opposed to that method, and chemical methods, potentially more acceptable to the stakeholders, could potentially be used instead.

Unlike earlier Army chemical demilitarization programs, ACWA has involved citizen stakeholders in every aspect of the program, including the procurement process. The ACWA program manager hired a nonprofit organization, the Keystone Center, to institute a unique public involvement process—the Dialogue on ACWA. In July 1997, DOD requested proposals from industry for complete technical packages for destroying assembled chemical munitions and, with assistance from the Dialogue, selected seven proposals submitted by industrial teams, called technology providers, for initial consideration. After two additional evaluation steps, three of the seven (those proposed by Burns and Roe, General Atomics, and Parsons-AlliedSignal) were chosen by DOD to proceed to demonstration. At the writing of this report, the demonstrations were under way.

When the committee began meeting and organizing this study, DOD had not selected the three technology packages for demonstration, and the program manager's stated policy was to demonstrate as many technologies as he could afford. In addition, the NRC's stringent peer-review process prior to publishing its reports usually requires about four months to complete from the time that the draft report is submitted for review. Thus, to publish this report in time for DOD's final report to Congress (September 30, 1999), the committee had to proceed without the test results from the demonstrations. Therefore, the committee has investigated and evaluated all technologies that passed DOD's initial screening. Our rationale was that any of these technologies could easily be resubmitted when requests for proposals for the Kentucky and Colorado sites are issued. This report provides the results of this NRC study based on the technical information that was available before March 15, 1999.

I wish to express my gratitude to the members of this committee. They served unselfishly, but they also provided their expertise in chemical processing, permitting and regulations, energetic materials, and public acceptance. Each member attended plenary meetings, visited the technology providers at various sites, visited the sites where the demonstrations were being held, and reviewed the extensive literature provided by the technology providers in the form of proposals, data-gap responses, and demonstration plans. We all served as volunteers.

The committee also recognizes and appreciates the assistance of the DOD ACWA team, which provided the committee with valuable information, and the members of the Dialogue on ACWA, particularly the Citizens' Advisory Technical Team, who attended all of our meetings and shared their opinions with us. We also appreciated the openness and the cordiality of the technology providers.

Thanks are also due to the reviewers who provided valuable and constructive comments on the draft of the report. These individuals, who were not known to the committee at the time of the review, also served as volunteers.

A study such as this requires extensive logistical, administrative, and technical support. We are all indebted to the NRC staff for their assistance. I would like to particularly acknowledge the program director for this study, Dr. Robert Bailey, with whom I developed a close working relationship. Robert and I worked as a team during this study. We spoke on the phone daily and e-mailed each other incessantly. Robert provided the technical, writing, and editorial skills that enabled us to complete this report. He also provided the organizational skills that I lacked. Harrison Panella, NRC research associate, also provided the committee and me with invaluable help. He took extensive minutes at all of our meetings, provided suggestions on how to best organize this report, and assisted in its development. In addition, Jenifer Austin, NRC senior project assistant, provided the logistical support that allowed us to concentrate on our task. Special thanks are also due to Carol Arenberg for her technical editing of the report, and Margo Francesco for her management of the publication process.

I would also like to acknowledge Sidney Cullipher, a graduate student in the Department of Political Science and the University of New Mexico, who assisted the committee in the development of the public acceptance portions of this report. Finally, I would like to thank one additional person who was never directly involved with this study, but whose assistance was essential to me. I am indebted to Professor David A. Dows, my colleague in the Chemistry Department at the University of Southern California, who willingly took over my teaching duties while I traveled on behalf of this study.

ROBERT A. BEAUDET, *Chair*
Committee on Review and Evaluation of
Alternative Technologies for Demilitarization of
Assembled Chemical Weapons

Acknowledgment of Reviewers

This report has been reviewed by individuals chosen for their diverse perspectives and technical expertise, in accordance with procedures approved by the National Research Council's Report Review Committee. The purpose of this independent review is to provide candid and critical comments that will assist the authors and the National Research Council in making the published report as sound as possible and to ensure that the report meets institutional standards for objectivity, evidence, and responsiveness to the study charge. The content of the review comments and draft manuscript remain confidential to protect the integrity of the deliberative process. We wish to thank the following individuals for their participation in the review of this report:

David Archer, Westinghouse Electric Company (retired), Pittsburgh, Pennsylvania
Gene Dyer, Bechtel Corporation (retired), San Rafael, California
Richard Magee, New Jersey Institute of Technology, Newark, New Jersey
Raymond McGuire, Lawrence Livermore National Laboratory, Livermore, California
Alvin Mushkatel, Arizona State University, Tempe, Arizona
Vernon Myers, U.S. Environmental Protection Agency, Washington, D.C.
Robert Olson, Consultant, Clinton, Tennessee
George Parshall, E.I. DuPont de Nemours & Company, Wilmington, Delaware
Janice Phillips, Lehigh University, Bethlehem, Pennsylvania
Cesar Pruneda, Lawrence Livermore National Laboratory, Livermore, California
Martin Sherwin, ChemVen Group, Inc., Boca Raton, Florida

While the individuals listed above have provided many constructive comments and suggestions, responsibility for the final content of this report rests solely with the authoring committee and the National Research Council.

Contents

APPENDICES

Figures and Tables

FIGURES

TABLES

Abbreviations and Acronyms

ACAMS	automatic continuous air monitoring system
ACFM	actual cubic feet per minute
ACTODEMIL	process developed by ARCTECH
ACWA	Assembled Chemical Weapons Assessment
AEA	a technology provider
a-HAX	solution containing potassium hydroxide base and humic acid (defined by ARCTECH)
AHR	agent hydrolysis reactor
AlTech Panel	Panel on Review and Evaluation of Alternative Chemical Disposal Technologies
BAA	broad agency announcement
BATNA	best alternative to a negotiated agreement
BIF	boiler and industrial furnace
CAA	Clean Air Act
CAC	Citizens Advisory Committee
CATT	Citizens' Advisory Technical Team
CLIN	contract line item number
CSDP	Chemical Stockpile Disposal Program
CSEPP	Chemical Stockpile Emergency Preparedness Program
CWA	Clean Water Act
CWC	Chemical Weapons Convention
CWWG	Chemical Weapons Working Group
DAAMS	depot area air monitoring system
DAD	decide, announce, and defend
DOD	U.S. Department of Defense
DPA	diphenylamine
DPE	demilitarization protective ensemble (suits)
DRE	destruction and removal efficiency
EMPA	ethyl methylphosphonic acid

EPA	Environmental Protection Agency
ERH	energetics rotary hydrolyzer
FEMA	Federal Emergency Management Agency
FTIR	Fourier transform infrared
FY	fiscal year
GAO	General Accounting Office
GB	a type of nerve agent
GC	gas chromatography
GC/MS	gas chromatography/mass spectrometry
GPCR	gas-phase chemical reduction
H	mustard agent
HA	humic acid
HAZOP	hazards and operability study
HD	distilled mustard agent
HEPA	high efficiency particulate air
HMX	cyclotetramethylene-tetranitramine
HRA	health risk assessment
HT	a mix of agents H and T
JACADS	Johnston Atoll Chemical Agent Disposal System
KOH	potassium hydroxide
LC/MS	liquid chromatography/mass spectrometry
LMIDS	Lockheed Martin Integrated Demilitarization System
MACT	maximum achievable control technology
MDC	material decontamination chamber
MPF	metal parts furnace
NEPA	National Environmental Policy Act
NMR	nuclear magnetic resonance
NRC	National Research Council
PAS	pollution abatement system
PCB	polychlorinated biphenyl(s)
PETN	pentaeryhritol tetranitrate
PIC	product(s) of incomplete combustion
PRH	projectile rotary hydrolyzer
PWC	plasma waste converter
QRA	quantitative risk assessment
RCRA	Resource Conservation and Recovery Act

RDX	cyclotrimethylenetrinitramine
RFP	request for proposal
SCFM	standard cubic feet per minute
SCWO	supercritical water oxidation
SET	solvated electron technology
TCLP	toxicity characteristic leachate procedure
TNT	trinitrotoluene
TSCA	Toxic Substances Control Act
UTS	universal treatment standards
VX	type of nerve agent
WHEAT	water hydrolysis of explosives and agent technology

Executive Summary

The U.S. Army is in the process of destroying the United States' stockpile of aging chemical weapons, which are stored at eight locations in the continental U.S. and on Johnston Atoll in the Pacific Ocean. The deadline for completing the destruction of these weapons, as specified by the Chemical Weapons Convention (CWC) international treaty, is April 29, 2007. The Army selected incineration as the preferred "baseline" destruction technology and currently has two operating incineration facilities—one on Johnston Atoll and another at the Deseret Chemical Depot near Tooele, Utah. Because federal law prohibits the cross-country shipment of these weapons for disposal, the Army originally planned to construct similar incineration facilities at the other seven storage sites. However, the incineration process has met with strong public and political opposition.

In response to this opposition, "neutralization" processes (based on the hydrolysis of chemical agent in water or in sodium hydroxide solution) are being developed as the primary treatment for destroying the chemical agents stored in bulk containers at Aberdeen, Maryland, and Newport, Indiana. For the remaining five sites, where weapons are stored that contain both chemical agent and energetic materials (assembled chemical weapons), incineration was still the planned process for destruction. However, in late 1996, Congress enacted Public Law 104-201, which instructed the U.S. Department of Defense (DOD) to "conduct an assessment of the chemical demilitarization program for destruction of assembled chemical munitions and of the alternative demilitarization technologies and processes (other than incineration) that could be used for the destruction of the lethal chemical agents that are associated with these munitions." The law required that the assessment be conducted by a program manager not associated with the Army's baseline incineration program.

Through Public Law 104-208, the new program manager was required to "identify and demonstrate not less than two alternatives to the baseline incineration process for the demilitarization of assembled chemical munitions." In addition, this law prohibited any obligation of funds for the construction of incineration facilities at two storage sites—Richmond, Kentucky, and Pueblo, Colorado—until the demonstrations had been completed and an assessment of the results had been submitted to Congress by DOD.

ASSEMBLED CHEMICAL WEAPONS ASSESSMENT PROGRAM

As a result of Public Laws 104-201 and 104-208, DOD created the Assembled Chemical Weapons Assessment (ACWA) program. To ensure public involvement in the program, the program manager for ACWA enlisted the Keystone Center—a nonprofit, neutral facilitation organization—to convene a group of 35 interested stakeholders, called the Dialogue on ACWA (or simply the Dialogue), who would be intimately involved in all phases of the program. The Dialogue included representatives of the affected communities, national citizen groups, state regulatory agencies, affected Native American tribes, the Environmental Protection Agency, and the DOD. All non-DOD members served as volunteers without remuneration from DOD (although travel expenses were paid by DOD).

The program manager established a three-phase program to bring at least two technologies to the demonstration stage[1] as mandated by the Congress:

- *Phase 1.* Develop evaluation criteria for assessing alternative technologies, and issue a request for proposals (RFP) from industry of technologies for destroying assembled chemical weapons without using incineration.
- *Phase 2.* Assess the proposed technologies, and select the most promising ones for demonstration. (The program manager's policy was to bring as many technologies to demonstration as possible if they met the evaluation criteria and if resources were available.)
- *Phase 3.* Demonstrate whether the selected technologies could destroy assembled chemical munitions.

In August 1997, after detailed evaluation criteria had been developed with extensive input from the Dialogue, the program manager for ACWA issued an RFP calling for "a total system solution" for the destruction of assembled chemical weapons. Twelve proposals were submitted in response to the RFP, and seven were selected for possible demonstration. These seven "technology packages" are summarized in Table ES-1. The companies primarily responsible for the technology packages (referred to as technology providers) are also listed. Because the demilitarization of assembled chemical weapons involves (1) the disassembly of the weapons and (2) the destruction of both agent and energetic materials, each technology package includes multiple technologies.

In July 1998, the technology packages proposed by Burns and Roe, General Atomics, and Parsons-AlliedSignal were selected by DOD for demonstration. Resource limitations precluded demonstrations of more than three technology packages.

[1]*Demonstration* was not defined by the Congress. Because the total program was to be completed in approximately two years (later extended to just under three years), the time available for demonstrations was limited. Eventually, the ACWA program manager defined a demonstration as "a series of tests conducted to prove [that] each critical process step meets the Program Evaluation Criteria; the results are repeatable; and each process step can be incorporated into an overall system—[i.e., a] 'total system solution.' " This is more limited than a traditional DOD demonstration, which involves complete integrated systems at near-full scale.

NATIONAL RESEARCH COUNCIL ASSEMBLED CHEMICAL WEAPONS COMMITTEE

Public Law 104-201 also required that DOD conduct the assessment "in coordination with the National Research Council." Accordingly, the program manager for ACWA asked the National Research Council (NRC) to perform an *independent* technical review and evaluation of the seven technology packages that had passed DOD's initial screening criteria. DOD plans to use this technical review as one factor in determining whether to recommend further development and implementation of any of the technology packages in its report to Congress on September 30, 1999. The NRC was not asked to compare the seven technologies to the baseline incineration system. Nor was it asked to rank the technologies or to recommend a technology. DOD asked that each technology be evaluated on its own merits, that its strengths and weaknesses be identified, and its potential for implementation be assessed.

To conduct this evaluation, the NRC formed the Committee on Review and Evaluation of Alternative Technologies for Demilitarization of Assembled Chemical Weapons (ACW Committee). This report contains the committee's findings and recommendations and details the factual data, the information supplied by the technology providers, and analyses and arguments that support the findings and recommendations. Chapter 1 describes the context for the committee's work, including the history of the Chemical Stockpile Disposal Program, the role of NRC committees in reviewing and advising that program, and the approach adopted by the ACWA program. Chapter 2 is a discussion of the broad evaluation factors assembled by the ACW Committee to organize information about the seven technology packages in terms of (1) the technical requirements of agent-destruction processes; (2) safety, health, and environmental considerations (including critical environmental permitting issues identified during informal discussions with state regulators about the proposed alternative technologies); and (3) the potential for public acceptance.

EVALUATING THE TECHNOLOGY PACKAGES

The committee found it necessary to evaluate much more than just the proposals submitted by the technology

TABLE ES-1 Descriptions of the Seven Technology Packages that Passed DOD's Initial Evaluation

Technology Provider	Access to Munitions	Treatment of Agent	Treatment of Energetics	Treatment of Metal Parts	Treatment of Dunnage
AEA Technology	Modified reverse assembly (high-pressure wash, new rocket shearing).	Electrochemical-oxidation using silver ions in nitric acid (SILVER II).	Treated with SILVER II process.	High-pressure acid wash; thermal treatment to 5X.	Shredded and treated with SILVER II process.
ARCTECH	Modified reverse assembly.	Hydrolysis with a-HAX (humic acid and strong base, KOH).	Hydrolysis with a-HAX.	Hydrolysis with a-HAX; shipped to Rock Island Arsenal for 5X treatment.	Hydrolysis with dilute a-HAX; shipped to landfill.
Burns and Roe	Modified reverse assembly.	Plasma arc.	Plasma arc.	Melted in plasma arc.	Shredded; processed in plasma arc.
General Atomics	Modified reverse assembly; cryofracture for projectiles.	Hydrolysis; supercritical water oxidation (SCWO).	Hydrolysis, SCWO.	Hydrolysis; thermal treatment to 5X.	Shredded; destroyed in SCWO.
Lockheed Martin	Modified reverse assembly (multiple lines, compact layout, new drain and wash).	Hydrolysis; SCWO; Eco Logic gas-phase chemical reduction (GPCR).	Hydrolysis, SCWO, GPCR.	Hydrolysis; GPCR to 5X.	Hydrolysis; GPCR to 5X.
Parsons	Modified reverse assembly (fluid-jet cutting and energetic wash-out for rockets).	Hydrolysis; biotreatment.	Hydrolysis, biotreatment.	Thermal treatment to 5X.	Thermal treatment to 5X.
Teledyne Commodore	Fluid-jet cutting; access and drain agent; wash out energetics with ammonia.	Solvated electron process in ammonia for reduction; chemical oxidation with sodium persulfate.	Solvated electron process in ammonia for reduction; chemical oxidation with sodium persulfate.	Wash in solvated electron solution; oxidation to 3X; ship to Rock Island Arsenal for 5X treatment.	Crushed or shredded; treated in solvated electron solution; shipped to landfill.

Source: Adapted from DOD, 1998.

providers. Because the ACWA program was moving forward rapidly, the technology packages were constantly being changed, and the committee had to stay abreast of the latest modifications. Working groups were formed to perform in-depth evaluations of each technology package and visit the facilities of the technology providers. Members of the working groups conducted probing interrogations of the technology providers through written or verbal questions, requests for further information, and face-to-face inquiries during site visits. Members of the ACW Committee also visited the demonstration sites (Aberdeen Proving Ground, Dugway Proving Ground, and Deseret Chemical Depot) in early 1999, just prior to the demonstration phase.

A freeze date of March 15, 1999 was set for accumulating data so the report could be submitted to external review by May 1, 1999, and published by September 1, 1999, in time for DOD's presentation to Congress. Thus, the committee did not have the benefit of the final demonstration results (which would not be available until June 21). However, interim progress reports were made available to the committee.

Chapters 3 through 9 summarize what the committee learned about the seven technology packages selected for review. Each chapter is devoted to a separate technology package, but the chapters are similar in format. After a short introduction, the package, as proposed by the technology provider (as of March 15, 1999), is described in some detail. This description is factual, focusing on the process flow and the supporting scientific and engineering principles and data. This description is followed by the committee's evaluation of the package in terms of the evaluation factors described in Chapter 2 and a discussion of the requirements for implementation. Each chapter ends with findings summarizing the committee's evaluation.

PUBLIC ACCEPTANCE

The most significant impetus for seeking alternative technologies to destroy assembled chemical weapons has been public opposition to incineration—and support for alternatives—by national activist groups and some members of the communities near the stockpile sites. These opponents have argued that the incineration of chemical weapons "presents an imminent threat to public health" in the event of a catastrophic failure, the release of small amounts of chemical warfare agent, and the emission of trace amounts of products of incomplete combustion, such as dioxin. Because the committee was aware of the emphasis on community involvement in previous stockpile-related reports by NRC committees and others, the views of these communities were important considerations in the committee's evaluations of the technology packages. The activities undertaken by the committee to solicit public perspectives of the alternative technologies included: (1) attendance at public meetings in Richmond, Kentucky; Anniston, Alabama; and Pueblo, Colorado; (2) discussions with many residents and concerned citizens who attended these meetings; (3) attendance at meetings of the Dialogue; and (4) discussions with Dialogue members and Dialogue facilitators (the Keystone Center).

Analyzing the information proved to be a complex undertaking for a number of reasons. First, the committee attempted to identify who is meant by "the public" and how "acceptance" of a chemical weapons destruction technology should be understood. The committee found that very little systematic data were available for evaluating public acceptance of the alternatives to incineration for chemical weapons destruction. Because the collection of new data was beyond the scope of this report, the committee was unable to assess how specific characteristics of alternative technologies affected public acceptance.

For these reasons, the discussion of public acceptance in this report does not follow the technology-by-technology approach of Chapters 3 through 9. Instead, the analysis of public acceptability is presented as a whole in Chapter 10, beginning with an overview of the processes by which public views of controversial policy options tend to be shaped and how these views are likely to affect public acceptance of alternatives to incineration for the disposal of chemical weapons. The focus of the discussion then shifts to the development of public views about incineration, the only technology for chemical weapons destruction that has received broad and sustained political attention. Finally, the prospects for public acceptance of alternatives to incineration are evaluated, with particular attention to the Dialogue, the innovative process for public involvement adopted by the ACWA program.

FINDINGS AND RECOMMENDATIONS

Based on 18 months of intensive information-gathering from the technology providers, the Dialogue, the affected communities, and DOD personnel, the committee generated findings and recommendations. Findings for the individual technology packages are contained at the end of each of the technology chapters (Chapters 3 through 9). Findings and recommendations that are broadly applicable to many or all of the proposed technology packages, together with supporting narrative, are presented in Chapter 11. These general findings and recommendations are also listed below.

General Findings

General Finding 1. The chemistries of all four of the primary technologies, (hydrolysis, SILVER II, plasma arc, and SET) as proposed, can decompose the chemical agents with destruction efficiencies of 99.9999 percent. However, each technology package raises other technical issues that must be resolved. One of the crucial issues is the identity and disposition of by-products.

General Finding 2. The technology base for the hydrolysis of energetic materials is not as mature as it is for chemical agents. Chemical methods of destroying energetics have only been considered recently. Therefore, there has been relatively little experience with the alkaline decomposition of ACWA-specific energetic materials (compared to experience with chemical agents). The following significant issues should be resolved to reduce uncertainties about the effectiveness and safety of using hydrolysis operations for destroying energetic materials:

- the particle size reduction of energetics that must be achieved for proper operation
- the solubility of energetics in specific alkaline solutions
- process design of the unit operation and the identification of processing parameters (such as the degree of agitation and reactor residence time) necessary for complete hydrolysis
- the characterization of actual products and by-products of hydrolysis as a function of the extent of reaction
- the selection of chemical sensors and process control strategies to ensure that the unit operation following hydrolysis can accept the products of hydrolysis
- development of a preventative maintenance program that minimizes the possibility of incidents during the cleanup of accumulated precipitates

General Finding 3. The conditions under which aromatic nitro compounds, such as trinitrotoluene (TNT) or picric acid, will emulsify in the aqueous phase and not be completely hydrolyzed are not well understood. Therefore, this type of material could be present in the output stream from an energetic hydrolysis step.

General Finding 4. The products of hydrolysis of some energetic materials have not been characterized well enough to support simultaneous hydrolysis of different kinds of energetic materials in the same batch reactor.

General Finding 5. The primary chemical decomposition process in all of the technology packages produce environmentally unacceptable reaction products. Therefore, all of the packages are complicated processes that include subsequent treatment step(s) to modify these products.

General Finding 6. The waste streams of all of the ACWA technology packages could contain very small amounts of hazardous substances (besides any residual chemical agent). These substances were not fully characterized at the time of this report; therefore, all waste streams must be characterized to ensure that human health and the environment are protected. If more than one phase (gas, liquid, or solid) is present in a waste stream and, each phase should be characterized separately.

General Finding 7. None of the proposed technology packages complies completely with the hold-test-release concept for all gaseous effluents (both process and ventilation effluents).

General Finding 8. Hold-test-release of gaseous effluents may not ensure against a release of agent or other hazardous material to the atmosphere. No evidence shows that hold-test-release provides a higher level of safety than current continuous monitoring methods for gaseous streams with low levels of contamination. Furthermore, none of the technologies provides for hold-test-release of effluents from ventilation systems that handle large volumes of gases from contaminated process areas.

General Finding 9. Solid salts will be hazardous waste, either because they are derived from hazardous waste (see Chapter 2) or because they leach heavy metals above the levels allowed by the Resource Conservation And Recovery Act Toxicity Characteristic Leaching Procedure. Stabilization—mixing waste with a reagent or reagents to reduce the leachability of heavy metals—will probably be required before the salts can be sent to a landfill. The potentially high chloride and nitrate content of these salts will make the waste difficult to stabilize, and treatability studies will be necessary to determine a proper stabilization formula.

General Finding 10. Testing, verification, and integration beyond the 1999 demonstration phase will be necessary because the scale-up of a process can present many unexpected challenges, and the ACWA demonstrations were limited in nature.

General Finding 11. Although a comprehensive quantitative risk assessment (QRA), health risk assessment (HRA), and ecological risk assessment (similar to assessments performed for the baseline process) cannot be completed at this stage of process development, these assessments will have to be performed and refined as process development continues.

General Finding 12. The "optimum" system for a particular chemical weapons storage depot might include a combination of unit operations from the technology packages considered in this report.

General Finding 13. Some of the ACWA technology providers propose that some effluent streams be used commercially. New or modified regulations may have to be developed to determine if these effluent streams can be recovered or reused.

General Finding 14. An extraordinary commitment of resources will be necessary to complete the destruction of the assembled chemical weapons stockpile in time to meet the current deadline using any of the ACWA technology packages. This would demand a concerted national effort. It is unlikely that any of the technology packages could meet this deadline.

General Finding 15. The Dialogue process for identifying an alternative technology is likely to reduce the level of public opposition to that technology. The committee believes that the Dialogue has been and continues to be a positive force for public acceptance of alternatives to incineration. Although the Dialogue process requires a significant commitment of time and resources, it has been a critical component of the ACWA program to date.

General Finding 16. Although the committee did not have access to scientific data on the attributes of a technology that would be most acceptable to the public, input from members of the active publics and previous research indicates that technologies with the following characteristics are likely to stimulate less public opposition:

- minimal emissions, particularly gaseous
- continuous monitoring of effluents to verify that the process is operating as designed (process assurance measurement)
- provisions for representatives of the local community to observe and participate in the process assurance measurement

General Recommendations

General Recommendation 1. If a decision is made to move forward with any of the ACWA technology packages, substantial additional testing, verification, and integration should be performed prior to full-scale implementation (see General Finding 10).

General Recommendation 2. The sampling and analysis programs at each phase of development should be carefully reviewed to ensure that the characterization of trace components is as comprehensive as possible to avoid surprises in the implementation of the selected technology (see General Finding 6).

General Recommendation 3. If a decision is made to move forward with any of these technology packages, health and safety evaluations should progress from qualitative assessments to more quantitative assessments as the process design matures. Quantitative (QRA), health (HRA), and ecological risk assessments should be conducted as soon as is practical. Early initiation of these assessments will allow findings to be implemented with minimal cost and schedule impact. (See General Finding 11.)

General Recommendation 4. Any of these technology packages, or any component of these technology packages, should be selected on a site-specific basis. (See General Finding 12.)

General Recommendation 5. Whatever unit operation immediately follows the hydrolysis of energetic materials should be designed to accept emulsified aromatic nitro compounds, such as TNT or picric acid, as contaminants in the aqueous feed stream. (See General Finding 3.)

General Recommendation 6. Simultaneous processing of different types of energetic materials should not be performed until there is substantial evidence that the intermediates formed from the hydrolysis of aromatic nitro compounds will not combine with M28 propellant additives or ordnance fuze components to form extremely sensitive explosives, such as lead picrate. (See General Finding 4.)

General Recommendation 7. The Department of Defense should continue to support the Dialogue throughout the current ACWA program and should seriously consider the participation of the Dialogue in any follow-on programs.

1

Introduction

THE CALL FOR DISPOSAL

The United States has maintained a stockpile of highly toxic chemical warfare agents for more than half a century. These chemical agents are lethal and, as components of obsolete munitions in storage, they present an unnecessary risk to surrounding communities.

Destruction of the U.S. chemical stockpile has been a concern for more than a decade. In 1985, Public Law 99-145 mandated an "expedited" effort to dispose of one particular type of chemical munition, the M55 rocket, because this munition could self-ignite during storage if the stabilizer were depleted. This mandate soon expanded into the U.S. Army's Chemical Stockpile Disposal Program (CSDP), whose mission is to eliminate the entire stockpile of unitary[1] chemical weapons. The CSDP developed the current "baseline" system, which uses incineration to destroy agents, energetic materials, and packing materials known as dunnage. The baseline system also uses incineration to decontaminate the residual metal parts. In 1997, after having set several intermediate goals and dates for completing the destruction of the U.S. chemical weapons stockpile, Congress ratified the President's signing of the Chemical Weapons Convention (CWC), which mandates that destruction be completed by April 29, 2007.

[1]The term *unitary* indicates a single chemical loaded in munitions or stored as a lethal material. More recently, *binary* munitions have been produced in which two relatively safe chemicals are loaded into separate compartments to be mixed to form a lethal agent after the munition is fired or released. The components of binary munitions are stockpiled in separate states. They are *not* included in the present CSDP, but they are being destroyed in a separate program.

THE CALL FOR ALTERNATIVES TO INCINERATION

The CSDP currently has two operating baseline incineration systems—one on Johnston Atoll in the Pacific Ocean and one at the Deseret Chemical Depot near Tooele, Utah. Together, these chemical agent disposal facilities are expected to destroy approximately one-half of the total U.S. stockpile, the remainder of which is dispersed among seven other storage sites in the continental United States. Because federal law prohibits the cross-country shipment of these weapons for disposal, the Army planned to construct similar incineration systems at the other sites. However, the choice of incineration as the disposal technology has met with strong public and political opposition. In response to this opposition, "neutralization" processes (based on the hydrolysis[2] of chemical agent in water or in sodium hydroxide solution) have been developed to destroy the chemical agents stored in bulk containers at Aberdeen, Maryland, and Newport, Indiana. The design of these

[2]*Hydrolysis* is a reaction of a target compound with water, an acid, or a base in which a chemical bond is broken in the target and OH^- or H is inserted into the bond cleavage. The destruction of chemical agent via hydrolysis is often referred to as chemical *neutralization*. This term is derived from the military definition of *neutralize*: to render something unusable or nonfunctional. Technically, neutralization is a chemical reaction between an acid and a base to form a salt and water. Chemical agents are neither acids nor bases, however, and the use of the term neutralization for two very different processes is somewhat confusing. Nevertheless, in the literature on chemical demilitarization in aqueous systems, the terms neutralization and hydrolysis have been used interchangeably. Therefore, unless otherwise specified, neutralization will be used to refer to the destruction of chemical agent via hydrolysis. The word *decontamination* is used to indicate the destruction of chemical agents that are impregnated into a matrix or adsorbed onto surfaces.

facilities is well under way. For the remaining sites, where explosively configured, "assembled" chemical weapons are stored, incineration is the planned approach for disposal. However, in late 1996, Congress enacted Public Law 104-201, which instructed the Department of Defense (DOD) to "conduct an assessment of the chemical demilitarization program for destruction of assembled chemical munitions and of the alternative demilitarization technologies and processes (other than incineration) that could be used for the destruction of the lethal chemical agents that are associated with these munitions." The law required that the assessment be conducted by a program manager not associated with the Army's CSDP.

Through Public Law 104-208, $40 million was appropriated by Congress for the program manager for Assembled Chemical Weapons Assessment (ACWA) to "identify and demonstrate not less than two alternatives to the baseline incineration process for the demilitarization of assembled chemical munitions." In addition, this law prohibited any obligation of funds for the construction of incineration facilities at two storage sites—Richmond, Kentucky, and Pueblo, Colorado—until the demonstrations had been completed and an assessment of the results submitted by DOD to Congress.

Public Law 104-201 also required that the program manager for ACWA conduct the assessment "in coordination with the National Research Council." The National Research Council (NRC) currently has a standing committee charged with oversight of the CSDP, including the proposed neutralization facilities in Aberdeen, Maryland, and Newport, Indiana. This standing committee, referred to as the Stockpile Committee,[3] could have been assigned to oversee the ACWA program, as well. However, in the spirit of PL 104-201, the program manager requested that the NRC establish a separate study committee to conduct an independent evaluation of the alternative technologies. In response, the NRC formed the Committee on Review and Evaluation of Alternative Technologies for Demilitarization of Assembled Chemical Weapons (ACW Committee). This report provides the results of the study performed by the ACW Committee.

[3]More formally, the Committee on Review and Evaluation of the Army Chemical Stockpile Disposal Program.

DESCRIPTION OF THE STOCKPILE

Agents

The principal unitary chemical agents in the U.S. stockpile are two nerve agents (GB and VX)[4] and three related forms of blister, or mustard, agent (H, HD, and HT). All of the agents are stored under ambient conditions primarily as liquids. Nerve agent VX is a nonvolatile, high boiling-point liquid that adheres to surfaces for days or weeks. Nerve agent GB (sarin) is a liquid that has a volatility similar to water and, therefore, evaporates relatively quickly. Blister agent (mustard) is a skin vesicant that evaporates very slowly, although somewhat faster than VX. The original stockpile contained 30,600 tons of unitary agents (U.S. Army, 1996) stored in a variety of munitions and containers. The stockpiles at Johnston Atoll in the Pacific and at Deseret Depot in Utah are in the process of being destroyed by incineration, so the amount currently in storage has been reduced somewhat.

The nerve agents GB and VX, which are highly toxic and lethal in both liquid and vapor forms, can kill in a matter of minutes by interfering with respiratory and nervous system functions. In pure form, these nerve agents are practically colorless and odorless. GB is relatively nonpersistent in the environment, whereas VX can persist for a long time under average weather conditions.

Bis(2-chloroethyl) sulfide is the proper chemical name of the blister agent mustard.[5] Mustard has a garlic-like odor and is hazardous both on contact and as a vapor. Because it is practically insoluble in water, mustard is very persistent in the environment. Table 1-1 lists some of the physical properties of GB, VX, and HD; Figure 1-1 shows the structural formulas of these chemical agents.

Containers and Munitions

Unitary chemical agents are stored in spray tanks, bulk storage (ton) containers, and a variety of munitions,

[4]Nerve agents are organophosphonate compounds. GB is O-isopropyl methylphosphonofluoridate; VX is O-ethyl-S[2-(diisopropyl amino) ethyl]-methylphosphonothiolate.

[5]Names such as mustard gas, sulfur mustard, and yperite have also been applied to this agent. The term mustard "gas" is often used, but the chemical is a liquid at ambient temperature.

TABLE 1-1 Physical Properties of Chemical Warfare Agents

Agent Characteristic	Nerve GB	Nerve VX	Blister H/HD
Chemical formula	$C_4H_{10}FO_2P$	$C_{11}H_{26}NO_2PS$	$C_4H_8Cl_2S$
Molecular weight	140.10	267.38	159.08
Boiling point, °C	158	298	217
Freezing point, °C	–56	<–51	14.45
Vapor pressure, torr (mm Hg)	2.9 @ 5°C	0.0007 @ 25°C	0.072 @ 20°C
Volatility, mg/m^3	22,000 @ 25°C	10.5 @ 25°C	75 @ 0°C (solid), 610 @ 20°C (liquid)
Surface tension, dynes/cm	26.5 @ 20°C	32.0 @ 20°C	43.2 @ 20°C
Viscosity, cS	1.28 @ 25°C	12.256 @ 20°C	3.95 @ 20°C
Liquid density g/cm^3 at 20°C	1.0887	1.0083	1.2685
Solubility, g/100 g of distilled water	100; soluble in organic solvents	5 @ 25°C; best solvents are dilute mineral acids	0.92 @ 22°C; soluble in acetone, carbon tetrachloride, methyl chloride, tetrachloroethane, ethyl benzoate, ether
Heat of vaporization, Btu/lb (cal/g)	144 (80)	141 (78.2)	190 (94)
Heat of combustion, Btu/lb (cal/g)	10,000 (5,600)	15,000 (8,300)	8,100 (4,500)

Source: Adapted from NRC, 1993; U.S. Army, 1988.

including land mines, rockets, bombs, and artillery and mortar projectiles. Some munitions contain fuzes, boosters, bursters, and/or propellants. These components, referred to collectively as energetic materials, or energetics, incorporate a variety of chemical compounds that must also be eliminated as part of the CSDP. The term *assembled chemical weapon* is used to describe munitions that contain both chemical agent and energetic materials in an assembled configuration. Rockets, mortars, and land mines are typically stored with their energetic components in place; projectiles may or may not contain bursters or fuzes. Alternative means for disposing of these four types of munitions are the focus of the ACWA program (and this NRC evaluation). Additional details on assembled chemical weapons are provided in Appendix A.

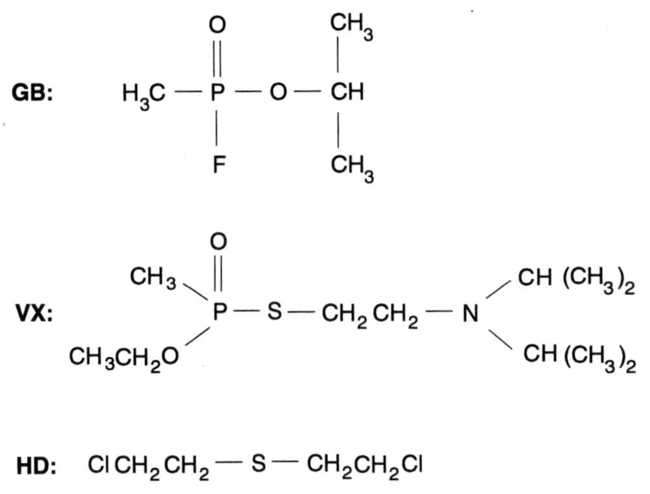

FIGURE 1-1 Structural formulas for GB, VX, and HD.

Geographical Distribution

The chemical stockpile is located at eight continental U.S. storage sites (see Figure 1-2) and at Johnston Atoll in the Pacific Ocean, about 700 miles southwest of Hawaii. Table 1-2 identifies the composition of the stockpile at each continental U.S. site by type of container or munition and by type of agent. All sites except Aberdeen, Maryland, and Newport, Indiana, contain assembled chemical weapons.

HISTORICAL ROLE OF THE NRC IN CHEMICAL DEMILITARIZATION

As the CSDP has evolved over the past 15 years, the NRC has provided extensive independent oversight and technical advice to the U.S. Army. This section provides a brief history of NRC activities related to this

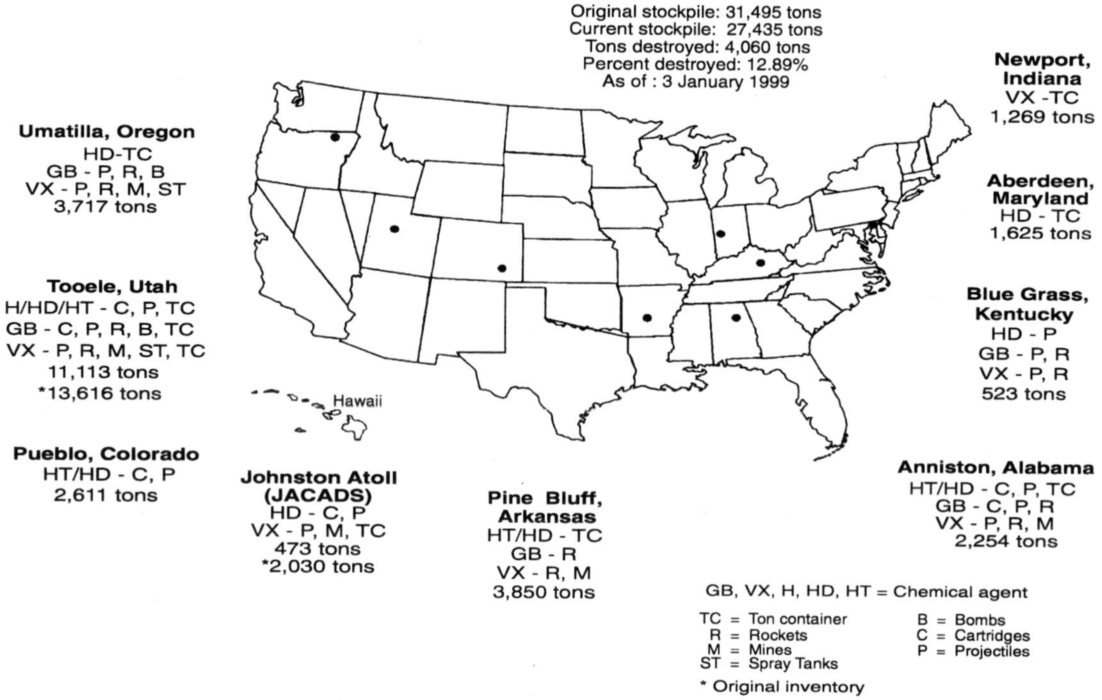

FIGURE 1-2 Types of agent, quantities of agent, types of munitions, and percentage of total agent stockpile at each storage site. Source: Pecoraro, 1999.

program, including previous evaluations of alternatives to incineration.

In August 1983, the NRC Committee on Demilitarizing Chemical Munitions and Agents was formed to review the status of the stockpile and technologies for disposal. The committee examined a range of technologies and, in its final report in 1984, endorsed incineration as an "adequate" technology for the safe disposal of chemical agents and munitions (NRC, 1984). The committee also concluded that the stockpile was well maintained and posed no imminent danger, but it expressed concern about future risk from stockpile deterioration.

In 1987, at the request of the Undersecretary of the Army, the Committee on Review and Evaluation of the Army Chemical Stockpile Disposal (Stockpile Committee) was established under the aegis of the NRC Board on Army Science and Technology to provide the Army with technical advice and counsel on specific aspects of the disposal program. Under this charter, the Army has to date received 18 reports from the Stockpile Committee evaluating progress and assessing specific aspects of the program.

In March 1991, growing public concerns and opposition to the baseline incineration system, as well as the rising cost of the CSDP, led the Stockpile Committee to suggest, with the Army's agreement, that a new study of alternatives to incineration for the destruction of the stockpile be undertaken. In January 1992, the NRC, at the request of the Army, established the Committee on Alternative Chemical Demilitarization Technologies (Alternatives Committee) to develop a comprehensive list of alternative technologies and to review their potential as agent and munitions disposal technologies. In June 1993, this committee published its report, *Alternative Technologies for the Destruction of Chemical Agents and Munitions* (NRC, 1993).

In February 1994, the Stockpile Committee, based on the report of the Alternatives Committee and on its own knowledge of the baseline system and disposal requirements, formulated recommendations regarding the destruction of munitions and the further investigation of potential alternatives to incineration in *Recommendations for the Disposal of Chemical Agents and Munitions* (NRC, 1994). The Stockpile Committee concluded that the baseline system was adequate for disposal of the stockpile and that the storage risk would persist until disposal of all stockpile materials is complete. The report recommended that the CSDP proceed expeditiously to minimize total risk to the public at each site.

TABLE 1-2 Chemical Munitions Stored in the Continental United States

Chemical Munitions (Agent)	APG	ANAD	BGAD	NECA	PBA	PCD	TEAD	UMDA
Mustard agent (H, HD or HT)								
105-mm projectiles (HD)		X				X		
155-mm projectiles (H, HD)		X	X			X	X	
4.2-in. mortars (HD, HT)		X				X	X	
Ton containers (HD)	X	X			X		X	X
Ton containers (HT)					X			
Agent GB								
105-mm projectiles		X						
155-mm projectiles		X	X				X	
8-in. projectiles		X	X				X	X
M55 rockets		X	X		X			X
500-lb bombs							X	X
750-lb bombs							X	X
Weteye bombs							X	X
Ton containers		X					X	
Agent VX								
155-mm projectiles		X	X				X	X
8-in projectiles							X	X
M55 rockets		X	X		X		X	X
M23 land mines		X			X		X	X
Spray tanks							X	X
Ton containers				X			X	
Miscellaneous								
Ton containers (L)							X	
Ton containers (GA)							X	

NOTE: APG, Aberdeen Proving Ground, Maryland; ANAD, Anniston Army Depot, Alabama; BGAD, Blue Grass Army Depot, Kentucky; NECA, Newport Chemical Activity, Indiana; PBA, Pine Bluff Arsenal, Arkansas; PUDA, Pueblo Chemical Depot, Colorado; TEAD, Tooele Army Depot, Utah; UMDA, Umatilla Depot Activity, Oregon

Source: NRC, 1996a.

The Stockpile Committee also found, after examination of all the technologies brought to its attention by the Alternatives Committee and others, that four neutralization-based systems showed the most promise for agent destruction (NRC, 1994). In view of the increasing risk associated with delays in the disposal program, and recognizing that public opposition to incineration technology might further delay the program, the Stockpile Committee recommended that alternative technologies be developed promptly. The committee also recommended that the Army continue to monitor research programs and developments that could lead to possible alternatives.

By the summer of 1995, the Army had decided that a reexamination was warranted of specific alternatives for use at two sites—the Aberdeen, Maryland, and Newport, Indiana, sites—where only bulk agent was stored. In August 1995, the Army advertised in the *Commerce Business Daily* for alternative disposal technologies other than the two already being evaluated by the Army. The purpose of this announcement was to determine if any other technologies were capable, within the CSDP schedule, of meeting chemical demilitarization requirements for the two sites where bulk agent was stored. The announcement requested proposals from industry on nonincineration technologies that were sufficiently developed to meet the needs of the CSDP. In November 1995, the Army selected three technologies for review and evaluation by the NRC—gas-phase reduction with molecular hydrogen, molten metal catalytic extraction, and electrochemical oxidation with silver(II)—in addition to the two processes, neutralization and neutralization followed by biodegradation, that were already being developed by the Army.

At the request of the Congress and the Army, the NRC formed the Panel on Review and Evaluation of Alternative Chemical Disposal Technologies (AltTech Panel) to evaluate the five technologies. The results of the study were published in *Review and Evaluation of Alternative Chemical Disposal Technologies* (NRC, 1996a). The panel recommended that aqueous neutralization of HD followed by biodegradation of the resulting hydrolysate be used at Aberdeen. The panel further recommended that neutralization of VX with sodium hydroxide followed by off-site treatment of the resulting hydrolysate be used at Newport. The Army is now developing hydrolysis processes to destroy the agents at Aberdeen, Maryland, and Newport, Indiana. However, the Army will not treat the VX hydrolysate off site as recommended by the NRC AltTech Panel. Instead, it is developing supercritical water oxidation (SCWO) technology to treat the hydrolysate on site prior to disposal.

More recently, the Stockpile Committee issued a report on SCWO technology entitled, *Using Supercritical Water Oxidation to Treat Hydrolysate from VX Neutralization* (NRC, 1998). The findings and recommendations in that report are pertinent to the ACWA program because two of the proposed ACWA technology packages include the use of SCWO.

EVOLUTION OF THE ACWA PROGRAM

Congress created the ACWA program (separate from the CSDP) to investigate alternatives for disposing of *assembled* chemical weapons (i.e., rockets, projectiles, and mines) in the face of continuing opposition to the baseline incineration system. The program manager for ACWA implemented two initiatives for identifying and demonstrating candidate alternative technologies. In the primary initiative, a request for proposal (RFP) for "a total system solution" was placed in the *Commerce Business Daily* (U.S. Army, 1997a). Proposals submitted in response to this RFP described technology packages that could replace the baseline system *in its entirety* for at least one munition/agent combination. The other initiative involved publication of a broad agency announcement (BAA) in the *Commerce Business Daily* (U.S. Army, 1997b). The BAA is seeking technologies that do not represent a total replacement of the baseline system but that accomplish one or more portions of the total process (e.g., the disposal of energetic material). By using this dual approach, DOD hoped to identify total solutions as rapidly as possible while still providing an avenue for furthering the development of promising component technologies. However, funding constraints have so far prevented the program manager from investigating any of the proposals submitted under the BAA.

ACWA Program Organization

The program manager for ACWA established a three-phase program to bring at least two technologies to demonstration as mandated by the Congress.

Phase 1. Evaluation criteria for assessing the technologies were developed and an RFP issued soliciting industry proposals for technologies to destroy the ACWs without using incineration.

Phase 2. The proposed technologies were assessed, and the most promising ones selected. This phase was performed in four steps with three sequential contractual awards, labeled Contract Line-Item Numbers (CLINs) 0001, 0002 and 0003. The program manager's policy was to bring as many technologies that met the criteria to demonstration as possible with the available resources.

Phase 3. The selected technologies had to demonstrate that they could destroy assembled chemical munitions by means other than incineration. However, the word *demonstration* was not specifically defined by the Congress. Because the total program had to be completed in approximately two years,[6] the time available for the demonstrations was severely limited. Eventually, the program manager for ACWA defined *demonstration* as "a series of tests conducted to prove [that] each critical process step meets the Program Evaluation Criteria; the results are repeatable; and each process step can be incorporated into an overall system—[i.e., a] 'total system solution' " (Richmond, 1998). The demonstrations would characterize the major feeds, intermediate materials, and final products and effluents of the processes, including any hazardous and/

[6]This deadline was eventually extended, but the schedule was still challenging.

or toxic compounds formed during the process. This would enable the evaluation of proposed environmental control systems and lay the groundwork for future permitting. During the demonstrations, the analytical methods proposed by the technology providers would also be validated.

The program manager established four teams to carry out the program. The *technical team* performed a detailed assessment of the proposed technologies based on the evaluation criteria. The *environmental team* identified the applicable environmental regulations and assessed the impact of these regulations on the technologies and on the demonstration schedule. The *business team* was responsible for procurement and legal activities. The *public outreach team* provided tools and information for communicating and interacting with the public.

To ensure public involvement in the program, DOD enlisted the Keystone Center, a nonprofit, neutral facilitation organization, to convene a diverse group of interested stakeholders, called the Dialogue on ACWA (or simply the Dialogue), who would be intimately involved in all phases of the program. The 35 members of the Dialogue included representatives of the affected communities, national citizen groups, state regulators, tribal representatives, the Environmental Protection Agency (EPA), and the DOD staff, including the program manager for ACWA, his deputy, and the deputy assistant to the Secretary of the Army for Chemical Demilitarization. All non-DOD members of the Dialogue were volunteers and received no remuneration from DOD (except for travel expenses). Ground rules were developed for the involvement of the Dialogue, which met regularly and participated in all phases of the assessment (Keystone Center, 1997). However, all decisions remained the responsibility of DOD.

In response to a request from the Dialogue for independent advice on technical issues throughout the program, the program manager for ACWA agreed to fund a consulting firm, SBR Technologies from South Bend, Indiana, to meet with a four-member liaison team. Together with representatives of SBR, these four Dialogue members formed the Citizen's Advisory Technical Team (CATT) to represent the Dialogue in procurement-sensitive matters. Once the members of CATT had signed nondisclosure agreements with all technology providers, they were given access to all proprietary information, and they participated as nonvoting members in DOD's procurement, evaluation, and selection processes.

Phases of the ACWA Program

Phase 1. The Development of Criteria and Preparation of the Request for Proposals

Prior to completing the RFP, the evaluation criteria for selecting the technologies at each of the three award steps and for evaluating them at the conclusion of the program had to be established and prioritized. The Dialogue was asked to help DOD develop and rank three sets of criteria:

1. *Threshold (go/no-go) criteria* established whether a technology package met the most basic requirements for supporting test data and could be accurately characterized as a nonincineration process. The threshold criteria were applied by DOD to determine the awarding of CLIN 0001 contracts.
2. *Demonstration selection criteria* focused on the potential of the proposed technology package for successful demonstration. Major factors included process efficacy, safety, human health and the environment, and business considerations. The criteria were applied by DOD prior to the award of CLIN 0002 contracts (with business factors omitted) and again before the award of CLIN 0003 contracts (with business factors included).
3. *Implementation evaluation criteria* focused on the potential of the proposed technology package to move into pilot-scale and full-scale implementation. These criteria were similar to the demonstration selection criteria but included more detailed process efficacy, safety, and human health and environment factors. The business factors were replaced by factors more relevant to implementation (e.g., life-cycle cost). DOD will use these criteria in its evaluations of the results of the demonstrations (CLIN 0003) and as a basis for recommendations to Congress at the end of the program.

All of these criteria were adopted with consensus from the Dialogue in July 1997. After three Dialogue meetings, DOD also held two technical workshops and a

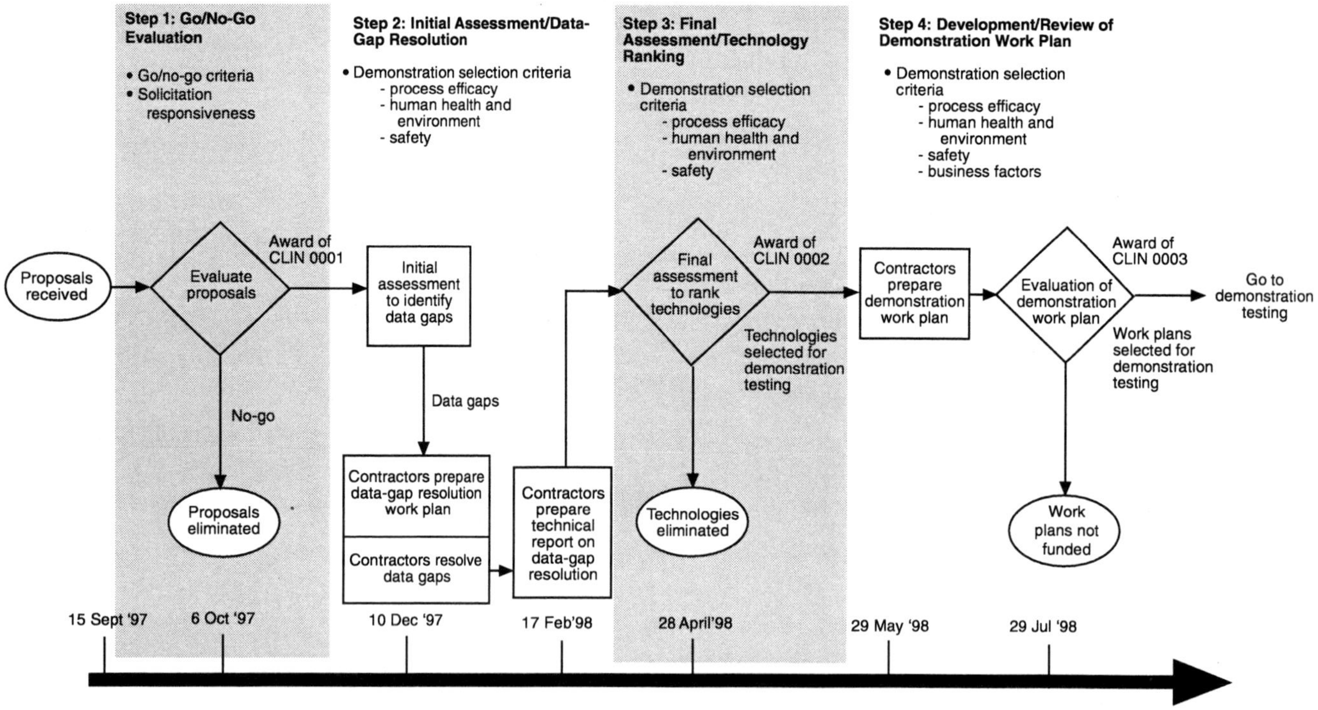

FIGURE 1-3 Schedule for the assessment and evaluation phase of DOD's ACWA Program. Source: Derived from U.S. Army, 1997a.

presolicitation conference for industry. Phase 1 culminated on July 28, 1997, when DOD issued the RFP. Twelve firms submitted proposals by the deadline, September 15, 1997.

Phase 2. Assessment and Evaluation

The ACWA assessment and evaluation phase was divided into a complicated four-step process during which technology providers would be selected and provided with funds at three stages, labeled CLIN 0001, CLIN 0002 and CLIN 0003. The steps in this process are summarized, with schedules, in Figure 1-3.

In Step 1, *the go/no-go evaluation*, the proposals were evaluated against the threshold criteria established by DOD and the Dialogue. On October 6, 1997, DOD announced that seven technology providers had passed the go/no-go evaluation. These technology providers and their teaming partners are listed in Table 1-3, and brief descriptions of their technology packages are listed in Table 1-4.

During Step 2, *initial assessment/data-gap resolution*, the technical team, together with the CATT, identified data gaps in the seven proposals and categorized them as mandatory, program, or demonstration gaps. Mandatory gaps had to be resolved in the data-gap resolution step (this step) before the provider could proceed to Step 3. Program gaps could be resolved at this time or during the demonstration phase, at the discretion

TABLE 1-3 Seven Technology Providers Selected in the ACWA Go/No-Go Evaluation Step

Team Lead	Partner(s)
AEA Technology	CH2M HILL
ARCTECH	ICF Kaiser, Inc.
Burns and Roe	Foster-Miller Startech
General Atomics	none
Lockheed Martin	SAIC Kvaerner John Brown Foster Wheeler Corporation Eli Eco Logic International El Dorado Engineering, Inc. Aerojet General Corporation Illinois Institute of Technology Research Institute
Parsons	AlliedSignal
Teledyne Commodore	Mason & Hanger Stone & Webster Southwest Research Institute University of Kentucky

TABLE 1-4 Descriptions of the Seven Technology Packages that Passed DOD's Initial Evaluation

Technology Provider	Access to Munitions	Treatment of Agent	Treatment of Energetics	Treatment of Metal Parts	Treatment of Dunnage
AEA Technology	Modified reverse assembly (high-pressure wash, new rocket shearing).	Electrochemical oxidation using silver ions in nitric acid (SILVER II).	Treated with SILVER II process.	High-pressure acid wash; thermal treatment to 5X.	Shredded and treated with SILVER II process.
ARCTECH	Modified reverse assembly.	Hydrolysis with a-HAX (humic acid and strong base, KOH).	Hydrolysis with a-HAX.	Hydrolysis with a-HAX; shipped to Rock Island Arsenal for 5X treatment.	Hydrolysis with dilute a-HAX; shipped to landfill.
Burns and Roe	Modified reverse assembly.	Plasma arc.	Plasma arc.	Melted in plasma arc.	Shredded; processed in plasma arc.
General Atomics	Modified reverse assembly; cryofracture for projectiles.	Hydrolysis; supercritical water oxidation (SCWO).	Hydrolysis, SCWO.	Hydrolysis; thermal treatment to 5X.	Shredded; destroyed in SCWO.
Lockheed Martin	Modified reverse assembly (multiple lines, compact layout, new drain and wash).	Hydrolysis; SCWO; Eco Logic gas-phase chemical reduction (GPCR).	Hydrolysis, SCWO, GPCR.	Hydrolysis; GPCR to 5X.	Hydrolysis; GPCR to 5X.
Parsons	Modified reverse assembly (fluid-jet cutting and energetic wash-out for rockets).	Hydrolysis; biotreatment.	Hydrolysis, biotreatment.	Thermal treatment to 5X.	Thermal treatment to 5X.
Teledyne Commodore	Fluid-jet cutting; access and drain agent; wash out energetics with ammonia.	Solvated electron process in ammonia for reduction; chemical oxidation with sodium persulfate.	Solvated electron process in ammonia for reduction; chemical oxidation with sodium persulfate.	Wash in solvated electron solution; oxidation to 3X; ship to Rock Island Arsenal for 5X treatment.	Crushed or shredded; treated in solvated electron solution; shipped to landfill.

Source: Adapted from DOD, 1998.

of the technology provider. Demonstration gaps were not expected to be resolved until the demonstration phase. Each technology provider was awarded $60,000 on November 3, 1997, to prepare a data-gap resolution report to be submitted to the program manager by February 17, 1998 (CLIN 0001). During Step 2, the technical team, with participation from the CATT, made site visits to all seven technology providers.

In Step 3, *final assessment/technology ranking*, the data-gap resolution reports submitted by the technology providers were reviewed in detail. Using the first three categories of the demonstration-selection criteria (process efficacy, safety, and human health and environment), DOD, with the help of the CATT, ranked the seven technology packages. DOD announced on April 28, 1998, that six of the seven had passed this step, and the six technology providers were awarded $150,000 each to prepare demonstration work plans and purchase items for demonstration that required long lead times.

During Step 4, *development of the demonstration work plan*, DOD met informally with the six providers to finalize the portions of the technology packages that required demonstration and to clarify the information that would be provided in the demonstration work plans. The work plans had to be submitted to DOD by May 29, 1998, at which time they were evaluated according to all of the demonstration-selection criteria, including business factors. The business factors included the demonstration schedule, past performance of the technology provider, socioeconomic factors, and technical and management approach. All six technology packages were found to be technically acceptable. Final selection for CLIN 0003 contracts was made on the basis of a technology provider offering the "best value" to the government within the available program resources. The announcement of the demonstration awards was delayed while the ACWA program manager attempted to obtain sufficient funds to award more than two contracts. On July 29, 1998, Burns & Roe, General Atomics, and Parsons-AlliedSignal were selected[7] to proceed to the demonstration phase. The program manager stated that if additional funds became available in fiscal year (FY) 99, more demonstrations might be funded. The awards were based on the requirements of the demonstrations and available funding.

Phase 3. Demonstrations

For demonstration, DOD identified the unit operations in each technology package that were the least mature and the most likely to be problematic. Detailed characterization of both major and trace constituents of the effluent streams was given high priority. The unit operations to be demonstrated and the quantities of agent and energetics involved are given in Tables 1-5 through 1-7. The deadline for completion of the demonstrations was May 7, 1999, and the deadline for submitting the reports describing and analyzing the results was June 21, 1999. The program manager for ACWA will submit his recommendations to Congress on September 30, 1999.

ROLE OF THE NATIONAL RESEARCH COUNCIL

The Congress instructed DOD to conduct its assessment of these alternative technologies "in coordination with the National Research Council." At the request of DOD, the NRC ACW Committee conducted an independent evaluation of the technology packages that had passed the go/no-go criteria.

Statement of Task

The NRC reached agreement with DOD on the ACW Committee's Statement of Task in March 1997, and the study officially commenced on May 27, 1997. The Statement of Task is reprinted below:

> The National Research Council (NRC) will provide an independent scientific and technical assessment to the Program Manager for Assembled Chemical Weapons Assessment (PMACWA) of alternative disposal technologies, identified through the Department of Defense (DOD) screening process, for assembled chemical weapons located at U.S. weapon storage sites. The NRC study will address the alternative technology research and

[7]The Army originally scheduled CLIN003 to be awarded on July 6, 1998. However, award announcements were delayed until July 29, 1998, while the program manager attempted to obtain additional funds to award more demonstration contracts. After the three contracts were awarded, one technology provider filed a protest with the General Accounting Office (GAO), and a stop-work order was announced. The protest was dismissed on September 15, 1998, and the stop-work order was rescinded two days later. The protest was subsequently reinstated on December 1, 1998, but no new stop-work order was issued. On March 5, 1999, after a detailed evaluation, the GAO again dismissed the protest. These events delayed the completion of the program from April 1, 1999, to September 30, 1999.

TABLE 1-5 Technology Demonstrations Performed by Burns and Roe

Location	Unit Operation	Test Material	Quantity
APG[a]	Energetic- deactivation chamber	Energetics	
		Composition B (Comp B)	0.2 lb
		Tetrytol	0.2 lb
APG	Plasma-waste converter	Energetics	
		Comp B	2.1 lb
		Tetrytol	2.1 lb
		M28 propellant	20 lb
		Simulated mortar with 10% GB heel	60 mortars
		Agent	
		HD	400 lb
		GB	400 lb
		VX	136 lb
		Dunnage	
		Wood pallets (spiked with PCP)	600 lb
		Carbon	600 lb
		DPE[b]	800 lb
		Fiberglass firing tubes	600 lb
		Decontamination solution	600 lb

[a]APG = Aberdeen Proving Ground
[b]DPE = demilitarization protective ensemble

Source: Adapted from Eissner, 1999.

TABLE 1-6 Technology Demonstrations Performed by General Atomics

Location	Unit Operation	Test Material	Quantity
CAMDS[a]	Energetics rotary hydrolyzer	Energetics	
		Fuzes	7 fuzes
		Comp B, tetrytol, and tetryl bursters	7 bursters each
		M28 propellant	7 sections
DPG[b]	Shredder/hydropulper	Dunnage	
		DPE, wood, carbon	5,000 lb shredded 4,600 lb hydropulped
DPG	Supercritical water oxidation	Agent Hydrolysates[c]	
		HD hydrolysate	1,760 lb
		VX hydrolysate	1,760 lb
		GB hydrolysate	1,760 lb
		VX simulant	4,400 lb
		Energetic/Dunnage Hydrolysates	
		Comp B/M28 propellant hydrolysate and dunnage slurry	1,760 lb
		Tetrytol hydrolysate and Dunnage slurry	1,760 lb

[a]CAMDS = Chemical Agent Munitions Disposal System
[b]DPG = Dugway Proving Ground
[c]Hydrolysate = by-product of the agent or energetics neutralization reaction

Source: Adapted from Eissner, 1999.

TABLE 1-7 Technology Demonstrations Performed by Parsons-AlliedSignal

Location	Unit Operation	Test Material	Quantity
DPG[a]	Rocket cutting and fluid mining	M60 115mm rockets (inert)	10 rockets
		M61 115mm rockets with agent simulant	10 rockets
APG[b]	Immobilized cell bioreactor	Agent/energetic hydrolysates	
		HD hydrolysate and tetrytol hydrolysate	20,100 lb
CAMDS[c]	Immobilized cell bioreactor	Agent/energetic hydrolysates	
		VX hydrolysate, Comp B hydrolysate, and M28 hydrolysate	22,400 lb
		GB hydrolysate and Comp B Hydrolysate	13,000 lb
CAMDS	Metal parts treater	Metal Parts	
		M2A1 4.2-inch mortar spiked with HD	64 lb
		M2A1 4.2-inch mortar spiked with VX	64 lb
		M2A1 4.2-inch mortar spiked with GB	64 lb
		Dunnage	
		DPE with butyl (spiked with HD)	64 lb
		Wood pallets (spiked with HD and PCP)	64 lb
		Carbon (spiked with HD)	64 lb
		Fiberglass firing tubes	64 lb

[a]DPG = Dugway Proving Ground
[b]APG = Aberdeen Proving Ground
[c]CAMDS = Chemical Agent Munitions Disposal System

Source: Adapted from Eissner, 1999.

development program for assembled chemical weapons, looking especially at ensuring maximum protection for the general public, the environment, and the personnel involved in the demilitarization program.

The specific tasks to be performed are to:

- Gather data and analyze information on stakeholder interests at the assembled chemical weapons storage site locations through site visits (U.S. and foreign) and meetings with residents and concerned citizens, state and federal regulators, and advocacy groups, in addition to necessary briefings by DOD officials and managers;
- Identify critical technology factors that should be considered in the selection criteria for advanced development and demonstration of selected alternative technologies;
- Perform in-depth, technical assessments of a set of technologies that are chosen by DOD and provided to the NRC by the PMACWA. These technical assessments will include: (1) conducting industrial site visits, as appropriate, to evaluate the status and maturity of the technologies; (2) assessing technical aspects, strengths and weaknesses, advantages and disadvantages of each technology; and (3) ensuring that all aspects of chemical demilitarization and disposal are considered, e.g., disposal of all waste products; and
- Provide to the PMACWA the results of the assessment in the form of an NRC approved report. This publicly released report will contain an evaluation of the current status of each technology (including advantages, disadvantages, knowns, unknowns, and potential for implementation). Information and data generated as part of DOD's ACWA efforts will be used to the fullest extent possible.

Scope and Approach of the Study

At the initial meeting, August 25–27, 1997, the committee found itself confronted with a unique task. Although previous NRC studies had focused on destruction technologies for chemical agents, the ACW committee had been asked to evaluate complete destruction systems, including the disassembly of a variety of munitions, the nonincineration destruction of energetic materials, and the decontamination of metal parts, dunnage, and other wastes. Moreover, the technologies were still evolving as technology providers were filling data gaps and improving their systems. In addition, four of the technology providers the committee evaluated did not proceed to the demonstration phase in 1999.

The committee, therefore, decided to evaluate the seven technology packages separately and not to compare them directly. The committee also agreed to develop criteria (see Chapter 2), identify gaps in the technologies, and comment on the maturity of the technologies and their potential for implementation. The committee decided not to rank technologies, recommend a best technology, or compare technologies to the baseline incineration system. There are two primary reasons for this approach:

- The Statement of Task called for the NRC to conduct an independent assessment of the technologies identified through the DOD screening process. To ensure independence, the committee decided that all seven technology packages that passed the basic threshold criteria should be evaluated, regardless of subsequent eliminations by DOD.
- The ACWA program was conceived in Public Law 104-201 as separate from the baseline incineration program. The goal of the ACWA was to demonstrate alternative technologies, not to compare those technologies with each other or with incineration. In keeping with the intent of the law, the committee decided to evaluate each technology package on its own merits and not to compare them with incineration.

A freeze date for accumulating data (March 15, 1999) was set so the report could be submitted to external review by May 1, 1999, and published by September 1, 1999. Although this meant the committee did not have the benefit of the final demonstration results, interim demonstration progress reports were available.

The committee established seven working groups, one for each technology provider, to perform in-depth evaluations of each technology package. The composition of the working groups was based on the expertise required for each technology package and included two technical experts, one expert on risk and one expert on energetic materials. Site visits by the groups proved to be very helpful in assessing the maturity of the technologies and the technical foundations of the designs. The committee also made site visits to the Aberdeen Proving Ground, the Dugway Proving Ground, and the Deseret Chemical Depot during the demonstration phase. (Dates and locations of these visits and of other committee meetings are given in Appendix B.)

Sources of Information

Because of the evolving nature of the ACWA program and the technologies, the committee had to be flexible about accepting information as it became available. The wide variety of information sources used in this evaluation can be grouped into the following 11 general categories:

- *Materials Submitted in Response to the RFP*. The committee reviewed, in detail, the documents submitted by the technology providers in response to DOD's RFP. These documents included proposals, data-gap reports, and demonstration work plans.
- *Briefings by DOD*. Representatives of the ACWA program provided briefings at each committee meeting. The briefings included general updates of the program, as well as detailed descriptions of certain aspects of the program (e.g., specifics of the demonstration plans, sampling and analysis methods for residual energetic material, etc.).
- *Briefings by Technology Providers*. The technology providers came to three committee meetings to give briefings and answer questions about their technology packages. These meetings took place in October 1997, December 1997, and March 1999. (See Appendix B for more information about these meetings.)
- *Other Briefings*. The committee also received briefings by individuals not directly associated with DOD or with the technology providers. Examples include briefings by the Dialogue CATT and a briefing by Dr. David Kosson, the chair of the NRC Stockpile Committee, on that committee's recent SCWO report (NRC, 1998).
- *Visits to Technology Providers*. Representatives of the working groups visited each technology provider to speak directly with experts about the technology packages and to examine existing equipment or systems. (See Appendix B for the dates and locations of these visits.)
- *Questionnaires for Technology Providers*. On several occasions, the committee sent written questions to the technology providers requesting additional information. A list of questions was sent before each site visit, and other questions were transmitted as they arose.

- *Visits to Storage Sites and Public Meetings*. The committee visited the Blue Grass Army Depot in Richmond, Kentucky, and the Pueblo Chemical Depot in Pueblo, Colorado, to meet with the personnel responsible for the storage of munitions and with local citizens to discuss the proposed alternative technologies. In Richmond, Kentucky, the committee attended a Citizens' Advisory Commission meeting; in Pueblo, members of the committee manned an information booth at the depot summer festival (part of their public outreach activities).
- *Discussions with Regulators*. The committee reviewed the existing federal regulations and conducted informal telephone discussions with state regulators in Colorado, Kentucky, and Utah. The purpose of these discussions was to identify major issues regulators thought might affect the permitting of each technology. The discussions were held with the full knowledge that these individuals were expressing personal opinions and could not render formal judgments.
- *Demonstration Visits*. Representatives of the committee visited each demonstration site to discuss the tests and to examine the equipment firsthand.
- *Dialogue Meetings*. Representatives of the committee attended all of the Dialogue meetings, and the committee chair updated the Dialogue members on the committee's activities. Committee members solicited the views of Dialogue members on the proposed alternatives and their interpretations of the public response to them. The ACW Committee and the Dialogue also held a joint meeting early in the process (October, 1997) to establish lines of communication.
- *Review of the Open Literature*. NRC staff and committee members reviewed the open literature on topics relevant to the evaluations (e.g., hydrolysis of agent and energetic materials, plasma technology, public response to waste-disposal technologies, etc.). Previous NRC reports on the chemical demilitarization program and alternatives to incineration were included in this review.

ORGANIZATION OF THIS REPORT

This report has been organized in the following fashion. Chapter 1 provided background on the CSDP, the ACWA process, and the NRC's role in that process. Chapter 2 describes the criteria the committee used to evaluate each technology package. Chapters 3 through 9 are devoted to the seven technology packages. Each chapter includes a description (without proprietary information); an evaluation in terms of the criteria described in Chapter 2; identification of information gaps; and a discussion of the requirements for implementing the package. Chapter 10 is a discussion of issues that affect public perceptions and acceptance of the technologies. Finally, Chapter 11 includes a summary of the technology packages in tabular form, as well as general findings and recommendations.

For the benefit of readers who may not be familiar with chemical weapons and the technologies used in demilitarization, the report includes eight appendices. The assembled chemical weapons are described in some detail in Appendix A. Appendix B lists the committee meetings and site visits. Because several technology providers proposed using (1) the baseline disassembly method, (2) hydrolysis to destroy chemical agents, and (3) hydrolysis to destroy energetic materials, a general description of these three technologies is given in appendices C, D, and E, respectively. Because two technology packages included SCWO, Appendix F contains a short description of SCWO, as well as the findings and recommendations from the recent NRC report on SCWO. Two technology packages also involve the use of fluid-jet cutting, so Appendix G provides a description of the fundamentals of jet cutting and a brief review of its use in the demilitarization of ordnance. Potential environmental permitting issues identified during informal discussions between the committee and state regulators are described in Appendix H. Biographical sketches of the committee members are provided in Appendix I.

2

Evaluation Factors

This chapter describes the factors the ACW Committee used to evaluate the technology packages for ACWA. The committee reviewed the criteria in DOD's RFP (U.S. Army, 1997a) and in the 1996 NRC report, *Review and Evaluation of Alternative Chemical Disposal Technologies* (AltTech Report), which focused on the disposal of bulk chemical agents stockpiled at the Aberdeen Proving Ground, Maryland, and the Newport Chemical Activity, Indiana (NRC, 1996a). The criteria in the RFP represented a consensus developed by DOD and the Dialogue.

After some deliberation, the committee concluded that the program-implementation criteria in the RFP, which are very similar to the criteria in the AltTech Report, incorporated all of the major factors required for this study. The committee made some slight additions to provide more detail in selected areas and eliminated cost as a factor, except as it was reflected in the maturity and complexity of the technology package. The committee did not attempt to estimate costs. The primary evaluation factors are:

- *Process Efficacy.* Does the system meet the requirements for the demilitarization of munitions (especially those covered by the CWC), including the destruction of agent and energetic material, the disassembly of munitions (if needed), and the decontamination of metal and other parts? Are the sampling and analysis methods well developed and appropriate? Is the system likely to operate in a stable and reliable manner under industrial conditions? Have the components and processes of the system been proven in similar applications? Is the system flexible enough to treat several munition types, and can it deal with anomalies in the munition feeds?
- *Process Safety.* Is the process safe, and does it include adequate protection for workers and the public in the event of an accident? (The definition of *safe* is the same one used by the Stockpile Committee and the AltTech Panel: minimization of total risk to the public and the workers.)
- *Human Health and the Environment.* During normal operation, does the system expose workers, the public, or the environment to excessive health risks? Are the waste streams adequately characterized, and can they be managed in accordance with regulatory limitations? What are the resource requirements? Is permitting relatively straightforward, or are there significant unknowns?
- *Public Acceptance.* Are there impediments to the acceptance of this technology package by the public? Will the package be perceived as too similar to incineration by concerned citizens? Is the public likely to accept the composition and disposition of the final waste streams?

Regardless of the technical approach, the destruction of assembled chemical weapons is a complex process involving many interrelated steps. For the purposes of the evaluation, the committee divided the overall demilitarization process into the following six major operations:

- *Munitions disassembly* involves the segregation of parts into chemical agent, energetics, other parts, and dunnage. The latter three may or may not be contaminated with chemical agent.

- *The treatment of chemical agent* involves detoxifying the agent and reducing it to environmentally acceptable products.
- *The treatment of energetics* involves decomposing energetic materials and reducing them to inert and environmentally acceptable products.
- *The treatment of metal parts* involves decontaminating the munition metal casings.
- *The treatment of dunnage* involves decontaminating munition packing materials and demilitarization protective ensemble (DPE) suits.
- *The disposal of waste* involves disposing of waste streams from all of the treatment systems.

The technology packages were evaluated for all of these operations in terms of the four primary evaluation factors (process efficacy, process safety, human health and environment, and public acceptance), each of which has several subfactors. A detailed discussion of the primary factors and their associated subfactors is provided below. Some of the subfactors could have been placed under more than one primary factor. Thus, the grouping of the subfactors under a particular factor is somewhat arbitrary. The groupings parallel the breakdown in the RFP whenever it was reasonable.

PROCESS EFFICACY

Process efficacy encompasses the effective demilitarization of the assembled chemical munitions and the reduction of the waste products to disposable materials. The proposed process must also be able to destroy *all* of the agents and *all* of the energetics in the stockpile at a given site. The process must also be controllable, reliable, and robust. That is, if some variation in the process conditions occurs, the process must be capable of continuing or returning to normal operation automatically and easily. Possible variations include changes in feedstocks, excursions in temperature or pressure, and changes in pH, electrical conductivity, or other process conditions.

In addition, the process must generate material streams that can be reliably sampled and analyzed in order (1) to control the process, (2) to obtain accurate mass balances, and (3) to verify the composition of waste streams. Finally, the waste streams must be well characterized to support health and environmental evaluations and to determine options for further waste management.

Efficacy includes the maturity of the process. During development, a process advances from simple laboratory bench-scale experiments to larger scale laboratory trials to pilot-plant scale and, finally, to full-scale operation. In general, the further a process is from the full-scale phase, the more likely unforeseen problems are to arise that will delay its development.

The subfactors under process efficacy are listed below and discussed in the following pages.

- effectiveness
  - — ability to disassemble the munitions
  - — capacity to decompose and detoxify chemical agents and to reduce the products to disposable waste streams
  - — capacity to decompose and deactivate energetic materials and to reduce the products to disposable waste streams
  - — ability to decontaminate munition parts and other materials
- sampling and analysis
- maturity
- robustness
- monitoring and control
- applicability

Effectiveness

Ability to Disassemble Munitions

Most of the technology providers proposed using the baseline disassembly process discussed in Appendix C. However, some proposed modifying it, and some proposed alternatives. If a system uses the baseline disassembly process, the committee considered the extent of the modifications and the impact of these modifications on the timing and operation of the disassembly procedure. In general, the more extensive the modifications, the greater the likelihood of delays and problems during development. The committee also considered the level of detail in the design and whether any tests had been performed to evaluate the proposed modifications.

When a technology provider selected another method for disassembly or pretreatment, the committee examined the past history and performance of the method in other applications and its suitability for the unique tasks at hand. The following areas were of particular interest:

- How were previous applications of the disassembly method similar to the proposed application? How were they different?
- At what throughput rate had the method been demonstrated, and for what period of time?
- What problems were encountered, and how were they solved?
- Had remote operation been demonstrated?

Capacity to Decompose and Detoxify Agent

An effective process must be capable of consistently destroying (i.e., decomposing and detoxifying) the chemical agent during operation under all conditions, including the presence of impure agent, gelled agent, contaminated explosives, and contaminated solids. To detoxify a chemical agent satisfactorily, the reaction must proceed until the concentration of agent is below a specific level, often specified in terms of a *destruction efficiency,*[1] defined as the percentage destroyed. The Army has specified a destruction efficiency of 99.9999 percent for chemical agent, based on the most stringent regulatory requirement under the Resource Conservation and Recovery Act (RCRA) for the destruction of dioxin—one of the most toxic regulated substances. The committee's evaluation factors include the ability of the technology packages to meet this Army requirement.

An acceptable process must, therefore, have an agent destruction efficiency of 99.9999 percent or greater.[2] In addition to the destruction efficiency, the Army sets limits on allowable contamination by chemical agent of materials to determine if the material (1) must be retained in an agent-controlled facility, (2) may be released to a hazardous waste treatment facility for further treatment, or (3) may be released to the environment or to the public sector. The contamination levels, which differ for gases, liquids, and solids, are given below.

Gases. The release of gases to the atmosphere is constrained by a health-based general population limit at the site boundary. The limit values for HD, GB, and VX are, respectively, 0.1, 0.003, and 0.003 μg (micrograms) per cubic meter of air.

Liquids. No standards have been established for the unconditional release of liquids containing chemical agents. The standard for the release of certain specified liquid wastes from incineration facilities to qualified disposal facilities is 200 ppb (parts per billion) for HD and 20 ppb for GB and VX.[3] These levels were taken from the standard for military drinking water in the field.

Solids. The Army has three primary classifications for solids contaminated with chemical agent. The first classification, 1X, refers to contaminated solid material that has not been subjected to decontamination or testing. This material cannot be released from Army-supervised agent-control areas. The second classification, 3X, is for solids that have been decontaminated to the point that the agent concentration in the head space above the encapsulated solid does not exceed the health-based, eight-hour, time-weighted average limit for worker exposure. These levels for HD, VX, and GB are, respectively, 3.0, 0.01, and 0.1 μg per cubic meter of air. A 3X material may be handled by qualified plant

[1]For incineration systems, the term *destruction and removal efficiency* (DRE) is often applied. DRE is a very specific term used by the Environmental Protection Agency (EPA) to evaluate the performance of incineration systems in destroying hazardous wastes. EPA defines DRE as $(W_{in} - W_{out})/W_{in}$, where W_{in} is the mass feed rate of hazardous waste to the incinerator, and W_{out} is the mass emission rate of hazardous waste present in the gaseous exhaust prior to release to the atmosphere (ASME, 1988). Thus, DRE has a very specific meaning and is a measure of the waste remaining in gaseous exhaust emissions. DRE does not take into account potentially hazardous constituents in the input streams that become part of the solid or liquid effluent phases. This committee believes the hazardous waste remaining in *all* effluent streams should be considered and, therefore, uses the more "generic" term *destruction efficiency* throughout this report to refer to the fraction of a particular material destroyed. For a treatment step, destruction efficiency will be defined as $(M_{in} - M_{out})/M_{in}$, where M_{in} is the mass feed rate of the particular material in the treatment step, and M_{out} is the mass emission rate of that material present in all effluent streams after that treatment step. The committee has found that DRE is sometimes inappropriately used by the Army and the technology providers.

[2]Destruction efficiencies are often expressed as the number of 9's in the percentage. Therefore, 99.9999 percent may be referred to as "six 9's."

[3]At the time of this writing, the Army was verifying that detection of VX down to 20 ppb in hydrolysate was possible with current analytical methods. Resolution of this issue is pending.

workers using appropriate procedures but is not releasable to the environment or for general public reuse (i.e., not releasable "to the public"). In specific cases in which approval has been granted, a 3X material may be shipped to an approved hazardous waste treatment facility for disposal in a landfill or for further treatment. The third classification, 5X, is defined as follows (Department of the Army, 1997):

> An agent symbol with five "Xs" (XXXXX) indicates an item has been decontaminated completely of the indicated agent and may be released for general use or sold to the general public in accordance with all applicable federal, state, and local regulations. An item is decontaminated completely when the item has been subjected to procedures that are known to completely degrade the agent molecule, or when analyses, submitted through MACOM (Major Army Command) and DA (Department of the Army) channels for approval by the DDESB (Department of Defense Explosives Safety Board), have shown that the total quantity of agent is less than the minimal health effects dosage as determined by The Surgeon General. 5X condition must be certified by the commander or designated representative. One approved method is heating the item to 538 degrees C (1,000 degrees F) for 15 minutes. This is considered sufficient to destroy chemical agent molecules.

In addition to meeting the Army requirements listed above, a process must meet the requirements of the CWC, which states that chemical weapons destruction must take place using "a process by which chemicals are converted in an essentially irreversible way to a form unsuitable for production of chemical weapons, and which in an irreversible manner renders munitions and other devices unusable as such." The requirement of irreversibility implies that both the chemical agents and any by-products that could be readily converted to chemical agent (CWC Schedule 2 [agent "precursor"] compounds) must be destroyed by the process. The CWC imposes no numerical requirements (e.g., the destruction efficiency) on the degree of agent or Schedule 2 compound destruction. It specifies only that the destruction should be irreversible, safe, and environmentally friendly.

Environmental regulations include specifications for the allowable quantities of some Schedule 2 compounds in process effluents. Because Schedule 2 compounds are much less toxic than chemical agents, the destruction efficiencies set by the Army for these compounds are less stringent than for agents. For example, the design-basis destruction efficiency for ethylmethylphosphonic acid ([EMPA]; a Schedule 2 precursor to VX) is 99.9 percent, compared to 99.9999 percent for VX (U.S. Army, 1997c).Thus, to evaluate the capacity of a technology package to decompose and detoxify agent, the committee considered its ability to achieve the required destruction efficiency for agents and Schedule 2 compounds and to reduce the agent contamination in other media to below the allowable levels.

Capacity to Decompose and Detoxify Energetics

Effective processes must be capable of consistently destroying energetic materials by decomposing them to nonenergetic compounds. The concentration of residual energetic materials or toxic by-products must not exceed established limits for release to the environment.

This is significantly different from merely rendering a material safe to handle or reducing the hazard classification of explosives or propellants. Energetic materials must not be simply diluted so that they will not react or propagate with explosive or propulsive violence. Standard laboratory sensitivity tests, such as impact, friction, electrostatic discharge, vacuum thermal stability, or differential scanning calorimetry cannot be used to measure acceptance or ensure quality because substantial explosive residues may be present but may not produce a positive response to standard tests.

The decomposition of energetic materials such as RDX (cyclotrimethylenetrinitramine), Composition B, tetrytol, or M28 double-base propellant requires that the process reaction(s) destroy the energetic chemical bonds (e.g., N-NO_2, C-NO_2, or C-ONO_2). The reaction is usually considered complete when the concentrations in liquid process effluents do not exceed the limits permitted for either a publicly owned treatment facility or by state environmental permits. For example, the limit typically established by local sanitation districts is less than or equal to 1 ppm (part per million) of energetic material in water. Lot acceptance testing is usually performed by either high-pressure liquid chromatography or a gas chromatograph/mass spectrometer. The vapor-phase process emissions (e.g., NO, NO_2, CO, CO_2, etc.)

must be managed to meet local or state ambient air quality standards. Any solids and/or hazardous wastes generated by the process must be sufficiently characterized so that appropriate landfill sites or other means of disposal can be identified.

The committee, therefore, considered the ability of the proposed system to decompose the energetic materials into nonenergetic compounds and achieve the residual concentrations required by environmental regulations.

Decontamination of Metal and Other Munitions Parts

Once the agent and energetic materials have been removed, the remaining parts of the munition must be decontaminated to either the 3X or (preferably) 5X level before disposal or release to the public sector (as allowed). If the parts were only at the 3X level, the technology provider was required by the RFP to describe how these parts would be disposed of.

Disposal of Other Contaminated Materials

The process must also dispose of a variety of other agent-contaminated wastes generated during demilitarization. These include decontamination solutions, used DPE suits, and spent activated carbon filters. The committee evaluated the ability of each technology to dispose of these contaminated materials while meeting the required decontamination standards.

Sampling and Analysis

To verify process performance, stream compositions must be sampled and analyzed at various stages. Sampling is required to validate monitoring and control of the process, to determine mass balances of the major constituents, and to characterize waste streams before release to the environment. Analytical techniques must be sensitive enough to determine the presence of and measure the levels of trace constituents in the waste streams. These trace components are often the constituents of greatest concern to the public and in health risk assessments (HRAs). Detection limits and sensitivities depend on the nature of the compound and the media in which it is contained. All phases (solid, liquid, and gas) within each waste stream must be analyzed (e.g., liquid plus particulates).

A process could, for example, simply dilute critical streams to below the sensitivity level of available sampling and analytical methods. In that case, it would be impossible to verify that the process had achieved its treatment objectives during operation. Although there is some dilution of toxic materials in most processes, this cannot be the primary mechanism for reducing the concentration of toxic material in effluent streams to acceptable levels. Therefore, the committee examined the proposed processes in the context of available sampling and analytical methodologies to establish whether their performance could be verified. The committee considered the detection limits of current analytical methods for chemical agents, energetics, and the major and minor products of destruction.

Process Maturity

The committee defined maturity of a technology as the stage to which the technology had progressed toward industrial operation and, hence, the level of confidence that the process would operate successfully at full scale. In general, chemical-process technologies can be located along a developmental continuum from laboratory-scale to proof-of-concept testing to pilot-plant demonstration and, ultimately, to full-scale operation. Laboratory-scale testing refers to the basic development of the treatment processes. Proof-of-concept refers to the testing at sufficient scale to demonstrate that the technology is a workable process. The earlier a process is on the continuum, the greater the uncertainty of its full-scale performance and the greater the likelihood that unanticipated problems will cause delays in its implementation.

Many considerations are involved in determining whether a technology is ready to move to the next stage or how close it is to being "successfully demonstrated" at a given stage. For instance, at the laboratory scale, assays and chemical analyses are important for establishing that the desired reactions predominate and that unwanted side reactions can be controlled or eliminated. During proof-of-concept testing, it is important that critical components of the treatment process be tested with either actual target chemicals or with realistic surrogates under process conditions that simulate the expected conditions under full-scale operation. At the pilot-plant stage, precise mass and energy balances become essential, along with quantitative characterizations of how key process variables affect outcomes. The documentation for a pilot design must be complete enough for a preliminary assessment of risks related to the hazard inventory (e.g., agent concentrations at each process step, reactive materials, pressure) and the safety features, such as process interlocks and safe means of releasing excess material or energy. A maturity status is, therefore, not a simple classification but a running checklist of what has been accomplished to date and what remains to be done.

The maturity level of a technology was based on the documentation and evidence submitted by the technology providers. Committee representatives visited sites suggested by the technology providers to confer with knowledgeable personnel and to observe experimental equipment the providers had designed and constructed. In this way, the committee was able to assess the current state of development of the proposed systems and technologies.

Process Robustness

Robustness, a significant factor in the evaluation, is defined as the ability of the total process to achieve its objectives even when the properties of the material being processed deviate from the nominal or average or when a process component does not behave as intended. For example, a process must be able to contain the explosion of a fuze or the ignition of rocket propellant during processing. If a more modest deviation occurs, the process must be capable of returning to normal operation without incident.

Determining the ability of a process to treat pure, uncontaminated (neat) chemical agents was the first step in the committee's evaluation. However, past experience with the baseline systems at Johnston Island and Tooele, Utah, has shown that unforeseen conditions occur frequently. Examples are listed below:

- A significant fraction of the chemical agents and energetics found in the munitions contain substantial impurities.
- The agent in many munitions may have become partially gelled or crystallized, making it difficult or impossible to drain the agent from the munition casing. (At Johnston Island, such munitions are treated in the metal parts furnace where the gelled or crystallized agent is vaporized and burned. Because the munitions are essentially undrained, throughput rates are greatly reduced to limit the amount of agent present in the furnace. For nonincineration processes, other ways of handling these munitions must be found.)
- Some of the energetics may have deteriorated to the point of potential instability, increasing the risk of deflagration or explosion during processing. (At Johnston Island and Tooele, explosion containment structures are used to house the energetics processing and destruction activities. A similar approach could be used with alternative technology packages.)
- Components of some munitions have rusted or corroded, making disassembly difficult.
- The attempted removal of the lifting lugs from projectiles during baseline disassembly has sometimes caused the lug to fuse to the shell body.

The proposed system must be capable of dealing with these conditions and should be flexible enough to respond effectively to other unanticipated occurrences.

The committee also considered the ability of the system to process multiple feeds (agent, energetics, metal parts, process wastes) simultaneously and the impact this could have on the overall robustness of the system.

Process Monitoring and Control

Each process must be monitored continuously at various stages and locations to ensure the destruction of agent and energetics and to verify that operating conditions are satisfactory. Proper operation requires that the process include built-in controls to maintain temperatures, pressures, flow rates, pH, and other key parameters within the necessary ranges.

All of the technology providers proposed using standard Army technology to monitor for chemical agent. They also proposed using state-of-the-art distributed monitoring and control systems linked to a central control room and data-collection system.

Because the overall proposed monitoring and control systems were standard industrial configurations, the committee focused its evaluation on aspects of monitoring and control that appeared to be unique or potentially difficult. The committee examined each process from the perspective of how effectively it could be monitored and controlled, whether the process was understood well enough to allow implementation of a sound monitoring and control strategy, and whether existing monitoring and control technologies could prevent or control process upsets.

Process Applicability

In the RFP, DOD allowed the technology providers to discuss the destruction of only one type of munition. Clearly, a practical process must be able to destroy all types of munitions at a given site. Thus, a process must be capable of treating *all* of the chemical agents and energetics found at a site.

PROCESS SAFETY

Process safety factors include risks to workers and risks to the nearby public from accidents. In this report, the term *risk* refers to the chance of adverse consequences (e.g., fatalities) from some event. Risk evaluations of processes for destroying chemical weapons should include the consequences of releases of chemical agent and of accidental detonations and conflagrations of energetic materials. Risks during the storage, transportation, handling, and disassembly of munitions should also be considered, as well as risks from the actual destruction process.

A comprehensive assessment of safety requires quantitative risk assessments (QRAs), which can only be done based on a detailed plant design. Because of the immature status of the systems reviewed in this report, quantitative evaluations at almost any level that would be consistent across all of the technologies could not be made. However, the committee has performed a qualitative evaluation of whether each technology can be operated safely. The qualitative evaluation focused on identifying intrinsic and probable safety issues and how technology providers propose to respond to these issues. Issues unique to a technology were emphasized over across-the-board issues (e.g., baseline munition unpacking and handling operations). The safety-related analyses of the committee are referred to as evaluations rather than assessments. The term assessment is reserved for actual risk assessments, which should be performed later (following the demonstration phase) when more complete descriptions will be available and a commitment to a particular design has been made.

The ACW Committee considered the following subfactors in the category of process safety:

- worker health and safety
  - normal facility operations
  - facility accidents
- public safety
  - facility accidents
- transportation accidents

Worker Health and Safety

In-plant safety and health risks depend on the nature and magnitude of the hazards inside the process facility. The committee's preliminary evaluation of each alternative technology included the following aspects of in-plant risks:

- major failure and agent release
- worker exposure to agent[4] without catastrophic failure
- worker exposure to other hazardous chemicals used or produced during the process
- worker exposure to other hazardous process conditions (e.g., high temperature, electromagnetic radiation, electrical energy, moving equipment, etc.)

The severity and likelihood of these risks are affected by the following factors:

- hazard characteristics (e.g., mass of agent and other toxic chemicals; stored thermal, mechanical, and electrical energy; mass of reactive chemicals)
- inherent safety of the process (e.g., limits on inventory, self-limiting characteristics of chemical reactions, etc.)
- the need for and feasibility of systems and procedures to prevent or mitigate accidents (e.g.,

[4]The Army and the Centers for Disease Control have established maximum allowable dose standards and human toxicity estimates for exposure to chemical warfare agents. These standards form the basis for protecting both workers and the public from exposure to chemical agent. The acute-exposure standards were recently reviewed by the NRC Committee on Toxicology, which recommended that the database supporting these standards be improved. In this report, the ACW Committee has not commented on the existing standards. Interested readers should consult NRC, 1997, for a complete discussion of the standards.

potential for energetics to carry over from primary to secondary treatment that might require increased containment)
- systems, equipment, and training to prevent worker exposure (e.g., in-plant monitoring for worker exposure, personal protective equipment, and well developed maintenance procedures)

Public Safety

An evaluation of public safety includes factors that characterize the effects of *accidental* releases. The propensity for uncontrolled low-level releases during *normal operations* and the latent health effects or gradual environmental damage caused by long-term exposure to these emissions are considered in the category of human health and the environment.

The committee reviewed handling and processing operations throughout the projected range of facility operations, from the removal of munitions from storage to the completion of munition destruction and the packaging of all waste streams for transport. Public exposure from accidents during the transportation of hazardous materials (including transport to waste-disposal sites) is considered in the next subsection on transportation accidents.

Specific components in the evaluation of public safety are:

- the likelihood and magnitude of agent release and exposure during the disassembly process
- the likelihood and magnitude of agent release and exposure during the agent and energetics destruction process
- the likelihood and impact of releases and exposures associated with storage, process chemicals, and unique forms of energy

Risks to the public and the environment from agent storage have been cited as a reason for prompt destruction of the stockpile (NRC, 1994), and storage risks have been the focus of ongoing debates in communities near the stockpile sites. The AltTech Report stated that reducing storage risk at individual sites, for the most part, is independent of the technology selected for stockpile destruction and that the critical factor affecting storage risk is the overall implementation/destruction schedule. The shorter the schedule, the lower the risk. Thus, the key evaluation factor for storage risk is the likelihood of a technology meeting or beating the required schedule. The implementation/destruction schedules for the alternative technology packages, however, are currently too uncertain for meaningful evaluations of storage risk. Therefore, storage risk is not considered further in this report.

Evaluating public safety during the destruction of the agent and energetics must also take into account the coexistence of energetics and agent and the possible presence of large amounts of stored energy. In addition, the use of technology-specific forms of energy must be considered, as well as the formation of intermediate reaction products that may be highly chemically reactive or unstable. Some of the proposed systems require unique forms of energy (e.g., high-energy plasma) that must also be evaluated for their risk to public safety.

Another factor considered in this evaluation is the risk of storage of hazardous or reactive chemicals required for the proposed technology. The risk in this instance does not involve exposure to agent or agent by-products but exposure to other hazardous chemicals.

Transportation Accidents

Technology-specific safety issues also may arise during the transportation required for each proposed technology. The issues or factors that have been evaluated include both worker and public exposure to accidental releases during the transport of assembled chemical weapons from storage to the disassembly area, hazardous chemicals transported to and on the processing site, and hazardous waste transported from the processing site to the disposal or post-processing site. The transport of weapons from storage to the disassembly area was essentially identical for all of the proposed systems (i.e., the Army's on-site container unit is used) and was not considered further in this study.

The risk from transporting hazardous process materials to, or hazardous waste from, the destruction facility is proportional to several parameters. At this stage of the ACWA program, the only parameters that one can reasonably evaluate are (1) the particular toxic, fire, or explosive hazards presented by the material; (2) the quantity of material transported per shipment; and

(3) the number of shipments. Uncertainties at this stage of facility design make even these few parameters difficult to evaluate (e.g., the concentration and/or physical form of some materials).

Therefore, the committee made several assumptions to establish a consistent basis for evaluating transportation risks across sites, technologies, and materials. First, all shipments to and from the facility were assumed to be by tractor-trailer truck rather than by a combination of truck and train. The number of shipments will be determined from weight limits, volume limits, schedules, and other practical considerations. The second assumption was that the truck gross weight limit was 80,000 lb (36,300 kg), even though the limit is higher in some states. It was also assumed that the tractor weight was 15,000 lb, and the trailer weight, whether a flatbed, enclosed trailer, or tanker, was another 15,000 lb. The maximum material weight would then be 50,000 lb/shipment. The average amount across all materials was assumed to be 35,000 lb (15,900 kg) per shipment, which is slightly conservative (i.e., results in a higher number of shipments). The last assumption was that the consequences of a release of toxic, flammable, or explosive material would be roughly equal; thus, the transportation risk would be proportional to the number of shipments to and from the site.

All shipments by railcar or heavy truck have the potential for public injury or fatality regardless of the cargo; however, only the shipments of process chemicals or process wastes are considered in this analysis.

Using the approach described above, it was estimated that the number of incoming truck shipments per week would be 50 or less for any of the technology packages. Similarly, the number of outgoing truck shipments was estimated to be less than 60. Even these maximum values do not appear unusual for a large industrial facility or a moderately busy highway. Therefore, the risk from the transportation of process materials to or from the site was not considered a significant criterion for the evaluation of the technology packages.

HUMAN HEALTH AND THE ENVIRONMENT

The human health and environment factor includes the impact of normal facility operations on the health of the public and the surrounding environment, including various ecosystems. All waste streams released from the facility as part of the proposed process should be characterized and their impacts evaluated. The following subfactors are considered:

- characterization of effluents and their impact on human health and the environment
- the completeness of effluent characterization
- effluent-management strategy
- resource requirements
- regulatory environmental compliance and permitting

Characterization of Effluents and Their Impact on Human Health and the Environment

Treatment processes generally have effluent streams from certain parts of the overall process. These effluent streams can be discharged into the air, water, or land. The committee considered not only the potential for agent release but also the potential for release of other hazardous constituents under normal (not accident) conditions. Normal conditions include typical types of process upsets under steady-state and unsteady-state operations, such as start-up, shut down, and normal process variabilities in treating older chemical weapons components. In addition, all waste streams that must be subsequently treated and disposed of were considered effluent streams that could pose risks to human health and the environment.

The level of risk associated with potential and actual effluents released into each medium (air, water, and land) is addressed separately. The goal is to gather enough information on the possible pathways of potentially harmful materials for comprehensive health and environmental risk assessments. For each discharge into each medium, the assessments should include the types, quantities, and duration of releases; the fate and transport of releases (particularly off-site); and the potential effects on humans, plants, and animals.

Another consideration is whether appropriate and proven methods for characterizing normal releases can be incorporated into the treatment processes. This is particularly important for public acceptance of the technology. Many citizens protested that process designs should include testing prior to the release of effluents (i.e., a hold-test-release sequence for all effluents). The committee considered whether existing monitoring techniques for detecting low levels of contamination within process effluents could be used. The process-specific contaminants of concern include those that could have long-term chronic effects. Ideally, the monitoring methods will be validated by comparison with standard sampling and analysis methods.

Completeness of Effluent Characterization

The systems under consideration have not yet been applied to the specific task of demilitarizing assembled chemical weapons. All of the proposed technology packages would generate effluents that must eventually be fully characterized in order to assess the overall impact to human health and the environment and to determine the best disposal process or control method for effluents.

The committee's assessment of the human health and environmental impacts is based on completeness of characterizations of the potential effluents in bench-scale or pilot-scale operations. The committee took into consideration the methods used to characterize the effluents. Determining the chronic effects to human health and the environment requires measuring very low levels of certain pollutants. Therefore, not only are mass balances on the major effluent streams important but also the completeness of the tests conducted to date on trace substances that might be of concern to human health or the environment.

The effluent characterization should include normal transient conditions. Often, the greatest environmental impacts of treatment processes are incurred during start up and shut down when process controls are not always optimal. Effluent characterizations should include both species at high concentration that would be considered in normal mass balances and trace species of environmental concern.

Effluent-Management Strategy

All of the waste streams generated by the proposed systems must be managed in an environmentally acceptable manner. Some systems would produce waste streams that are particularly difficult to manage. This evaluation factor addresses whether the proposed system has a well developed effluent waste management plan that is compatible with applicable laws and regulations. Because these technologies are all rather immature, management strategies may still have significant unknowns. The committee defined these unknowns in its assessment of the requirements for moving toward implementation.

The committee paid particular attention to the materials of potential health and environmental concern that might be left in the waste stream or created by the technology. The residuals and materials that remain after treatment can have significant environmental impacts, as well as permitting challenges. For example, RCRA regulated hazardous wastes are subject to stringent permitting requirements for treatment, storage, and disposal. Thus, any RCRA-regulated material generated must be clearly identified. If the plan proposes off-site treatment or disposal, it is important to ensure that existing facilities will accept the waste. If the plan calls for on-site waste management, it is important to evaluate current experience for treating the material. The committee evaluated special requirements for the disposal of waste streams that might be problematic. The committee also evaluated whether the waste could be separated into batches, which would allow for testing prior to release of the waste stream.

Resource Requirements

The resources required for a technology, such as energy, water, and land, can have a significant impact on the implementation of a technology at a specific site because some resources are limited at some sites. In addition, if excessive resources are required, the economic viability of the process could be decreased. For this reason, it is important to assess the projected demand for water (including quantity and quality); the requirements for energy, such as electricity and fuel; and the requirements for land, particularly for large equipment and storage areas.

Environmental Compliance and Permitting

The committee evaluated each technology to determine whether its inherent features might create problems with environmental compliance or permitting. Although a complete analysis of the required permits is well beyond the scope of this study, the committee tried to identify potential problems. Two specific issues were examined:

- the quantity and hazardous nature of the air emissions similar to those from other processes that have encountered permitting problems
- discharges that may be difficult to dispose of at existing waste disposal facilities (e.g., solid waste that might not be acceptable to a hazardous waste disposal facility)

The RCRA permits (40 CFR §264, 266, 270) for all of the technology packages are likely to be issued under one of the following categories: 40 CFR§264 Subpart X,[5] Miscellaneous Units or 40 CFR§264 Subpart J, Tank Systems. In general, permitting for tank treatment (Subpart J) processes is much simpler than for Subpart X processes, and the permitting period for Subpart J processes has historically been much shorter than for Subpart X.

Because the technologies evaluated in this study are relatively new, few precedents have been established for permitting. In all likelihood, some aspects of the process(es) (e.g., the hydrolysis of agent or energetics) would be permitted under Subpart J; other parts (e.g., SCWO or thermal treatment units) would be permitted under Subpart X.

The RCRA permitting process is generally administered (with minor exceptions) by the state. In addition to RCRA standards, most states also have unique requirements. Assessing the requirements of each state that may be impacted by one of these technology packages is also beyond the scope of this study; however, the committee did consider the state-specific issues discussed below.

RCRA permits are usually comprehensive and include conditions that restrict the ranges of operating conditions (e.g., temperature, pressure, flow rates), limit waste feed rates, require specific maintenance procedures, and require monitoring of specific parameters. The permit conditions generally require that the system have an automatic interlock that shuts off the feed of hazardous material under certain specified excursions from the acceptable operating conditions. The permit is highly site-specific and process-specific and is issued on the basis of a complex process of engineering evaluations and testing.

The EPA has issued guidance documents for common waste-treatment units, such as incinerators, cement kilns, and boilers. These peer-reviewed documents are based on consensus opinions of many permit writers around the United States. Although the committee reviewed these documents, the technologies considered in this study are new, and established documents are not directly applicable.

Because the environmental regulatory agencies of each state will have to accept the technologies, the committee contacted representatives of state agencies where ACWA treatment facilities exist or are planned. Although regulators cannot pass judgment on any technologies until they receive a specific application, the committee was able to identify initial general concerns through informal discussions with permit writers. The issues that arose during these discussions are described in Appendix H.

The committee evaluated the technologies in the light of the above considerations. If no major stumbling blocks were found, (e.g., if the agent treatment aspects of the process could be permitted as a Subpart J system and if the wastes produced appeared to be amenable for ultimate disposal at commercially available sites), the committee concluded that the process did not appear to have any unusual permitting issues. If, however, potential permitting issues were identified (e.g., the process required Subpart X permits for its agent treatment system, its air emissions were similar to some processes that have encountered permitting problems, or the process created a waste that was different from typical hazardous waste), then these issues are discussed within the environmental compliance and permitting section of the technology chapter.

PUBLIC ACCEPTANCE

This committee was asked to "gather data and analyze information on stakeholder interests at the assembled chemical weapons storage site locations..." The committee gathered data from the following sources:

[5]Subpart X is a general category that covers treatment systems that do not fit any given category. Quoting 40 CFR §264 Subpart X, "A miscellaneous unit must be located, designed, constructed, operated, maintained, and closed in a manner that will ensure protection of human health and the environment. Permits for miscellaneous units are to contain such terms and provisions as necessary to protect human health and the environment, including, but not limited to, as appropriate, design and operating requirements, detection and monitoring requirements, and requirements for responses to releases of hazardous waste or hazardous constituents from the unit. Permit terms and provisions shall include those requirements of other rules that are appropriate for the miscellaneous unit being permitted."

- attendance at public meetings in Richmond, Kentucky; Anniston, Alabama; and Pueblo, Colorado
- private discussions with many of the residents and concerned citizens who attended the public meetings
- attendance at meetings of the Dialogue held during the preparation of this report
- private discussions with participants in the Dialogue
- interviews with Keystone facilitators
- discussions with state regulators in Colorado, Kentucky, and Utah
- briefings by DOD officials and managers

Analysis of the information proved to be a complex undertaking for a number of reasons. First, the committee had to determine who was meant by "the public" and how "acceptance" of a chemical weapons destruction technology should be understood. Furthermore, the committee found that very little systematic data were available with which to evaluate public acceptance of the alternatives to incineration for chemical weapons destruction. Because the collection of new data was beyond the scope of this study, the committee was unable to assess how the characteristics of the alternative technologies would be related to public acceptance.

For these reasons, the discussion of public acceptance in this report does not follow the technology-by-technology approach used to evaluate other factors (efficacy, safety, and human health and environment). The analysis of public acceptance is presented for all technologies in Chapter 10, which includes an overview of the processes by which public views of controversial policy options tend to be shaped and a discussion of how these views are likely to affect public acceptance of the alternative technologies.

Second, the focus of the discussion is on public views of incineration, the only technology for the destruction of chemical weapons that has received broad and sustained political attention. The discussion includes implications of public attitudes toward incineration for the chemical weapons destruction program.

Third, the committee evaluated the prospects for the public acceptance of alternatives to incineration, especially the innovative process for public involvement being used by the ACWA program. The discussion focuses on the development and workings of the ACWA Dialogue Group, which has directly involved interest groups, regulators, and citizens in the process of identifying and selecting alternative technologies. The committee also attempted to identify the characteristics of the alternative technologies that are likely to influence public acceptance.

Finally, the committee's findings are summarized and very general recommendations are offered for increasing public acceptance of alternative chemical weapons disposal technologies.

CLOSING REMARKS

The factors described in this chapter were used to evaluate the seven technology packages described in the next seven chapters. Three aspects of the evaluations that merit further explanation are mentioned below.

First, all of the technology providers proposed using the baseline structures, support systems, and support equipment as much as possible. The committee considers this a very positive indication. However, because significant experience has been gained with these systems, the technology providers did not provide detailed descriptions of how the baseline systems would be used, and the committee did not consider them in its evaluations.

Second, DOD required that the technology providers design their systems to achieve the throughput rates shown in Table 2-1. The overall system availability was set at 38 percent, and a five-year operating period was specified. In some proposals, however, the providers chose to optimize their processing rates to handle mixes

TABLE 2-1 Throughput Rates Prescribed in the ACWA RFP

Munition	Agent	Processing Rate (munitions/hr)	Processing Rate (lb agent/hr)
105 mm projectile	HD	100	300
155	HD	100	1,170
155	VX	80	600
155	H	80	1,170
4.2-in mortar	HD	50	300
4.2-in mortar	HT	50	290
8-in projectile	GB	20	280
M55 rocket	GB	20	214
M55 rocket	VX	20	200
M23 land mine	VX	30	315

Source: U.S. Army, 1997a.

of the munitions at actual storage sites. They also assumed higher system availabilities than those prescribed in the RFP. Thus, the technology providers included mass balances for a variety of munition feeds in their proposals and other documentation. The committee examined these mass balances to determine whether the providers had a solid understanding of their processes. The committee decided to include mass balances in this report so that the reader could get a feel for the mass flows and materials involved in the full-scale processes. Although the assumptions and levels of detail in the mass balances vary, the committee did not attempt to standardize the information.

Third, the committee was asked in the statement of task to evaluate each technology's "potential for implementation." In response, the committee included the process maturity factor described earlier in this chapter. To address this issue more fully, the committee also included a section in each technology evaluation chapter entitled "Steps Required for Implementation."

3

AEA SILVER II Technology Package

INTRODUCTION AND OVERVIEW

The technology provider team formed by AEA Technology, Inc., and CH2M HILL, Inc., has proposed a system based on the SILVER II chemical oxidation technology developed by AEA. (SILVER II is an electrochemical oxidation process that operates at relatively low temperatures [90°C; 194°F] and near atmospheric pressure.) A previous NRC committee evaluated the SILVER II technology for the destruction of chemical agents HD and VX stored in bulk (NRC, 1996a). This report evaluates SILVER II as part of a total system for destruction of the agents (HD, VX, and GB) and energetics in stored chemical munitions.

The first stage of the total system uses the Army's existing baseline disassembly system. The output streams from the disassembly lines are either oxidized directly by SILVER II or are rendered ready for disposal. The technology provider proposes three separate SILVER II units: one for the destruction of chemical agent, one for the destruction of energetics, and one for the destruction and/or decontamination of dunnage (also, small metal parts, decontamination solutions, and other mixed wastes). Large metal parts (i.e., munition bodies) are decontaminated using a combination of high-pressure jetting with recycled dilute nitric acid and thermal treatment in a heated decontamination chamber. Figure 3-1 shows an overview of AEA's package; Figure 3-2 shows a more detailed block flow diagram of the process. Table 3-1 summarizes how the AEA technology package accomplishes the six primary demilitarization operations listed in Chapter 1.

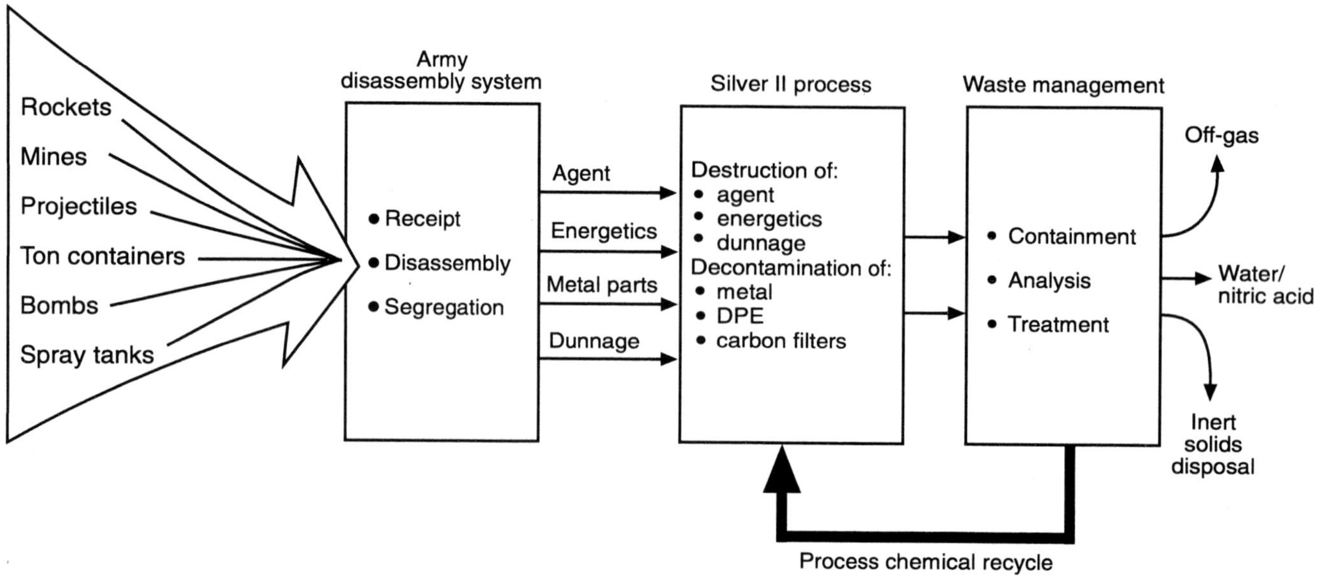

FIGURE 3-1 Overview of the AEA technology package. Source: AEA, 1997.

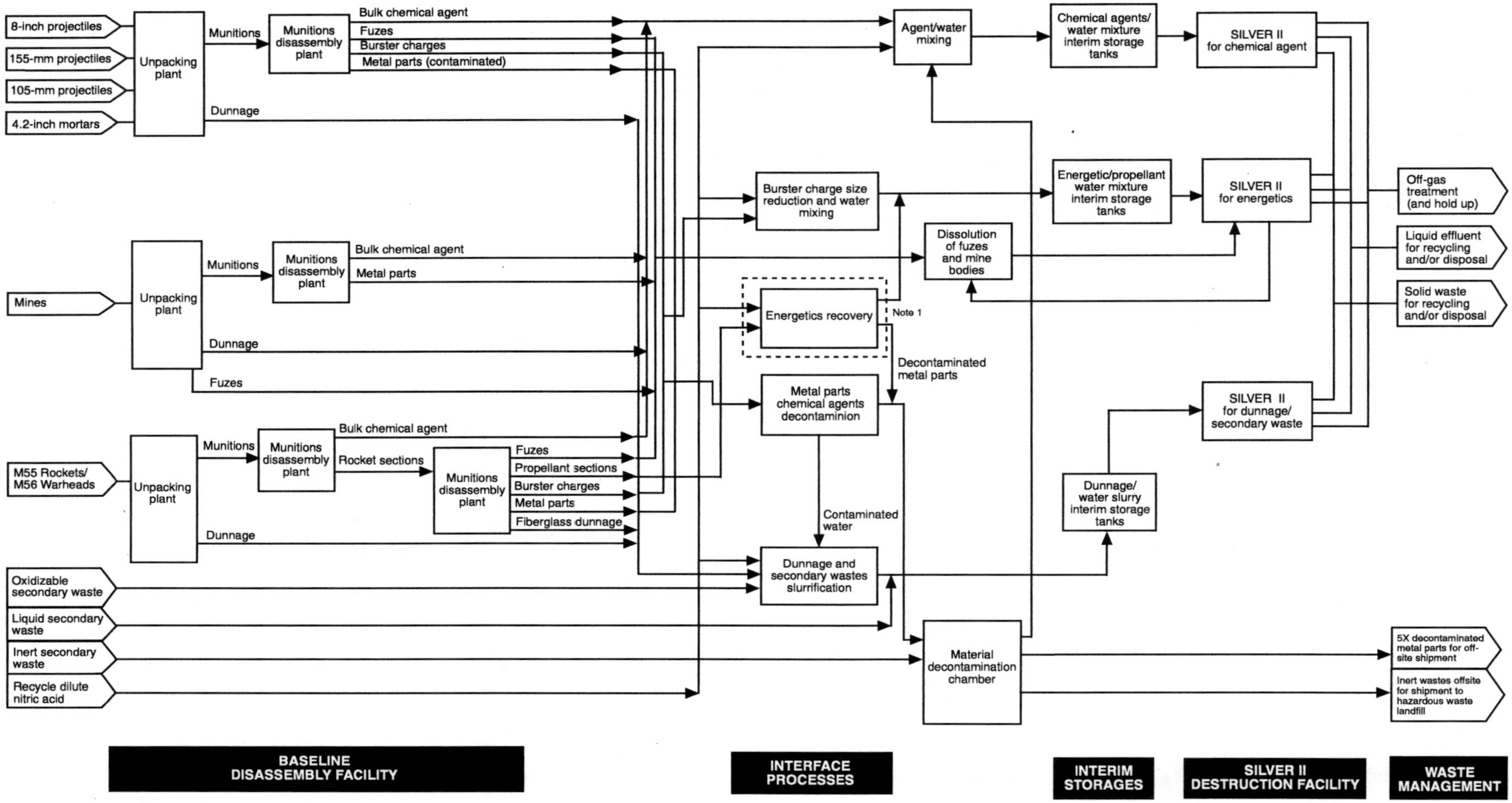

NOTE:
1. Although not specifically required in the RFP, this stage is required to remove propellant from M55 rocket motor sections.

FIGURE 3-2 SILVER II block flow diagram.

TABLE 3-1 Summary of the AEA/CH2M HILL Approach

Major Demilitarization Operation	Approach(es)
Disassembly of munitions	Modified version of Army's baseline disassembly method; processing of outputs from disassembly into pieces less than 1-inch in diameter; water-jet washout added; metal dissolution in nitric acid.
Treatment of chemical agent	Destruction using SILVER II electrochemical oxidation.
Treatment of energetics	Destruction using SILVER II electrochemical oxidation.
Treatment of metal parts	Heated in materials decontamination chamber to 5X.
Treatment of dunnage	Destruction using SILVER II electrochemical oxidation.
Disposal of waste	**Solids.** Various solids from SILVER II process, caustic scrubber salts, and assorted dunnage (uncontaminated or 5X) sent to permitted landfill; silver chloride sent off site for recovery of silver as silver nitrate; 5X metal parts sent to a recycling facility. **Liquids.** Excess dilute nitric acid neutralized and released to local wastewater treatment facility; excess concentrated nitric acid sent off site for reuse. **Gases.** Off-gas from process treated in caustic scrubber, held and tested, and released to the atmosphere if the test results are acceptable.

DESCRIPTION OF THE TECHNOLOGY PACKAGE

SILVER II Process Chemistry

The SILVER II Process is based on the highly oxidizing nature of silver (II) (Ag(II)) ions in a nitric acid solution (Lehmani et al., 1996). Ag(II) is one of the strongest oxidizing agents known; nitric acid also makes a significant contribution to the oxidizing process. In this process, a solution of silver nitrate in nitric acid is electrolyzed to produce the Ag(II) cation. The standard electrode potential of Ag(II) and other reactants involved in the process are given in Table 3-2. The standard half-cell reduction potentials are for all reactants and products having an activity of 1.0 M and all gases at 1.0 atmosphere and 25°C (77°F). For dilute concentrations (< 1.0 M), the activity can be approximated by the molar concentration. Above 1.0 M, the activity will deviate progressively from the value of the concentration.

Thus, the standard electrode potentials, $E°$, of the half-cell reactions for the desired electrolysis under standard conditions at 25°C (77°F) would be:

$$Ag^+ \rightarrow Ag^{+2} + e^- \qquad E° = -1.98 \text{ V}$$

$$HNO_3 + 2\,H^+ + 2\,e^- \rightarrow HNO_2 + H_2O \qquad E° = 0.94 \text{ V}$$

The overall reaction and standard potential for the electrolysis of the silver nitrate solution are therefore:

$$2\,Ag^+ + HNO_3 + 2\,H^+ \rightarrow 2\,Ag^{+2} + HNO_2 + H_2O \qquad E° = -1.04 \text{ V}$$

The electrode potentials at other concentrations are given by the Nernst equation:

$$E = E^o + \frac{RT}{\Im n}\ln(Q)$$

where Q represents the standard expression for the Law of Mass Action,

$$Q = \frac{[Ag^{+2}][HNO_2]}{[Ag^{+1}][HNO_3][H^+]^2}$$

Also, T is the temperature in Kelvin, R is the universal gas constant, $\Im$ is the Faraday constant (charge on one mole of electrons), and n is the number of electrons transferred in the reaction.

The high concentration of nitric acid used in the SILVER II process prevents the electrolysis of water that would occur at a lower potential (–1.23 volts[V]) than the electrolysis of silver nitrate in aqueous solution. For a simple 1.0 M silver nitrate solution, the following reactions would take place:

$$2\,H_2O \rightarrow O_2 + 4\,H^+ + 4\,e^- \qquad E° = -1.23 \text{ V}$$

$$HNO_3 + 2\,H^+ + 2\,e^- \rightarrow HNO_2 + H_2O \qquad E° = 0.94 \text{ V}$$

For an overall reaction:

$$2\,H_2O + 2\,HNO_3 \rightarrow O_2 + 2\,HNO_2 + 2\,H_2O \qquad E° = -0.29 \text{ v}$$

These standard electrode potentials are valid only at 25°C (298 K) and at unit activity of all reactants and products. However, the silver (II) process is normally

TABLE 3-2 Standard Electrode Potentials of Reactions Related to the SILVER II Process. All Concentrations are 1.0 M

Reaction	E° (volts)
$Ag^{+2} \rightarrow Ag^+ + e$	1.98
$O_2 + 4H^+ + 4e^- \rightarrow 2H_2O$	1.23
$O_2 + 4H^+ (pH = 7) + 4e^- \rightarrow 2H_2O$	0.82
$NO_3^- + 4H^+ + 3e^- \rightarrow NO + 2H_2O$	0.96
$HNO_3 + 2H^+ + 2e^- \rightarrow HNO_2 + H_2O$	0.94
$Ag^+ + e^- \rightarrow Ag$	0.80
$O_2 + 2H_2O + 4e^- \rightarrow 4OH^-$	0.40
$O_2 + 2H_2O + 4e^- \rightarrow 4OH (pH = 7)$	0.81
$2H^+ + 2e^- \rightarrow H_2$	0.00
$2H_2O + 2e^- \rightarrow H_2 + 2OH^- (pH = 7)$	–0.42

carried out at about 90°C (194°F) and at 8.0 M HNO_3. The potentials at this temperature and at these high concentrations of nitric acid are not well known. Moreover, the reaction is aided by the formation of a brown complex, $Ag(NO_3)^+$, which lowers the required potential for the oxidation of Ag^+. The association constant for its formation is unknown.

Although the electrochemistry for the conditions used is not known, there is no doubt that Ag(II) is formed and that it reacts only slowly with water to produce intermediates, such as OH and other radicals. The Ag(II) formed in the electrolytic cells is then transported into a reaction vessel. The generation of Ag(II) ions depends entirely on the electrical current to the cells, and it stops rapidly when the power is switched off. However, because Ag(II) is such a strong oxidizing agent, it will corrode most metal systems that are not glass-lined or coated with a noble metal. Concentrated nitric acid is also a strong oxidizing agent.

In practice, an overvoltage must be applied for the reaction to proceed rapidly. The technology provider states that oxidation of the silver proceeds rapidly and requires an overpotential of only 120 mV at 5 kA/m^2. The technology provider intends to use 2.0 V.

Ag(II) can oxidize all elements to their highest oxidation state if allowed to reach equilibrium. Thus, the following general theoretical reaction should occur:

$$\textit{Organic} + O_2 \rightarrow CO_2 + \textit{inorganic salts/acids in water}$$

The NO_x formed is converted to NO_3^-, but in practice, amines (RNH_2) are only converted to elemental nitrogen.

Table 3-3 shows the reactions of Ag(II) with chemical agents and energetic species and lists the number of electrons required. Table 3-4 shows the electrical energy requirements for the process in terms of energy per gram of material for a voltage of 2.0 V applied to the electrochemical cell. The technology provider claims that the total process will have 80 percent electrochemical efficiency (the fraction of current passing that produces useful oxidation to CO_2). The committee's calculations in Table 3-4 agree with the technology provider's claims.

SILVER II Process Arrangement

Figure 3-3 is a block flow diagram that shows a standard industrial electrochemical cell used in the SILVER II process. Ag(II) ions are generated by passing an Ag(I) nitrate/nitric acid solution past a platinum anode. A semipermeable membrane (to cations) separates the anode and cathode compartments of this cell, thus preventing bulk mixing of the anolyte and catholyte solutions and allowing the transport of cations and water (but not anions) across the membrane. The cell is operated in the following configuration: packs of individual electrodes are connected in monopolar format (electrically in parallel), and the packs are then connected in bipolar format (electrically in series). The

TABLE 3-3 Anode Reactions of Ag(II) with Chemical Agent and Energetic Materials

Compound	Reaction	Number of Electrons
GB	$C_4H_{10}PFO_2 + 10\ H_2O + 26Ag^{2+} \rightarrow 4CO_2 + H_3PO_4 + HF + 26H^+ + 26Ag^+$	26
H	$C_4H_8SCl_2 + 12H_20 + 28Ag^{2+} \rightarrow 4CO_2 + H_2SO_4 + 2Cl^- + 30H^+ + 28\ Ag^+$	28
VX	$C_{11}H_{26}SNPO_2 + 31H_2O + 82Ag^{2+} \rightarrow 11CO_2 + H_3PO_4 + H_2SO_4 + HNO_3 + 82H^+\ 82Ag^+$	82
Comp B	$C_4H_{5.2}N_{1.1}N'_{3.3}O_{5.6} + 5.7H_20 + 15.5Ag^+ \rightarrow 4CO_2 + 1.1HNO_3 + 1.65N_2 + 15.5H^+ + 15.5Ag^+$	15.5
Tetryl	$C_7H_5N_4N'O_8 + 18H_20 + 37\ Ag^{2+} \rightarrow 7CO_2 + 4HNO_3 + 0.5N_2 + 37H^+ + 37Ag^+$	37
TNT	$C_6H_3N_3O_6 + 15H_2O + 32Ag^{2+} \rightarrow 6CO_2 + 3HNO_3 + 32\ H^+ + 32\ Ag^+$	32

TABLE 3-4 Energy Required for the Destruction of Chemical Agents and Energetics

Compound	Electrons per Mole	Molecular Mass (grams)	Charge Required (Coul/mole)	Energy (kW-hr/mol)	Moles Destroyed (per kW-hr)	Mass Destroyed (per kW-hr)	Grams Destroyed at 80% Efficiency
GB	26	140.1	2,509,000	1.394	0.717	100.5 g	80.40
H	28	159.1	2,702,000	1.501	0.666	106.0 g	84.78
VX	82	267.4	7,913,000	4.396	0.227	60.8 g	48.66
Comp B	15.5	204.5	1,495,750	0.831	1.203	246.1 g	196.90
Tetryl	37	287.2	3,570,500	1.984	0.504	144.8 g	115.81
TNT	32	213.1	3,088,000	1.716	0.583	124.2 g	99.38

process liquids are fed to each cell in parallel via an external manifold, with each individual anode/cathode pair within the cell being fed in parallel by internal manifolds. The anolyte and catholyte solutions are circulated around separate closed loops between the cell and the anolyte and catholyte reaction vessels (see Figure 3-3). The organic-containing material is continuously metered into the anolyte tank to match the rate of destruction.

According to the technology provider, Ag(II) ions generated at the anode of the electrochemical cell react with the water and nitric acid of the anolyte solution to form a range of other oxidizing radicals (OH, NO_3^-). The Ag(II) ions and other oxidizing species then react with the organic material delivered into the anolyte vessel and are reduced to Ag(I) ions, nitrate ions, and water. The organic material itself is completely oxidized to carbon dioxide, oxides of nitrogen (NO_x) and traces of carbon monoxide, protons (H^+), and inorganic salts. No hydrogen is produced in the process. Off-gas from the reaction passes from the anolyte tank via a chiller (to condense nitric acid vapors) to an NO_x reformer.

To balance the electrochemical reaction in the anolyte vessel, a cathode reaction reduces nitric acid to nitrous acid and water, while other reduction reactions generate NO_2. The evolved gases pass from the catholyte tank to the NO_x reformer. The overall process is operated at a temperature of 90°C (194°F) and at atmospheric pressure.

During the electrochemical reaction, nitric acid is consumed in the catholyte circuit, and water is transferred across the semipermeable membrane in the electrochemical cell from the anolyte to the catholyte. Some Ag(I) ions are also transferred across the cell membrane. In order to maintain steady-state operating conditions, proportions of the anolyte and catholyte circuits are bled to a nitric acid and silver recovery unit, which includes the previously mentioned NO_x reformer. Here, a combination of evaporation and fractional distillation is used to recover the NO_x as nitric acid and to generate streams for the return of (1) silver ions together with 16 M nitric acid to the anolyte circuit and (2) 4 M nitric acid without any silver ions to the catholyte circuit. A dilute nitric acid stream is also produced, which is recycled within the plant and used to prepare organic materials before they are fed into the SILVER II process.

After leaving the NO_x reformer, all off-gas passes through a caustic scrubber to remove residual NO_x. Treated gas from the scrubber is pumped into one of the two tanks of the hold-test-release system. The tanks have sufficient capacity to hold approximately 15 minutes of generated off-gas at a nominal maximum pressure of 1 atm. When one tank is full, the second tank is switched on line, and the gas in the first tank is analyzed. The gas entering each tank is continuously analyzed for chemical agent vapor content by a standard system, such as the Army's automatic continuous air monitoring system (ACAMS). The proposed concept is to hold the gas and only release it to the atmosphere upon confirmation that it is free of chemical agent.

During the development of the SILVER II process, its applicability to many different organic compounds (including explosives, propellants, rocket fuels, pyrotechnics, and industrial solvents) was investigated.

Disassembly of Munitions and the Removal of Agent/Energetics

The transport, receipt, and pre-processing of munitions prior to processing with SILVER II are done

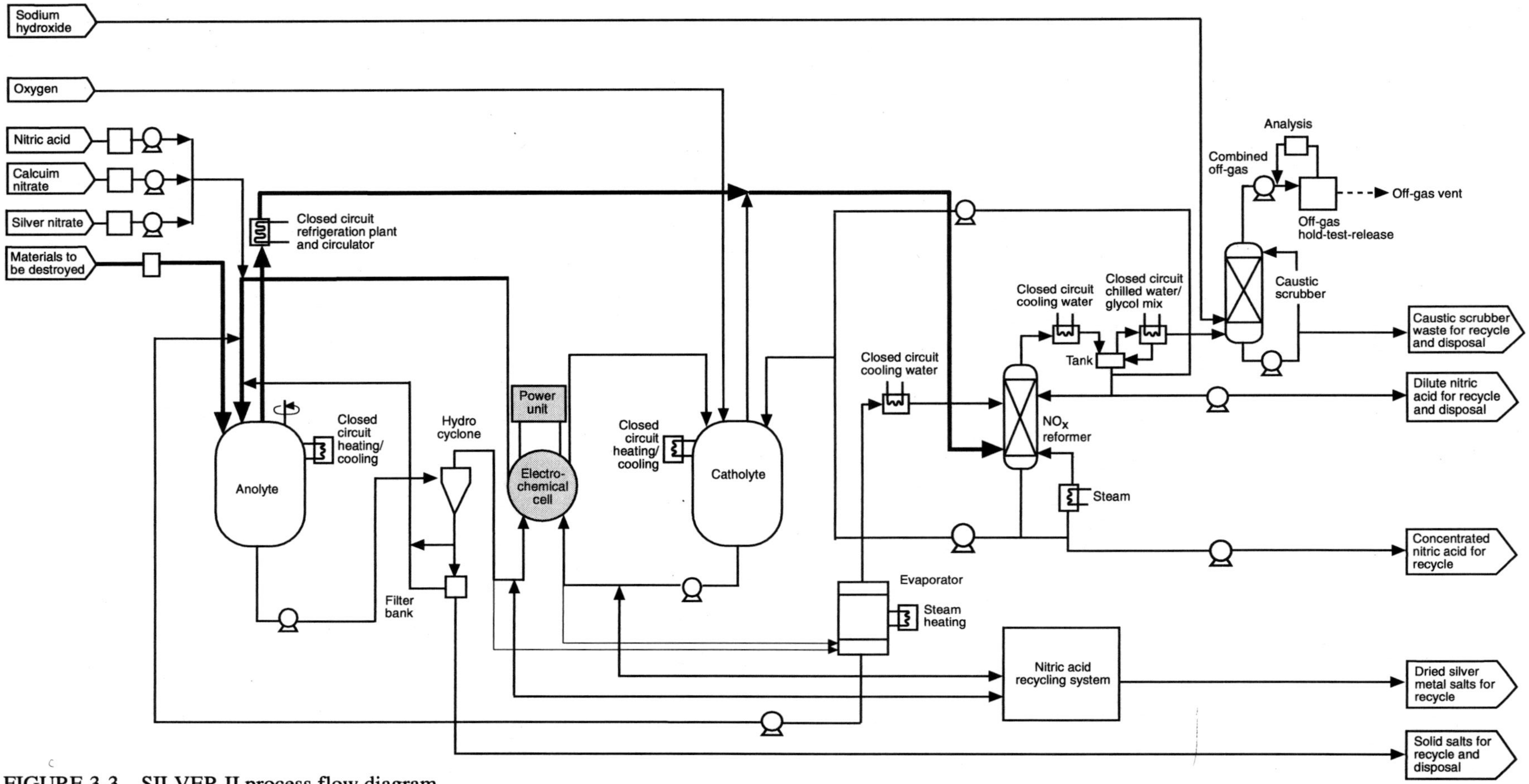

FIGURE 3-3 SILVER II process flow diagram.

using the Army baseline technology modified with additional material-handling stages at the end of the process. These stages reduce material size and segregate material for SILVER II processing. The baseline munitions disassembly process is described in Appendix C.

Rocket Disassembly

M55 rockets are sheared into six pieces to generate sections suitable for handling. The six pieces are different from those obtained in the baseline process. Subsequently, a sheared-parts handling machine performs a number of operations on the sheared sections of the rockets in order to segregate materials for subsequent processing, either in SILVER II reactors or by other treatment technologies. The segregated outputs from the rocket disassembly include:

- drained chemical agent, as already provided by baseline disassembly
- a warhead fuze section
- warhead burster sections, which are further reduced in size in a rocket shearer to small sections (< 1 inch) for subsequent processing
- agent warhead sections that have been drained of agent but are still agent contaminated
- a rocket motor section containing propellant
- fiberglass dunnage consisting of shipping/firing tube pieces
- miscellaneous metal parts, including end caps from the shipping/firing tubes, tail-fin sections from the rockets, and motor nozzle plates

These individual parts are treated separately in subsequent process units.

Prior to treatment, the energetic material is separated from the associated casings. The propellant from M55 rocket motor sections is removed by high-pressure jetting with hot dilute nitric acid recycled from the SILVER II process. High-pressure jetting is directed from below to a vertically oriented rocket motor section, and the propellant is washed into a sump. The process is carried out in batches in an enclosed chamber and is remotely controlled. The remaining metal parts are visually examined for residual contamination, and parts found to contain residual contamination are recleaned. The energetic material that has been removed is then mixed with recycled dilute nitric acid to make up a 20 percent (by weight) energetic-in-water slurry. This material is transferred to a continuously stirred interim storage tank from which it is continuously fed to the energetics SILVER II reaction circuit.

Burster charges and smaller energetic components are reduced in size by shearing to achieve a particle size of less than one inch in diameter. This material is then mixed with recycled dilute nitric acid and fed to the SILVER II reaction process. Whereas AEA originally proposed storing the resultant energetic/metal/water slurry in an interim storage tank, later discussions with the committee indicated that this material would be fed directly to the SILVER II reactor.

Projectile/Mortar Disassembly

Projectiles and mortars are disassembled using the Army's baseline systems without modifications. Burster charges and other energetic components are then reduced in size to less than one inch in diameter and fed to the SILVER II process.

Land Mine Disassembly

For M23 mines, the Army baseline disassembly system is used to punch and drain the agent and push out the burster charge. In addition to this baseline processing, the mines are sheared into four parts to open the radial initiator charge and expose the energetic components.

Treatment of Chemical Agent

The chemical agent drained from the various components is transferred to an interim storage tank where it is mixed and continually stirred with dilute nitric acid recycled from the SILVER II process. The dilute nitric acid provides the necessary water balance for the electrochemical oxidation reactions in the agent SILVER II reaction circuit. It also disperses the agent, thereby increasing the surface area/volume ratio of any gelled agent and provides partial destruction via hydrolysis of the chemical agent. The interim storage provides a buffer volume for the feed of materials to the SILVER II reactors. The design calls for interim storage capacity for up to 48 hours of SILVER II operation. The chemical agent/dilute acid mixture is fed continuously to the SILVER II reaction circuits for agent destruction during normal operation.

A schematic drawing of the anolyte and catholyte circuits is shown in Figure 3-3. The anolyte circuit consists of a glass-lined anolyte vessel in which the reagents and the agent solution are mixed to start oxidation. The vessel is maintained at 90°C (194°F) through the use of a thermal fluid circulating through a jacket surrounding the vessel. The agent material hydrolyzed by the dilute nitric acid is oxidized in the anolyte vessel by the Ag (II) ions, which are reduced to Ag (I) in the process. Carbon dioxide (CO_2), oxygen (O_2), carbon monoxide (CO), nitrous oxide (N_2O), and water are generated in the anolyte vessel. According to the technology provider, NO_x is not evolved from the anolyte, but under certain conditions, excess NO_x can be present. These gaseous products then flow to the NO_x recovery system. The anolyte solution is circulated to the electrochemical cell where the Ag(II) ions are regenerated. Other reagents added to the anolyte vessel include calcium nitrate to precipitate $CaSO_4$ and calcium fluoride, silver nitrate solution to maintain silver concentration (when chloride is present, silver chloride will precipitate), and nitric acid. A hydrocyclone between the anolyte vessel and the electrochemical cell prevents solids from entering and potentially blocking the narrow passages in the cell. A continuous bleed of anolyte solution takes place at the hydrocyclone for recovery of silver chloride and nitric acid.

The vented gases from the anolyte vessel are passed through a water/glycol-cooled condenser at 0°C (32°F) to condense nitric acid vapor, water vapor, and trace volatile organic compounds. The condensate is returned to the anolyte vessel, and the remaining off-gas is sent to the NO_x reformer.

The catholyte circuit also consists of a stirred glass-lined vessel from which the catholyte solution is circulated to the cathode side of the electrochemical cell. The catholyte vessel also has a thermal jacket to maintain the temperature at 90°C (194°F). In the cathode side of the cell, nitric acid is reduced to nitrous acid, NO_x, and water. The NO_x is vented from the catholyte vessel to the NO_x reformer. A slipstream of the catholyte solution is continuously bled to the silver and nitric acid recovery system. A make-up input stream of dilute acid to the catholyte vessel maintains the necessary concentrations of nitric acid.

Treatment of Energetics

A mix of sized-reduced propellant material and other energetics and dilute nitric acid is fed directly to a separate SILVER II reaction circuit via the anolyte vessel. This SILVER II process is similar to the process for treating agent slurries.

Fuzes are treated separately from other energetic material. The fuzes are first pretreated in small batches in a reaction vessel with nitric acid, which has been obtained from spent anolyte and catholyte solutions from the SILVER II process for energetics. The technology provider claims that the spent anolyte and catholyte solutions are rich in nitric acid, which attacks and completely dissolves the metal of the fuzes and neutralizes the energetics. The resultant liquid material is then evaporated and the dry salt extracted. Sheared mine-body sections are treated in a manner similar to the fuzes, but the effluent from the reaction vessel is directed to the SILVER II circuit for energetics.

Treatment of Metal Parts

After disassembly, munitions can still contain agent on internal surfaces and in the form of gelled heels. Therefore, metal parts associated with munition disassembly are oriented vertically and cleaned by high-pressure liquid jets of hot dilute nitric acid. This process is carried out remotely in batches in an enclosed chamber. In subsequent discussions, the technology provider indicated that some sort of agitated bed or submerged bath is under consideration as an alternative to high-pressure liquid jetting.

The metal parts are then decontaminated in batches inside a material decontamination chamber (MDC). The MDC is part of a sealed heat-treatment circuit and is used to treat batches of metal parts. Air circulates through the MDC, which is heated either electrically or inductively, and the off-gas is extracted, cooled to condense any trace agent liquid, and passed through an activated carbon filter prior to recycling back into the MDC. The condensate from the condensers is fed to the agent interim storage tank. The chamber operating temperature is in excess of 1,000°F (538°C), and the metal parts are maintained at this temperature for at least 15 minutes to ensure 5X decontamination.

Treatment of Dunnage

Both contaminated and agent-free dunnage are shredded, pulped, and mixed with recycled dilute nitric acid. This slurry is temporarily stored in a stirred storage tank. Other liquid secondary wastes, such as hydraulic fluids and decontamination solutions, are mixed with this material. These slurry mixes are then fed into the SILVER II reaction circuit for dunnage.

Other solid wastes, such as DPE suits and carbon filters, are also shredded and processed in batches. These materials are treated in the MDC similar to the way metal parts are treated.

Process Instrumentation, Monitoring, and Control

The technology provider proposes holding and analyzing gaseous and liquid effluents prior to release from the facility. Liquid streams (e.g., from the evaporator and NO_x reformer) are first accumulated in holding tanks to allow sampling and analysis for the presence of agent prior to recycling or prior to shipment off site for disposal. The off-gases vented from the anolyte and catholyte vessels (after scrubbing) are accumulated in evacuated tanks. The tanks are filled to atmospheric pressure; each tank has the capacity to hold the gases produced during approximately 15 minutes of operation. Multiple tanks are proposed to allow for continuous operation. As each tank is filled, the gas is sampled and analyzed for agent. If the gas is determined to be agent free, it is vented to the atmosphere. If the gas is determined to contain traces of agent, it is vented to the atmosphere through a dehumidifier and a multistage carbon bed. The following on-line chemical analyses are proposed:

- anolyte off-gas composition using gas chromatography/thermal conducting detector
- NO_x concentration in off-gas streams
- acidity (density and temperature for simple nitric acid/water mixtures)
- concentrations of chemical agent vapor using ACAMS or another certified analysis

Feed Streams

The requirements for the feed streams to the SILVER II reaction circuits are stringent. The feeds must be liquids, and the particle size of suspended solids must be smaller than one inch. Suspended solids larger than this could require excessive reaction times in the anolyte vessel. These stringent requirements entail more physical processing of the munitions than is routinely done in the Army's baseline disassembly system. In addition, materials that contain halogens will increase silver precipitation and necessitate more rapid changeout of reagents.

Effluent Streams

The process effluents and treatment/disposal strategies proposed by the technology provider are listed in Table 3-5. The effluent streams include solids (such as agent-free dunnage and 3X decontaminated trash, DPE suits, metal parts, filter candles, and spent carbon), liquids (spent anolyte and catholyte solutions and excess nitric acid), and gases from anolyte and catholyte vessels.

Liquid bleed streams from the anolyte and catholyte vessels are directed to a nitric acid evaporator in which water, nitrogen oxides, and some nitric acid are evaporated. The vapors from the nitric acid evaporator are then passed to the NO_x reformer. The evaporator bottoms, consisting of a silver nitrate/nitric acid solution, are returned to the anolyte circuit. The NO_x reformer receives the vapor effluent from the nitric acid evaporator and the off-gas streams from the anolyte and catholyte vessels. Pure oxygen is added to the reformer as a reagent. The reformer consists of a reactive distillation column designed to generate nitric acid by reacting NO_x, oxygen, and water. The nitric acid is accumulated in holding tanks where it is sampled and analyzed for agent prior to recycling to the SILVER II process. From the top of the distillation column, gases consisting primarily of CO_2, CO, N_2O, NO_x and O_2 are directed through a packed-column caustic scrubber (sodium hydroxide/carbonate/nitrate solution). The scrubber liquor blow-down is accumulated, analyzed for chemical agent, and evaporated. The salts are discharged as a solid waste. Part of the bottom product from the distillation column (a concentrated nitric acid stream) is recycled, and the remainder is prepared for off-site disposal. The net generation of nitric acid is attributed to the addition of the calcium nitrate and to the nitrogen in the energetics.

TABLE 3-5 The Process Effluents and Treatment/Disposal Strategies Proposed by AEA for the Silver (II) Process

Process Effluent	Management Strategy
Nonprocess-related inert dunnage (glass, plastic, metal band)	Shipped off site for disposal at a hazardous-waste landfill after decontamination using standard decontamination solutions, followed by polishing in the material decontamination chamber.
DPE suits	Shipped off site for disposal at a hazardous-waste landfill after decontamination as a part of personnel egress procedures, followed by treatment in the material decontamination chamber.
Spent carbon	Shipped off site for disposal at a hazardous-waste landfill after treatment in the material decontamination chamber.
Trash	Shipped off site for disposal in a solid-waste landfill.
Munition-related metal	Decontamination by high-pressure washing followed by treatment in material decontamination chamber to 5X. Decontamination solution treated in SILVER II Unit.
Nonmunition miscellaneous metal	Treatment in material decontamination chamber. Decontamination solution treated in SILVER II Unit.
Anolyte solution	Nitric acid and water recovered through an evaporator/condenser unit. Remaining solid mixture shipped off site for recovery of silver as silver nitrate. Ultimate disposal of waste from recycling at a hazardous-waste landfill.
Catholyte solution	Nitric acid from the catholyte solution recovered through an evaporator/condenser unit. Remaining solid material (mostly silver nitrate) shipped off site for recovery of the silver as silver nitrate. See anolyte solution above.
Off-gas	Treated by caustic scrubber and emitted to the atmosphere.
Dilute (1% wt) nitric acid	Dilute nitric acid recovered on site from NO_x reformer; most reused in plant operations. Excess dilute nitric acid treated on site in a tank treatment system to neutralize it prior to discharge to the local publicly-owned treatment works. Nitrate removal may also be required.
Concentrated (71% wt) nitric acid	Concentrated nitric acid recovered on site from NO_x reformer and reused in plant operations. Acid is generated during treatment of items containing nitrogen (e.g., energetics, mustard, and GB). Excess nitric acid reused off site.
Calcium fluoride filter candles	Solids from anolyte solution generated during treatment of GB shipped off site for disposal.
Silver chloride and calcium sulfate/filter candles	Solids from anolyte solution generated during treatment of mustard shipped off site for recovery of silver as silver nitrate.
Silver chloride/filter candles	Solids from anolyte solution generated during treatment of decontamination solution shipped off site for recovery of silver as silver nitrate.
Caustic scrubber waste	Dried salts from caustic scrubber shipped off site for disposal at a hazardous-waste landfill.
Glass fibers captured on filter candles	Treated in materials decontamination chamber and transported off site for disposal at a hazardous-waste landfill.

Source: Adapted from AEA, 1997.

Another effluent from the process consists of spent anolyte and catholyte solutions. The anolyte solution becomes saturated in metal ions and halogen salts from metal impurities and halogens in the agent. The spent anolyte and catholyte solutions are discharged into holding tanks, and after they are confirmed to be agent free, they are evaporated. The nitric acid and water are recycled, and the dry metal salts are sent off site for recovery of the silver. Wastes from silver recovery ultimately require disposal in a hazardous-waste landfill.

The MDC (material decontamination chamber) is used to desorb traces of agent thermally from a range of solid materials. Munitions-related metals are high-pressure washed and then treated in the MDC to a 5X decontamination level. Nonmunitions-related metals are also treated to 5X in this chamber. The treated metals are then suitable for release to the public sector. Spent carbon is also treated in the MDC and then shipped off site to a hazardous-waste landfill. Other solid materials shipped to off-site hazardous-waste landfills are: the glass fiber filtrate captured on filter candles, decontaminated DPE suits, residues from off-site recycling of silver nitrate, calcium fluoride/calcium sulfate solids collected from anolyte solutions on filter candles, and dried salts from the caustic scrubber.

AEA included a mass balance in the original

TABLE 3-6 Process Inputs for SILVER II per 155-mm Projectile

Description	Total Flow (lb/munition)	Details/Comments
Organic feed	20.7	Liquid mustard feed to agent plant; solid burster feed to energetics plant; "chipped" wood feed to dunnage plant.
Calcium nitrate	12.1	Feed to anolyte vessel (dissolved in recycled water to produce the feed solution).
Oxygen	26.6	Low-pressure gaseous feed.
Sodium hydroxide	2.4	Feed to caustic scrubber (dissolved in recycled water to produce a 20 wt% solution feed to the scrubber).
TOTAL[a]	61.8	Metal parts not included because they are only decontaminated and not destroyed.
Contaminated metal parts	*86*	*Typically the munition casing, etc.*

[a]Assumes no net makeup of silver or nitric acid.

Source: Adpated from AEA, 1997, and AEA, 1998b.

proposal (AEA, 1997) and revisions in the subsequent data-gap report (AEA, 1998a). The technology provider supplied the committee with additional information on the mass balances for an integrated facility consisting of three separate SILVER II units—one each for the destruction of chemical agent, energetics, and dunnage. Tables 3-6 through 3-9 provide the process inputs and net (after recycling) outputs for 155-mm projectiles filled with mustard and M55 rockets filled with VX. These estimates assume that regeneration and recycling of spent anolyte and catholyte will be conducted on site and, therefore, recycled acid will be reused. The precipitated silver chloride is assumed to be recycled off site to regenerate silver nitrate with an efficiency of 100 percent. The silver loss is projected by the technology provider to be less than 1 percent of the silver recycling rate.

The technology provider was also asked to supply an estimate of the amount of silver anticipated to be inventoried for operations at Bluegrass and Pueblo, including the total amount of silver circulating in the process, being reprocessed, stored as reserve, and

TABLE 3-7 Process Outputs for SILVER II per 155-mm Projectile

Description	Total Flow (lb/munition)	Details/Comments
Nitric acid	14.3	Produced as both a concentrated and dilute acid stream (mass quoted is for pure acid and does not account for any water of solution, which is detailed separately).
Calcium sulfate	9.9	Solid precipitate removed.
Water	8.8	Produced from several sources from the plant (e.g., as dilute nitric acid) (mass quoted is for pure water).
Sodium nitrate	2.6	This mass of sodium nitrate will be contained in the caustic scrubber waste stream.
Sodium nitrite	2.2	Contained in the caustic scrubber waste stream.
Carbon dioxide	27.3	Off-gas from scrubber.
Nitrogen dioxide	0.4	Off-gas from scrubber.
TOTAL Solid and liquid[a]	37.8	
TOTAL Off-gas	27.7	
Decontaminated metal parts	*86*	*Decontaminated to 5X standard.*

[a]Silver chloride is recycled as silver nitrate.

Source: Adapted from AEA, 1997 and AEA, 1998b.

TABLE 3-8 Process Inputs for SILVER II per M55 Rocket

Description	Total Flow (lb/munition)	Details/Comments
Organic feed	62.9	Liquid VX feed to agent plant; solid Comp B burster feed and grain propellant from water jetting to energetics plant; "chipped" wood and fiberglass feed to dunnage plant.
Nitric acid	0.4	Produced by destruction of agent and energetic; when wood is processed no nitric acid is generated; because one of the off-gas components is nitrogen, nitric acid must be added to maintain a constant "nitrogen" balance.
Oxygen	86.0	Low pressure gaseous feed.
Sodium hydroxide	9.2	Feed to caustic scrubber; dissolved in recycled water to produce a 20 wt% solution feed to the scrubber.
TOTAL[a]	158.5	
Contaminated metal parts	*44.8*	

[a]Assumes no net makeup of silver nitrate or nitric acid.

Source: Adpated from AEA, 1997 and AEA, 1998b.

maintained in any way for use in the process. The information supplied to the committee is included in Tables 3-10 and 3-11 for 155 mm projectiles filled with mustard and M55 rockets filled with VX, respectively. During the operation, the sources of silver-rich output streams are precipitated silver chloride (due to the presence of chloride in the mustard gas) and silver nitrate sludge in the spent anolyte and catholyte solutions.

The technology provider has estimated that 52,000 pounds of silver will have to be inventoried at Pueblo and 8,000 pounds of silver at Blue Grass. This estimate is based on an assumption of a 10-day supply of stored silver nitrate. The silver requirement for Pueblo is much greater because of the increased production of silver chloride from treatment of larger amounts of mustard-filled weapons, from which there is continuous

TABLE 3-9 Process Outputs for SILVER II per M55 Rocket

Description	Total Flow (lb/munition)	Details/Comments
Phosphoric acid[a]	3.7	Produced as a result of VX destruction (mass quoted is for pure acid and does not account for any water of solution, which is detailed separately).
Sulfuric acid[a]	3.7	As above.
Lead nitrate	0.2	Small amounts produced as a result of destruction of the lead stearate in the propellant.
Water	32.8	Produced from several sources from the plant (e.g., as dilute nitric acid); mass quoted is for pure water.
Sodium nitrate	9.9	Contained in the caustic scrubber waste stream.
Sodium nitrite	7.9	Contained in the caustic scrubber waste stream.
Carbon dioxide	98.3	Off-gas from scrubber.
Nitrogen dioxide	10.6	Off-gas from scrubber.
TOTAL Solid and liquid	58.2	
TOTAL Off-gas	108.9	
Decontaminated metal parts	*44.8*	

[a]Probably disposed of by conversion to calcium salts

Source: Adapted from AEA, 1997 and AEA, 1998b.

TABLE 3-10 Estimate of Spent Silver for Mustard-Filled 155-mm Projectiles

Source	Form	Mass Per Munition	Silver Content
Silver from precipitation	Silver chloride precipitate	21 lb of silver chloride	75% by weight
Silver from spent anolyte and catholyte solutions	Concentrated "sludge" following evaporation	1.1 lb of silver nitrate contained within the "sludge"	64% by weight

Source: AEA, 1998b.

precipitation of silver chloride from the chlorine in the mustard. At Pueblo, the technology provider estimates that the silver contained in the processing systems is 8,850 lb and that the flow rate of silver metal to the plant is 4,340 lb/day. At Blue Grass, the silver contained in the processing systems is only 2,040 lb, and the flow rate is 600 lb/day.

Start-up and Shutdown

Generation of Ag(II) ions depends entirely on the electrical current and stops immediately when the power is switched off. The technology provider claims that the ability to turn the process off rapidly by switching off the electricity provides a safety benefit because it ensures that the reaction is easily controllable. Electrical power to the cell can be shut off at any time—for example, from safety interlocks at other stages of the overall process. However, the committee notes that oxidation reactions will continue even without electricity until Ag(II) concentrations decline, and reactions with nitric acid will proceed even in the absence of electricity.

EVALUATION OF THE TECHNOLOGY PACKAGE

Process Efficacy

Effectiveness of Munitions Disassembly

Although the munitions disassembly process is based on the Army's baseline disassembly system, some significant differences and additions will require further development, particularly for disassembling rockets and mines. These differences are largely attributable to the need (1) to segregate munition parts and components and treat them in separate SILVER II reactors, and (2) to further reduce the size of components prior to treatment in SILVER II reactors. Solids must be reduced to a particle size of less than 1 inch in diameter to dissolve completely in nitric acid and to oxidize in a reasonable amount of time. The committee found significant barriers to implementing the additional processing because of the increased complexity of the mechanical processes and the need to ensure proper feed characteristics to the SILVER II reactors. The sophisticated segregation, material-handling, and size-reduction processes will require either the development of robotics equipment or manual segregation in DPE suits. Manual segregation would be undesirable because of the high hazard level and because performing lengthy work assignments in DPE suits is extremely uncomfortable. The added complexity of robotic equipment would require substantial development, which would certainly delay the overall development schedule.

The technology provider indicates that visual inspection would be used to determine whether residual energetic materials remained after wash out, but the committee has reservations about determining residual contamination by visual inspection.

Effectiveness of Agent Destruction

This process can be viewed as an oxidation of organic materials with Ag(II), with electrolytic regeneration of

TABLE 3-11 Estimate of Spent Silver Sent for Recycling for VX-Filled M55 Rockets

Source	Form	Mass Per Munition	Silver Content
Silver from spent anolyte and catholyte solutions	Concentrated "sludge" following evaporation	4.2 lb of silver nitrate contained within the "sludge"	64% by weight

Source: AEA, 1998b.

the silver. Mustard contains volatile low molecular weight chlorinated hydrocarbons that are expected to be oxidized by Silver (II), but this will have to be demonstrated. The committee has identified three potential problems that could interfere with agent destruction in a full-scale system.

Plugging of the Electrolytic Cell. A hydrocyclone is used in the anolyte circuit to prevent solids from entering the SILVER II cell, but the committee believes that the hydrocyclone may not remove solids to the level necessary to prevent plugging of the electrolytic cell used to regenerate the Ag(II). The plugging can be caused by (1) solids in the agent-contaminated materials and (2) precipitates generated in the anolyte vessel that migrate from the vessel to the electrolytic cell. The recirculation stream leaves substantial amounts of solids and precipitates in the anolyte vessel, and the electrolytic cell has very small flow channels that appear to be prone to plugging.

Impact of Metals on Electrolytic Efficiency. Other metal ions may reduce the electrolytic efficiency of the desired reaction by migrating through the membrane and diminishing the current via useless reactions. The tests conducted to date have been on relatively pure organic chemicals that did not contain iron, copper, aluminum, or other metals found in munitions. These metals will all react vigorously with the nitric acid of the SILVER II reagent, causing very large quantities of metal ions to dissolve. The technology provider states that the added metal ions should not affect the reduction/oxidation reactions that destroy the agent and energetics; however, no data have been presented to show this. In fact, iron and other metals found in munitions can exist in many oxidation states. The complex mixture of many metals that will be present during SILVER II treatment of chemical munitions will make predicting the behavior of electrolytic solutions under reduction/oxidation conditions very difficult.

Rapid Reaction with Organics. Rapid reactions of nitric acid with organics in the agent and energetic solutions could cause process upsets. Rapid reactions were observed by members of the ACW Committee during a bench-scale demonstration of the SILVER II process. That is, the acid by itself is a strong oxidizer, and in the absence of Ag(II) ions, it may initiate or sustain strong exothermic reactions causing the evolution of large quantities of gases (e.g., oxides of nitrogen).

Effectiveness of Metals Decontamination

In discussions with the committee subsequent to submitting the proposal, AEA indicated that some sort of agitated bed or submerged bath would be used for decontaminating metal parts. The committee did not have sufficient information to evaluate this process. The proposed thermal treatment of metals parts in the MDC to decontaminate the metal to a 5X condition appears to be adequate.

Effectiveness of Energetics Destruction

The technology provider proposed chemical dissolution of fuzes. However, no data are currently available on the efficacy of this dissolution process. All fuzes are complex units made up of multiple components and materials. Some contain metals that have been in contact with azides and heavy metals, including lead. The committee concluded that (1) the proposed approach is more complex than conventional techniques of separation and detonation, and (2) a strategy and demonstration of the plan for dealing with small amounts of lead azide, antimony, copper, and aluminum during the dissolution of fuzes in nitric acid would have to be developed.

The committee also had concerns about the dissolution of metal parts in an aqueous environment. For example, aluminum parts in the M55 rocket will generate hydrogen gas when dissolved in acid. The amount of hydrogen generated is uncertain but could be estimated based on the known reactions and thermodynamic properties of reactants and products. The reaction vessels must be designed to vent the anticipated amount of hydrogen. Moreover, the cumulative heat release associated with the exothermic reactions of both metals and energetics must be carefully managed for the fine-particle mix of metals and energetics.

Tests completed to date have indicated that very small particle sizes, long reaction times, and intense mixing are required for the destruction of energetics, such as trinitrotoluene (TNT). In initial laboratory tests, flake TNT was not completely destroyed in five hours. However, tests with larger amounts indicated much higher destruction efficiencies, and the carbon content remaining was at ppm levels. The committee did not have sufficient data to assess the efficacy of energetics destruction at larger scale. The strategy for scaling up the process to full-scale operation on mixtures of energetics and metals was not demonstrated to the committee.

Sampling and Analysis

The technology is amenable to a hold-test-release scheme for liquid and gaseous effluents. Liquid streams are accumulated in holding tanks, where they are sampled and analyzed for the presence of agents prior to recycling or to shipment off site for disposal. Vented off-gases from the anolyte and catholyte vessels are first directed to the nitric acid recovery system and then scrubbed. The off-gas is then collected in an evacuated tank that can hold gases from approximately 15 minutes of operation. The tank is sampled and analyzed for agent using ACAMS or another certified technique. If the gas is determined to be agent free, it is vented directly to the atmosphere. If agent is present, the gases are vented through a dehumidifier and a multistage carbon bed to the atmosphere without further analysis. The committee had reservations about obtaining a representative sample from the tank because revaporization of condensed material on the internal walls of the tank could occur when the pressure is lowered for venting.

Maturity

Technology Status. The SILVER II process has yet to be operated on a commercial scale for waste treatment. The electrochemical cell in the SILVER II process is used commercially in the chlor-alkali industry. However, the largest pilot-scale tests for waste treatment have been conducted using a 4-kW cell consisting of a single anode-cathode pair. (The technology provider has proposed a full-scale system made up of 60 cells in parallel.) The most extensive tests have been conducted with spent tributyl phosphate dissolved in kerosene, from the Purex process, as the feed material. These tests, which were run continuously, 24 hours per day for up to 14 days, destroyed a total of 150 liters of the feed material. The technology provider has successfully completed laboratory tests on 10-gram batches of agent and has constructed a pilot plant at Porton Down, United Kingdom, that is suitable for tests on 15-liter batches of agent. All of the tests prior to startup of the Porton Down plant had been conducted with only the electrochemical cell component of the agent-destruction system. The Porton Down facility also includes anolyte and catholyte feed circuits, an anolyte off-gas condenser, an NO_x reformer system, and a modified version of the combined off-gas treatment circuit, including a sodium hydroxide scrubber. The silver management system was scheduled to be tested at Dounreay with the effluent generated from the Porton Down plant.

The NRC AltTech report (NRC, 1996a) summarizes the results of a test conducted by the technology provider at Porton Down on 14.62 kg of "as supplied VX," which contained 12.7 kg of agent. The test consisted of a single continuous run of 6.5 days. At the end of the run, no agent was detected in the catholyte solution or in the process residuals. The lower detection limits for VX were 7.6 mg/m^3 in the anolyte, 9.2 mg/m^3 in the catholyte, and 1.7 mg/m^3 in the residuals discharged during the trial. The corresponding volumes were 0.0724 m^3 of anolyte, 0.0854 m^3 of catholyte solution, and 0.0929 m^3 of process residuals. The total residual VX was, therefore, less than 1.5 mg out of an input of 12.7 kg of VX, corresponding to an agent destruction efficiency of greater than 99.99998 percent.

The technology provider calculated that the 14.62 kg of "as supplied VX" contained 7.21 kg of organic carbon. At the end of the run, the total organic carbon remaining in the anolyte and catholyte circuits was 0.816 kg. Therefore, the destruction efficiency for conversion of organic carbon to CO_2 and CO was 88.7 percent. The technology provider suggests that further removal might have been possible by continuing the operation of the cell after the organic feed was ended.

The data-gap report (AEA, 1998a) included additional bench-scale tests (1/35 of full scale) involving the treatment of 100 g of GB at the Boscombe Laboratory. A 60-W cell was used, and small batches of material were added. The reaction time was 17 hr, after which the destruction efficiency was greater than 99.9999 percent based on detection limits. The oxidation efficiency based on CO/CO_2 in the off-gas as compared to the carbon feed was relatively poor (70 percent). There was no measure of total hydrocarbon levels in the off-gas. Energetic materials were tested in an isolation laboratory for a 20 percent slurry of energetic material in water. Separate runs were conducted with 100 g of TNT in the form of fine particles, 100 g of RDX, 100 g of tetryl, and 100 g of double base propellants. In two tests, TNT slurry was run for 42 hr, after which the residual TNT was below detection levels. This corresponds to a destruction efficiency of greater than 99.9999 percent.

Scale-up. The committee identified several issues that need to be addressed during the scale-up of the process into a fully integrated system for the wide diversity of assembled chemical weapons. The key scale-up issues are discussed below.

First, the reaction rates and processes are sensitive to temperature, which can be more difficult to control at larger scales. In the committee's opinion, temperature will be difficult to control to prevent localized boiling in the cells. The set point for the process is 90°C (194°F). Portions of the process are exposed to high electric currents that can raise temperatures, however, and any blockage that restricts flow could result in localized boiling. This difficulty would add complexity to the scale-up.

Second, reaction times for larger sized particles will limit the feed rate. Tests conducted to date have been limited to very small particles. Third, the conversion efficiency of the NO_x reformer will depend on the feed rate and composition. Process operation and feed variations cause fluctuations that could lead to performance problems. Fourth, a 240-kW system requires 60 cells in parallel. The committee is concerned about the problems with managing the feed-stream flows to that many parallel cells because of the potential for plugging and localized boiling, which could cause flow imbalances among the cells. A substantial increase in maintenance requirements would increase the risk of worker exposure.

Fifth, the tolerance of cells to particulate matter, such as very fine silver chloride precipitate and metals that pass through the hydrocyclone, is unknown. However, particles must be kept out of the small passages in the cells. If the hydrocyclone efficiency in removing particles declines at larger size, a series of smaller cyclones in a parallel arrangement (multicyclone) may be necessary. There is also a significant chance of corrosion/erosion of the hydrocyclone unless the materials for liners are carefully selected.

The committee concluded that all five of these scale-up issues will have to be addressed directly in larger-scale units and precommercial integrated-system prototypes.

The SILVER II process requires a very high recirculation rate of the reagent solution between the anolyte reactor vessel and the electrolytic regeneration cell. Interruptions in the recirculation stream caused by plugging or corrosion, for example, will prevent the complete oxidation of agent and other organic materials.

The proposed full-scale plant represents a significant scale-up from the existing system. The largest scale system operated to date uses a 4-kW cell, consisting of a single anode-cathode pair and a 15-liter reactor. In 1996, this system was used to destroy VX at rate of approximately 94 grams per hour. The test cell was operated at currents of between 600 and 1,400 A. Operation at the design current of 2,000 A was not successful because the pressure increased in the anolyte compartment when VX was added. AEA traced the problem to lower than expected efficiency of the NO_x reformer, which resulted in the passage of more than expected unreacted O_2 and NO_x gas through the condenser and into the scrubber. This increased the pressure drop across the scrubber, causing an increase in pressure in the anolyte gas stream. These results illustrate the types of scale-up problems the SILVER II process could encounter.

In the committee's opinion, these scale-up issues could create serious processing problems in larger scale systems. The uncertainty of scale-up of the SILVER II process units into an integrated treatment facility are serious enough to challenge the ultimate ability of this process to perform at the required level.

Robustness

The committee has several concerns about the ability of the process to handle nonoptimal feed streams, especially (1) the potential for plugging caused by the accumulation of solids, (2) the effects of metals on the electrolytic reaction efficiency, and (3) the reaction of nitric acid with organics in the feed. Until these issues can be thoroughly investigated, the committee considers the robustness of the process questionable.

Monitoring and Control

The monitoring and control system involves a relatively straightforward application of normal industrial techniques. The transducers for temperature, pressure, flow, and liquid level are commercially available. A centralized computer is used to achieve automatic closed-loop control via control algorithms and operator interface. The complexity of the control systems is consistent with chemical process plants in general. However, the committee is concerned about the ability to determine whether holes have been created in cell membranes.

Applicability

If the integrated SILVER II reactor system operates as envisioned by the technology provider, it would be applicable to the destruction of all of the assembled chemical weapons in the Army's inventories. Although limited data on the destruction of pure agent and energetics in a SILVER II reactor are available, no data are available on treating components from disassembled chemical weapons components in a fully integrated treatment process using SILVER II-based technologies. For this reason, that committee was not able to determine its applicability at this time. The committee did identify several issues (listed below) that should be addressed to ensure that the integrated technology could treat the full array of assembled chemical weapons:

- behavior of the SILVER II reaction circuit with high concentrations of metals
- effects of various metals on the electrochemical potential
- potential poisoning of electrodes by trace elements
- reaction times to destroy energetics and agent components completely
- precipitation of silver chloride and recycling of silver from silver chloride materials

Process Safety

The SILVER II process requires the following unique equipment:

- a machine for handling sheared rocket parts
- SILVER II cells for the generation of Ag(II) ions for the destruction of agent and energetics
- anolyte and catholyte tanks in which most of the agent and energetic destruction occurs
- hydrocyclones to separate solids from liquid flowing to the SILVER II cells
- nitric acid recovery equipment (NO_x reformer and HNO_3 recovery evaporator)
- MDC for metal parts not dissolved in the SILVER II process and inert secondary wastes (e.g., DPE suits and spent activated carbon)
- shredder for dunnage and secondary waste
- off-gas scrubbers using NaOH and H_2O_2
- off-gas tanks for hold-test-release of process gases
- oxygen supply system for the SILVER II process

The SILVER II processes operate at the relatively low temperature of 90°C (194°F)[1] and low pressure (essentially atmospheric pressure), thus minimizing stored energy and reducing process hazards. Nitric acid concentrations are expected to be within safe operating ranges of 4M–12M for anolyte and 2M–6M for catholyte. Higher temperatures (1,000°F [538°C] or more) occur in the MDC, which uses air in a closed-loop recirculation system to evaporate agent from metal parts and transfer any residual agent to a condenser where it is removed as a liquid and returned to the agent feed tank. The MDC operates at near atmospheric pressure. Liquid process effluent streams are sampled for completeness of reaction before they are released for acid recovery or other treatment or disposal.

A unique hazard for the process is the use of electrolytic cells to generate the Ag(II) ions. A unit size for these cells is 4 kW (2,000 amps at 2 V per cell). The plant size is estimated to be four 64-electrode pair cells operating in series with a total operating voltage of 8 V.

Worker Health and Safety

The SILVER II technology package operates in a batch or semi-continuous mode with verification of agent and energetic destruction prior to release of the process effluent. Electrical energy and the presence of

[1]The safe operating range is 40°C–120°C (104°F–248°F).

large quantities of nitric acid (and, to a lesser extent, the very corrosive nitric acid/hydrofluoric acid mixture [HNO_3/HF]) are the major process hazards. The latter material, which is derived from the decomposition of GB, is highly corrosive.

The technology provider intends to select corrosion-resistant materials that will not require increased maintenance and will not increase worker exposure to these substances or to agent. Although corrosion was identified as a potential hazard on the hazard and operability study (HAZOP) of the concept design, it was assigned a low-risk ranking by the technology provider (AEA, 1998a). To justify this ranking, the design must ensure that the selection of materials is compatible with HNO_3 and anticipate the HNO_3/HF mixture in areas where it is not normally expected. Even small amounts of this material in the presence of unprotected or unsuitable materials will result in rapid corrosion and leaks.

The electrolytic cells are similar to those used in many industrial applications and should be compatible with safe operation and maintenance; however, the potential presence of chemical agent will require that maintenance workers wear DPE suits or other protective gear, which will complicate the procedures. Because of the large number of electrode pairs and the potential for plugging and leakage, cells will have to be designed for ease of maintenance and with extremely reliable protection from electrical shock. Since the voltage is low, the electrical hazard is not unusual, and many electrolytic cell configurations of similar size are being used elsewhere. However, none of them has the DPE protection requirement of this application; hence, easy maintenance will be critical in this case.

As members of the ACW Committee observed during a bench-scale demonstration of the SILVER II process, the nitric acid has the potential for causing process upsets from its rapid reaction with organics in the agent and energetic solutions. The acid by itself is a strong oxidizer, and even if no Ag(II) ions are present, it could still initiate or sustain strong exothermic reactions in the presence of large quantities of unreacted organic material. These reactions could result in the evolution of large quantities of gases (e.g., oxides of nitrogen). Because the equipment containing these solutions has relief valves and would be installed in a controlled ventilation cell, upsets would probably be contained. The technology provider has stated that "there is no possibility for any of the conditions on the plant to 'run away,' all control is on a negative and not positive feedback basis" (AEA, 1997). The committee believes that a demonstration of the ability to limit or control reactions between nitric acid and agent and energetics or a demonstration that these reactions will be self-limiting will be necessary before this technology moves to full development.

The technology provider has addressed the problem of electrolytic cell plugging by providing hydrocyclones to remove particulates down to 100 mm in size. Although limited experience (approximately 156 hours) has shown that the design is effective, the increased maintenance for the hydrocyclones and the cells may increase the chances of worker exposure to agent. Therefore, the design must ensure ease of maintenance without undue worker exposure. This applies to the design of the dunnage slurrifier and the stirred catholyte and anolyte tanks, as well.

The proposed disassembly processes are basically derivatives of those used in baseline disassembly and do not present new or increased hazards for rockets, projectiles, or land mines during operation. These processes will take place in vessels or structures designed to withstand explosive overpressure, should an initiation occur. A similar approach is used in baseline disassembly (see Appendix C). However, the proposed modifications would change the requirements for the design of machine tools and could require high maintenance, thus increasing the possibility of agent exposure by workers.

Sheared parts will be cleaned of agent and energetic materials by a robotic water-jet washout system. Similar systems are commonly used for the destruction of conventional munitions and should not pose unique safety problems. One possible area of concern is the interfacing of this system with the sheared parts. This step will require that new tools and handling concepts be designed and tested. A safety concern is that higher than normal maintenance will be required in DPE suits to sustain process throughput rates and schedules.

In the MDC, air circulates over metal parts; the heat is applied externally to the MDC. The air leaving the MDC is cooled, and condensable gases are removed in a condenser and returned to the SILVER II process. Subsequently, the dried cool air passes through an activated carbon filter and is returned to the MDC. The

MDC operation does not appear to present any unusual process safety problems.

The SILVER II process uses a caustic scrubber similar to the baseline process for treatment of all process off-gases. No unusual process hazards are believed to exist in this system, and the system can be designed to handle all anticipated upsets that could increase the evolution of off-gas. The full range of possible off-gases (not just those associated with stoichiometric reactions) should be characterized to identify hazardous species that may not be readily absorbed by the caustic scrubber design. The technology provider has indicated that no liquid acid solution is expected to reach the scrubber because nitric acid condensers in the off-gas lines will carry acid gases to the scrubber. The design must ensure that a failure to remove liquid from the condensers will not result in the passage of accumulated liquid to the scrubbers, which could damage the scrubbing system. If this scenario were to develop, it could possibly overload the hold-test-release system and result in unmonitored releases to the atmosphere. Of course, the off-gas agent monitoring system should shut down the process if agent is found in the off-gas. But, if the nitric acid continues to react with agent, the evolution of gas will also continue and could lead to an upset in the condenser that would overload the hold-test-release system. This scenario might also result in the accumulation of N_2O in activated carbon filters.

Another scenario addressed by AEA is a possible explosion in the off-gas system because oxygen will be present at a concentration of 34.5 percent (or higher in an upset), and some CO will be present in the gas (approximately 5 percent under process design conditions). The technology provider has concluded that the CO-O_2 reaction is not likely because the lower flammability limit for CO in air is 12.5 percent, and the CO_2 will suppress the reaction or raise the lower flammability limit.

Public Safety

Releases of agent or other regulated substances to the atmosphere or to the groundwater system are expected to be below detection limits. Hold-test-release systems are applied to all effluent streams except the containment ventilation system, which uses baseline air-cleaning technology (carbon filtration). All processing equipment operations will take place in confined air spaces with air flowing from clean to contaminated areas and finally to baseline air-cleaning units.

The primary cause for the release of material containing agent or other regulated substances would be an explosion. The likelihood of such an event is expected to be extremely small at the conclusion of the design process for the full-scale facility. However, the design process will have to minimize the opportunity for explosive gas mixtures or nitrate salts.

Human Health and the Environment

Effluent Characterization and Impact on Human Health and Environment

The effluent streams produced by the process include solids (e.g., nonprocess-related inert dunnage and trash, DPE suits, munition- and non-munition-related metals, filter candles, and spent carbon), liquids (spent anolyte and catholyte solutions and excess nitric acid), and gases (cleaned and scrubbed off-gases from anolyte and catholyte vessels).

Solid Waste. The process will generate numerous solid waste streams, including several hazardous waste streams, such as filter solids (14 lb/ton of material processed), caustic scrubber waste (20 lb/ton), decontaminated DPE suits (17 lb/ton), and treated process dunnage (10 lb/ton). Currently, very little data are available on the characteristics of these solid wastes, and the committee has some concerns as to whether proper disposal methods would be available for these materials. The filter cake produced from filtering the effluent from the anolyte hydrocyclone stream will contain nitrates, chlorides, and hydroxides of many metals. Some of the metals are not regulated, but the cake will contain significant quantities of regulated metals, such as nickel, chromium, and lead—the vast majority of which will come from the munition bodies and the materials used to construct the reaction system. In any event, the waste will be hazardous under the "derived from" rule, and will have to be handled in compliance with hazardous-waste regulations. For wastes not subject to the "derived from" rule, a waste is considered hazardous if the filtrate from a prescribed toxicity characteristic leachate procedure (TCLP) contains certain metals at concentrations above specified limits. Very little data are available on the characteristics of the solid wastes generated in the process, and the committee is not convinced that proper disposal methods would be available. For example, the high silver content may cause the filter cake to fail TCLP, and the high nitrate content of the filter cake may complicate stabilization.

There are several types of effluent streams from the NO_x reformer. Some are recycled to the primary reactor cycle; others are subsequently processed or must be disposed of. In addition, the scrubber liquor blow-down is accumulated and, after being analyzed for chemical agent, is evaporated and the salts discharged as a hazardous solid waste. Some of the middle products from the reformer condenser are recycled to the process, but others (e.g., a portion of the dilute nitric acid) require further treatment. Finally, the bottom product from the NO_x reformer is a concentrated nitric acid stream, some of which is recycled and some of which must be disposed of.

Another effluent is spent anolyte and catholyte solutions, which are discharged into holding tanks. After they are confirmed to be agent-free, they are evaporated. The nitric acid and water are recycled, and the dry metal salts are sent off site for recovery of silver. Wastes from recycling will ultimately require disposal in a hazardous-waste landfill.

The MDC is used for thermal desorption of a range of solid materials. These 5X metals should be suitable for release to the public sector. Other materials will require off site disposal. The gas circulating in the MDC is directed through a condenser and carbon filter. The spent carbon filters will be shipped off site to a hazardous-waste landfill. The fiberglass captured on filter candles after they have been decontaminated in the MDC will also be shipped off site to a hazardous-waste landfill.

Air Pollution. Off-gases are generated in a number of locations in the process. The anolyte and catholyte vessels are continuously vented as the organic carbon is oxidized in the anolyte vessel and NO is formed in the catholyte. The process also produces particulates consisting mainly of droplets of liquids and vapor emissions from the handling of the munitions. Gases generated from the top of the reformer column are primarily CO_2, CO, N_2O, NO_x, and O_2. These gases are directed through a packed-column caustic scrubber (sodium hydroxide/carbonate/nitrate solution). The SILVER II process forms significant quantities of NO_x, which will have to be either captured or destroyed during operation. Technology for controlling NO_x in process gas streams of the size proposed is commercially available and would be efficient for this application, although the technology provider does not propose it for the full-scale facility.

Although the NO_x that is collected can be adequately removed, the very large recycle streams will also entrain NO_x and transfer it to parts of the process that are not vented to the NO_x control equipment. Releases of NO_x at unexpected points in the process were observed by members of the ACW Committee during a laboratory demonstration.

Completeness of Effluent Characterization

No characterization data is currently available except from small pilot-scale tests or theoretical mass-balance calculations. Agents and limited decomposition products were sampled and analyzed using gas chromatography/mass spectrometry (GC/MS) in some tests. However, no tests have been conducted to date to define the potential for secondary oxidation and nitration by-products. The amount of carbon monoxide produced is an important indicator of oxidation efficiency and should be monitored. Based on potential chemical synthesis, the committee identified several other potential intermediates and incomplete oxidation products that should also be investigated. These include nitropolyaromatic hydrocarbons, alkyl nitrates, intermediate dinitro compounds, hexavalent chromium, nickel carbonyl, and zinc. No data are available on the characteristics of the solid and liquid effluents of actual treatments of the components of assembled chemical weapons.

Effluent Management Strategy

The process effluents and treatment/disposal strategies proposed by the technology provider are listed in Table 3-5. The committee has several concerns about effluent management strategy, especially the recycling of 59 lb of concentrated nitric acid per ton of munitions treated. According to AEA, one option for recycling excess concentrated nitric acid is the production of ammonium nitrate fertilizer. The committee has reservations about the acceptability of this option because of potential contamination and the liability of commercial users of these waste materials.

The issue of the stabilization of high nitrate solutions will also have to be addressed prior to land disposal. The committee also doubts that commercial silver reclaimers would accept the spent material. Most silver recyclers in commercial operation handle photographic wastes and may not accept this material.

Resource Requirements

The major resources required are electricity and chemical reagents, such as silver nitrate, nitric acid, oxygen, calcium nitrate, and sodium hydroxide. Although water is used extensively, very little water will be needed for make-up because water is produced and recycled in the process. The resource requirements of most concern are silver and electricity. The use of silver is projected to be neutral in the mass balances as long as silver nitrate can be recycled off site by commercial reclaimers. The committee has some reservations about this recycling because (1) silver will inevitably be lost in the recycling process and (2) off-site recyclers may not be willing to accept the potentially contaminated material. The technology provider claims that more than 52,000 lb of silver will have to be inventoried at the Pueblo site. This represents a 10-day supply of material, which does not allow sufficient time for recycling, which could take more than 30 days. Thus, the silver inventory may actually be much larger, and an on-site silver recycling plant may be necessary.

The power requirements for the full-scale facility are projected to be 9.15 MW peak power, with the total energy consumption per year of 33,600 MW-hr for a full-scale facility capable of processing all of the munitions listed in the RFP in a five-year period (188,000 individual munitions in a single year). Thus, the average requirement per munition is 170 kW-hr. In fact, the energy requirements for different types of munitions can differ significantly from this average value. For example, the energy requirement for M55 rockets filled with VX was estimated by AEA to be 420 kW-hr per munition. Meeting this electricity requirement could be a problem at some remote sites but is otherwise not a significant concern.

Environmental Compliance and Permitting

The technology provider is considering using a tank treatment unit as the permitting protocol (see Chapter 2), which appears to be appropriate for the technology. Nonetheless, plants built for the SILVER II process are complex chemical plants and have never been permitted for waste treatment. Therefore, the permitting process is not well established and will require significant time and resources. Because there will be gaseous, liquid, and solid effluents, a broad array of permits must be secured, including RCRA, Clean Water Act (CWA), and Clean Air Act (CAA) [and possibly the Toxic Substances Control Act (TCSA) because of the presence of polychlorinated biphenyls (PCBs)]. Eventually, data on a similar full-size unit will be required to demonstrate that no adverse impacts to health or the environment would be caused by facility operations.

STEPS REQUIRED FOR IMPLEMENTATION

The committee identified several key development and demonstration factors that must be addressed for this emerging treatment process to move toward implementation:

1. Modified shearing locations for M55 rockets and a new shearing machine must be tested to show routine segregation of components and reduction in particle sizes to less than 1 inch in diameter.
2. The modified mine shearing approach must be tested.
3. The dissolution of fuzes and mine bodies in nitric acid and SILVER II solution must be evaluated.
4. All effluents must be characterized in detail when treating agents contaminated with metals from disassembled chemical weapons (i.e., potential trace species and reaction by-products, such as nitrated hydrocarbons, partially oxidized products, and metals, must be identified) and their environmental impacts evaluated.
5. Demonstrations of the scale-up, development, and integration of hardware with real materials of construction must focus on the robustness of the parallel flow in multiple-cell reactors. The issues of cell blockage, hydrocyclone performance, and NO_x reformer performance must be addressed.
6. The efficacy of high pressure-jet wash-out of

agent and gelled agent from M55 sheared pieces must be tested.

7. The treatment of burster charges and M28 propellant in the SILVER II reactor must be tested, and the material preparation required to ensure reasonable treatment times with no energetic events must be evaluated. This testing must also determine what happens to the lead stearate in the propellant during SILVER II treatment.
8. The process must be developed and tested for the efficacy of submerged-bath dilute nitric acid treatment for metals parts, including the effects of agitation and temperature.
9. The treatment of shredded dunnage material must be tested in a prototype-scale SILVER II reactor.
10. Techniques for controlling particulate matter to prevent plugging of SILVER II electrolytic cell channels must be developed and demonstrated.
11. Materials of construction must be evaluated under corrosive and oxidizing conditions.
12. The realistic potential for off-site recycling/reuse of silver salts and concentrated nitric acid must be evaluated, including recyclers' ability to accept, handle, and treat these materials.

FINDINGS

Finding AEA-1. Significant barriers to the development of the sophisticated equipment and processes for segregation, materials handling, and size reduction (to reduce materials to less than 1 inch in diameter) must be overcome.

Finding AEA-2. Potential problems associated with plugging of the passages in the electrolytic cells will have to be addressed.

Finding AEA-3. The proposed chemical dissolution of fuzes is a complete unknown because no data on this process are available. This approach is complex in comparison to more conventional techniques of separation and detonation.

Finding AEA-4. Data are not available to assess the efficacy of the treatment of energetics at larger scales.

Finding AEA-5. The ability to obtain a representative sample of gaseous effluents retained in the holding tanks prior to release has not been demonstrated.

Finding AEA-6. Several issues need to be addressed during the scale-up of the process into a fully integrated system, including temperature control, reaction times, efficiency of the NO_x reformer, cell flow management, efficiency of the hydrocyclone, and the tolerance of cells to particulate matter. These issues are potentially serious enough to create processing problems in larger scale systems.

Finding AEA-7. Identifying which corrosion-resistant materials would be compatible with HNO_3 and HNO_3/HF will require a significant development program.

Finding AEA-8. Limiting or controlling reactions between nitric acid and agent and energetics must be demonstrated.

Finding AEA-9. Conversion of excess nitric acid into ammonium nitrate fertilizer may be complicated because of the potential for contamination and the liabilities this would entail.

Finding AEA-10. A large inventory of silver is required for processing of chemical weapons, and finding an off-site recycler to accept the potentially contaminated materials could be a problem.

4

ARCTECH ACTODEMIL Technology Package

INTRODUCTION AND OVERVIEW

ARCTECH, Inc., has offered its ACTODEMIL technology for the destruction of assembled chemical weapons (ICF Kaiser is a member of ARCTECH's team). The ACTODEMIL technology uses potassium hydroxide (KOH) for base hydrolysis of energetic materials, and base hydrolysis with KOH followed by oxidation with Fenton's reagent to destroy VX and GB. For H, HT, and HD, destruction is done by hydrolysis at neutral or slightly acidic pH, followed by oxidation with hydrogen peroxide. Humic acid is added to the reaction solution to adsorb some of the organic compounds. ARCTECH has designated a solution that contains the KOH base and humic acid as "a-HAX." After the hydrolysis step, the humic acid is precipitated by lowering the pH with nitric acid, producing a solid residue that contains adsorbed organic compounds. The solid residue and remaining reaction liquor are then treated with Fenton's reagent or hydrogen peroxide to complete the destruction of agent and energetics. The solid residue is separated from the liquor, stabilized in a solid matrix, and disposed of at a commercial hazardous-waste landfill. To maintain water balance, the stabilization/solidification of large quantities of reaction liquor may also be required.

The technology package incorporates the Army's existing baseline disassembly process to separate the agent, energetics, and metal parts. All materials from the process, including agent, energetics, metal and plastic parts, dunnage, DPE, laboratory wastes, and miscellaneous wastes, are put in contact with a-HAX to destroy agent and energetics. Metal parts and dunnage are treated to a 3X decontamination condition and shipped to the Rock Island Arsenal for final decontamination to 5X and disposal. DPE suits and other nonmetal dunnage are treated by hydrolysis and shipped to a hazardous-waste landfill for disposal. Table 4-1 summarizes the demilitarization operations, and Figure 4-1 is a schematic drawing of the overall technology package.

Background on Humic Acid

The use of humic acid to treat agent/energetics is unique to the ARCTECH process. Humic acid is a high molecular weight (1,000–3,000 Daltons) polyelectrolyte derived from coal, peat, humate, and soil. The material proposed for treating assembled chemical weapons is produced from coal and is manufactured by ARCTECH at Sterling, Virginia. It is a complex aromatic macromolecule containing various functional groups and is dark brown to black in color. The compounds involved in the linkages include amino acids, amino sugars, peptides, aliphatic acids, and other aliphatic compounds. The functional groups include carboxylic, phenolic, aliphatic, enolic-OH, and carbonyl (C=O) structures of various types. The ARCTECH actosol® humic acid product is a Lewis acid that is manufactured with a high pH.

The interaction of humic acid with chemicals has been studied extensively; a large body of literature has been accumulated on the nature of bonding or binding between humic acids and various organic compounds. Rice and coworkers studied the binding of organic contaminants to the humic substances in soil (summarized

TABLE 4-1 Summary of the ARCTECH ACTODEMIL Approach

Major Operation	Approach(es)
Disassembly of munitions	Army's existing baseline disassembly method except that rockets are sheared in 10 places instead of seven; water jets may be used for washout of energetics.
Treatment of chemical agent	**Mustard.** Hydrolysis at near-neutral pH followed by oxidation with hydrogen peroxide, both steps in the presence of humic acid. **GB and VX.** Base (KOH) hydrolysis followed by oxidation with Fenton's reagent, both steps in the presence of humic acid.
Treatment of energetics	Base (KOH) hydrolysis in presence of humic acid.
Treatment of metal parts	Base (KOH) hydrolysis in presence of humic acid.
Treatment of dunnage	Base (KOH) hydrolysis in presence of humic acid.
Disposal of waste	**Solids.** Stabilized with cement and sent to appropriately permitted landfill; metal parts shipped to Rock Island Arsenal for 5X treatment. **Liquids.** Stabilized with cement and sent to appropriately permitted landfill. **Gases.** Offgas from process treated in acid scrubber, filtered through activated carbon, and discharged to the atmosphere; no hold-test-release step.

in Erickson, 1997) and concluded that the "humin" present in the soil, which is distinct from the humic acid, binds with the contaminants. Studies by Bollog (1992), Pennington et al. (1992), and Caton et al. (1994) concluded that the biodegradation of TNT in soil stops short of mineralization and that reaction products containing reduced nitro groups bind strongly to the humic fraction of the soil.

According to the technology provider, humic acid is an association of molecules that form aggregates of elongated bundles of fibers at low pHs and open flexible structures perforated by voids at high pHs. The voids can trap and adsorb both organic and inorganic compounds if the charges are complementary. Humic acid combines with organic compounds by electrostatic bonding (i.e., attraction of a positively charged organic cation to an ionized carboxylic or phenolic group), hydrogen bonding, and ligand exchange. In addition, the high concentrations of stable free radicals in humic acid are capable of binding organic compounds that can ionize or protonate to the cation form. The mechanisms that have been postulated for the adsorption of organic compounds include (1) Van der Waal's attractions, (2) hydrophobic bonding, (3) hydrogen bonding, (4) charge transfer, (5) ion exchange, and (6) ligand exchange. Van der Waals forces, which are involved in the adsorption of non-ionic and nonpolar compounds, are created by short-range dipole-dipole interactions. Van de Waals forces are additive in nature, and these forces between the atoms of the adsorbate and the adsorbent can create considerable attraction for large molecules. However, the data provided by ARCTECH indicate that the potential separation effect of this binding action is negligible for the chemical agents.

DESCRIPTION OF THE TECHNOLOGY PACKAGE

Disassembly of Munitions and the Removal of Agent/Energetics

Rocket Disassembly

The baseline disassembly process (described in Appendix C) is used except that rockets are sheared in 10 places to allow better access to agent and energetics. The normal baseline process shears the rocket in seven places. ARCTECH also stated that water jets could be used to aid in the removal of energetics from munitions, although details were not provided. After the agent is drained, the rocket parts are sent to a decontamination reactor containing a-HAX solution, which is caustic and dissolves the aluminum parts, helping to eliminate residual agent. The solution also attacks the energetics and hydrolyzes them.

Projectile/Mortar Disassembly

For the demilitarization of the 155-mm, VX-filled M121A1 projectile, ARCTECH uses the baseline disassembly process up to the point at which munitions

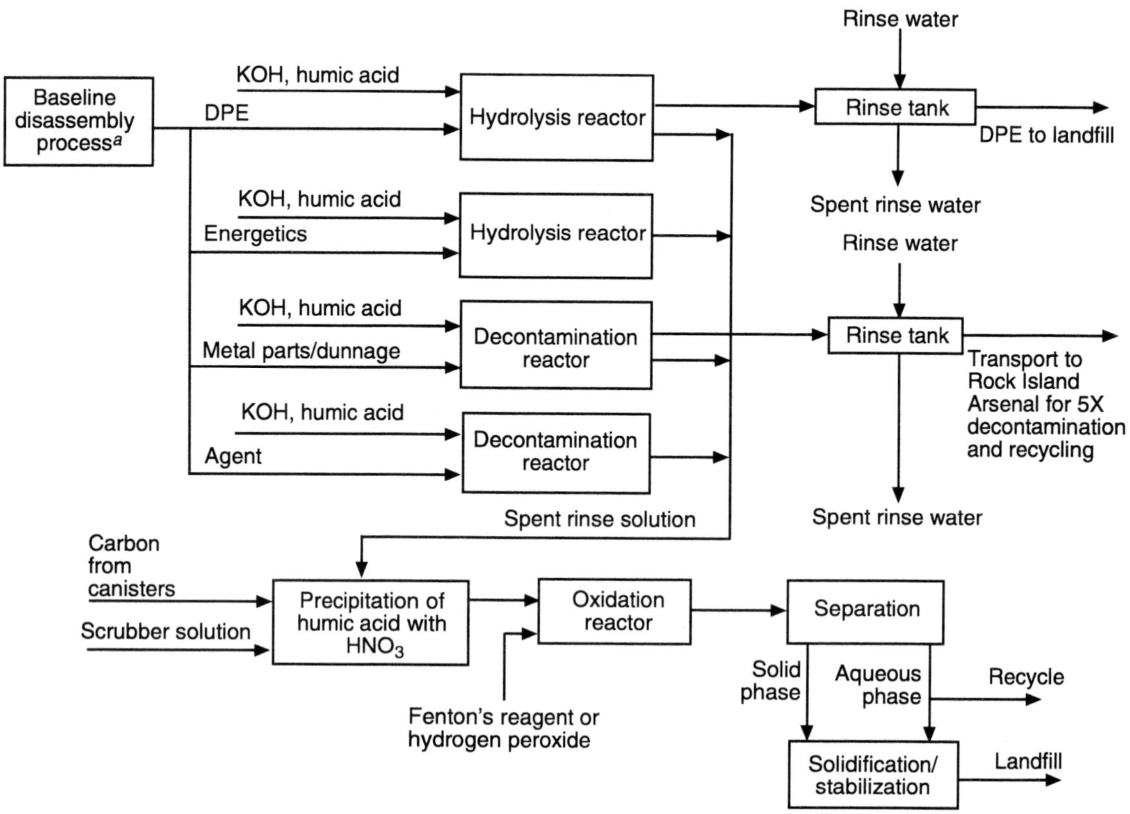

[a]Rockets are sheared in 10 places and water-jet cutting or washout may be included.

FIGURE 4-1 Schematic drawing of ARCTECH's ACTODEMIL process.

are conveyed to the projectile/mortar disassembly machine where the nose closure, supplementary charge, and burster are removed. Here, the first change from the baseline disassembly process occurs. The nose closure, supplementary charge, and burster parts drop down a chute equipped with a blast gate into the energetics reactor (filled with a concentrated a-HAX solution), rather than into the deactivation furnace system. The solution is caustic and reacts with the aluminum parts (e.g., the burster casing) to dissolve them, which eliminates the need to open them. The solution then attacks the energetics and hydrolyzes them.

The munitions leave the explosion containment room through another blast gate and enter the multipurpose demilitarization machine. Here, the burster wells are pulled from the munitions, and the chemical agent is drained through a suction tube. The agent is pumped to the agent reactor where it is treated. After the agent is drained, the burster well is crimped to prevent it from becoming reseated into the shell. The assembly is then sent to the metal parts/dunnage reactor.

Land Mine Disassembly

The baseline disassembly process is used up to the point at which the main explosive charge is removed. The mine is then sheared into four parts, which opens the remaining burster charge for access by the a-HAX solution. The mine parts, along with the fuze parts removed earlier, are fed to the metal parts washing tank.

Treatment of Chemical Agent

Destruction of GB and VX is carried out by alkaline hydrolysis using a starting reaction mixture containing 20 percent KOH. Mustard is hydrolyzed at a neutral pH or in slightly acidic conditions, again using KOH for pH control. Humic acid is present in the reaction mixture during agent hydrolysis.

The technology provider claims that humic acid has several properties that are favorable for the treatment of agents and energetics: it promotes hydrolysis in a reducing environment; it adsorbs organic compounds;

it chelates metals; and it reduces oxidized metal species. Humic acid can easily be precipitated by lowering the pH to less than 2.0. ARCTECH believes that the humic acid exists as a colloid at a pH of less than 2.0, and reducing the pH to below 2.0 causes the colloidal particles to agglomerate.

ARCTECH claims that the a-HAX solution effectively and irreversibly hydrolyzes chemical agents and energetic materials, converting them into nontoxic compounds. In the case of VX, the reaction is expected to destroy (or convert to other materials) 99.9999 percent of the agent in six hours. EA-2192 may be present temporarily as an intermediate but is not present in the product stream.

Neutralization, or hydrolysis, of chemical agents with a strong base has been documented extensively in the open literature and is the method the Army has selected for the destruction of HD and VX stored in bulk containers at Aberdeen, Maryland, and Newport, Indiana (see Appendix D). The most significant study of agent hydrolysis using ARCTECH's approach was done by GEOMET in early 1998 to assist ARCTECH in responding to the Army's list of data gaps. The study includes tests of the hydrolysis of GB, VX, H, HD, and HT. Two-liter glass flasks were used as reactor vessels. For GB and VX, initial experiments involved 100 g of agent treated in the presence of large quantities (240 g) of KOH, 280 ml of a-HAX solution, and 650 ml of added water. Temperatures of 90°C (Run 1) and 70°C (Run 2) were used. In a third set of experiments, the 650 ml of added water was replaced with 280 ml of reaction liquor from the first set of reactions and 280 ml of reaction liquor from the hydrolysis of energetics. The purpose of this replacement was to simulate steady-state plant operations in which the residual reaction liquor from one batch would be used for the next batch. In a final set of experiments (Run 4), only KOH and fresh water were used (no humic acid).

Based on results from previous studies, the hydrolysis reactions for mustard, including H, HD, and HT, were run at near neutral conditions. KOH was added only to keep the pH above 5. For H, only one reaction, at a temperature of 90°C, was carried out.

For each run, agent was continuously pumped into the reaction flask at a rate of 8.6 ml/hr until a total of 100 g of the agent had been added over a period of 12 hours. The reactions were allowed to continue for a maximum of 24 hours. The experimental conditions for the tests at 90°C are shown in Table 4-2 and the results are shown in Table 4-3. The appearance of an organic phase in the GB and VX hydrolysates has been reported by others (see Appendix D) but was not noticed by GEOMET. Destruction of all agents was complete after 24 hours to below detection limits shown in Table 4-3.

ARCTECH precipitates the humic acid by adding nitric acid to reduce the pH to less than 2.0, then destroys the reaction products using oxidation with Fenton's reagent (i.e., hydrogen peroxide plus ferrous ion) for GB and VX, and hydrogen peroxide for mustard. Some of the humic acid, perhaps 10 percent, is destroyed by the hydrogen peroxide during this step. ARCTECH claims that the presence of humic acid inhibits the formation of gaseous reaction products during the destruction of GB and VX.

In the full-scale process, two jacketed, 4,000-gallon stainless steel reactors are used in a two-day cycle. The temperature is held at 90°C, and the operating pressure is atmospheric. The reactors are fitted with internal and external mixing systems, with the external system using a static mixer. The reactors also have external cooling loops and reflux condensers. Vented gases from the

TABLE 4-2 Summary of Experiments by ARCTECH with Agents at 90°C. The agent addition rate was 8.6 ml/hr.

Agent	Reactants	Mass or Volume
GB		100 g
	water	650 ml
	dilute a-HAX (with 240 g KOH)	280 ml
VX		100 g
	water	650 ml
	dilute a-HAX (with 240 g KOH)	280 ml
H		100 g
	KOH	75 g (to maintain pH > 5)
	water	280 ml
	dilute humic acid	560 ml
HD		100 g
	KOH	66 g (to maintain pH > 5)
	water	280 ml
	dilute 3% humic acid	560 ml
HT		75g
	KOH	50 g (to maintain pH > 5)
	water	280 ml
	dilute 3% humic acid	750 ml

Source: ARCTECH and ICF Kaiser, 1998.

TABLE 4-3 Results of Analysis for Residual Agent during ACTODEMIL Neutralization Process at 90°C

Agent	Sample Designation	Residual Agent Concentration (mg/L)	Residual Hydroxide Concentration	Gas Evolved	Conclusions
GB	L1 at t = 0 hr	< 20	4.43 M	None (agent < 0.5 ppb)	Reaction complete by 12 hours; no gas evolved; all agents destroyed; KOH left at end of reaction.
	t = 6 hr	< 20	3.63 M		
	t = 12 hr	< 20	2.61 M		
	t = 24 hr	< 20	2.50 M		
VX	L1 at t = 0 hr	< 20	4.95 M	None (agent < 0.5 ppb)	Reaction complete by 12 hours; no gas evolved; all agents destroyed; excess KOH left at end of reaction.
	t = 6 hr	128	4.54 M		
	t = 12 hr	145	3.97 M		
	t = 24 hr	< 20	4.29 M		
HD	L1 at t = 0 hr	< 20	No hydroxide used in the tests except for pH adjustment.	None (agent < 0.5 ppb)	Complete destruction of agent at end of 24-hour period; no gas evolved.
	t = 6 hr	200,000			
	t = 12 hr	140,000			
	t = 24 hr	< 20			
HT	L1 at t = 0 hr	< 20	No hydroxide used in the tests except for pH adjustment.	None (agent < 0.5 ppb)	Complete destruction of agent at end of 24-hour period; no gas evolved.
	t = 6 hr	230,000			
	t = 12 hr	130,000			
	t = 24 hr	< 20			
H	L1 at t = 0 hr	< 20	No hydroxide used in the tests except for pH adjustment.	75 ml (agent < 0.5 ppb)	Complete destruction of agent at end of 24-hour period; minimal gas evolved; no unreacted agent in gas stream.
	t = 6 hr	420,000			
	t = 12 hr	310,000			
	t = 24 hr	< 20			

Source: ARCHTECH, and ICF Kaiser, 1998.

reactors and ventilated air from the reaction area are directed to a gas scrubbing system that contains both acid and caustic scrubbers. From these scrubbers, the gases are directed to activated carbon filters for further treatment.

The reactors are initially filled with an aqueous reaction solution prepared by mixing humic acid with either fresh water or recycled reaction liquor from which the humic acid has been precipitated. KOH is added for the hydrolysis of GB and VX. Agent is fed into one of the reactors as it is drained from the munitions. At the end of the shift, the feed is secured, and the reaction is allowed to proceed to completion. A six-hour reaction time will ensure the destruction of any residual EA-2192 from VX treatment. On the following day, the contents are tested. If the reaction is complete, the reactor contents are drained into a spent reaction solution tank, and small plastic and metal parts are removed. The drained reactor is then filled with fresh reaction solution. Thus, on any given day, one of the two reactors is being used for agent destruction, while the other is being tested, drained, and refilled.

Treatment of Energetics

The technology provider has conducted substantial tests on bulk energetic materials, and test results indicate that all of the energetic materials can be completely destroyed. Hydrolysis of nitrocellulose, cyclotetramethylene-tetranitramine (HMX), RDX, TNT, nitroglycerin, and other aliphatic nitrate esters, nitramines, and nitoaromatic type propellants and explosives has been demonstrated.

In laboratory-scale studies, complete denitration of the nitrocellulose was obtained in a matter of minutes at optimum operating conditions. No solid residue remained. The product was analyzed for a broad range of volatile and semivolatile compounds, and no hazardous products were found at levels of concern. The product was also characterized and formulated into a fertilizer product, which was evaluated in tests of plant growth.

Pilot-scale tests using 100 lb batches have been carried out on single, double, and triple base propellants. All of the energetic compounds were destroyed, yielding products that complied with RCRA, TCLP, and Universal Treatment Standards (UTS) levels. ARCTECH conducted these pilot-scale tests at the Hawthorne Army Depot using a stirred tank reactor under a contract with the U.S. Army Industrial Operations Command. The jacket of the reactor was used to circulate either steam or cold water, and the reactor was equipped with an anchor agitator.

The double-base propellant tested was type M26 from 105-mm shells.[1] The M26 propellant was successfully denitrated, with greater than 99.9999 percent removal of nitro groups within 24 hours. Tests of the products for TNT, RDX, dinitrotoluene, nitroglycerine, and nitroguanadine were negative. Ammonia was evolved during the reaction.

Propellants tested in the laboratory included IMR (a single-base, small-arms propellant), WC 830 (double-base, small-arms propellant), Hercules 2400 (double-base, small-arms propellant), Hercules RS nitrocellulose (neat powder), and M30 (triple-base propellant). Laboratory studies showed that the reaction was complete in a matter of minutes for nitrocellulose; all of the nitrocellulose was consumed, leaving no solid residue. The cyanide concentration in the product was below the detection limit of 100 mg/kg. Similar results were achieved for other explosives and propellants, including RDX, HMX, nitroglycerine, TNT, dinitrotoluene, and tetryl.

Energetics are removed from the munitions and handled differently depending on the particular munition. If the energetic material is in an aluminum container (e.g., some bursters), the container and energetic material are fed to the hydrolyzer. The concentrated KOH solution dissolves the aluminum and then attacks the energetic material. (See Appendix E for possible problems with burster material, such as tetrytol, where the reaction could be slow unless the energetic material is cut up further.) If the energetic material is in a steel container, it is washed out as a slurry, using KOH solution. The slurry is then fed to the hydrolyzer.

In the proposed full-scale process, two double-walled stainless steel reactors are used in a two-day cycle. The temperature is held at 80°C, and the operating pressure is atmospheric. The reactors are fitted with mixers and reflux condensers. One reactor is initially filled with concentrated a-HAX solution, and energetics are fed into it as they are removed from the munitions. At the end of the shift, the feed is secured, and the reaction is allowed to proceed to completion.

On the following day, the contents are tested and drained into the spent a-HAX tank, and small plastic and metal parts are removed. The reactor is then refilled with fresh concentrated a-HAX solution. On any given day, one of the two reactors is being used for the destruction of energetics, while the other is being emptied and refilled.

Treatment of Metal Parts

Decontamination of metal parts, along with the dunnage (but not DPEs), is carried out in double-walled, jacketed, stainless-steel reactors. High pH (20 percent KOH) a-HAX solution is used for parts contaminated with GB and VX. A near-neutral aqueous solution is used for parts contaminated with mustard. The processing temperature is 90 to100°C, and the operating pressure is atmospheric. The reactors are fitted with mixers, cleaning lances, and reflux condensers. Two process trains comprising three reactors each are installed, with each train sized for one days output of metal.

First, one tank is filled with the appropriate decontamination solution. For GB- and VX-contaminated materials, concentrated a-HAX solution is used for rocket parts, and dilute a-HAX solution is used for other parts. Water/humic acid is used for mustard-contaminated parts. Metal parts from munitions are conveyed to and then immersed in the tank using charge cars in a remote operation. At the end of the shift, the feed is secured, and the reaction is allowed to proceed to completion. A reaction time of 6 to 12 hours is planned. The parts are removed after treatment and combined with the small metal and plastic parts removed from the agent and energetic destruction tanks and rinsed with an unspecified amount of clean water. In concept, the treated metal parts/dunnage should comply with the 3X standard for decontamination and will be transported to the Rock Island Arsenal for final treatment to the 5X level. Small-scale tests carried out by ARCTECH indicate that this approach is technically feasible.

[1]The shells were described as follows: extruded cylindrical grains, 7 perforations, 0.219″ diameter, 0.518″ long; nitrocellulose 67.25 percent; nitroglycerin 25 percent; ethyl centrallite 6 percent; barium nitrate 0.75 percent; graphite 0.3 percent.

ARCTECH is considering using water-jet cutting for accessing mine initiator energetics. Fuzes are aluminum-bodied and will dissolve in high pH solutions. The technology provider was not certain whether or not azide would be destroyed by a-HAX solution.

Treatment of Dunnage (and Protective Suits)

Contaminated dunnage is treated to a 3X level using the metal parts decontamination lines. ARCTECH shreds or cuts DPE suits and contaminated pallets and treats these materials by immersion in a reactor containing dilute a-HAX solution. The reactor is a double-walled stainless-steel tank held at 90°C and operated at atmospheric pressure. It is fitted with an agitator and a reflux condenser. Nonmetal materials (including DPE suits) are sent to a hazardous-waste landfill for disposal, and cleaned metal is sent to the Rock Island Arsenal for treatment to 5X.

Process Instrumentation, Monitoring, and Control

Operating conditions, such as temperature, pH, pressure, and gas and liquid flow rates will be monitored using standard industrial equipment.

Feed Streams

In addition to the chemical munitions, which contain agent, energetics, and metal parts, several other major chemical feed streams are involved in the process: KOH for hydrolysis; humic acid; nitric acid for pH control (to precipitate humic acid); Fenton's reagent, an aqueous solution of hydrogen peroxide and ferrous sulfate, for final oxidation of most of the organics; and cement for stabilization of most of the final product. The required weight of these chemicals greatly exceeds the weight of agent treated.

For example, the chemicals required to treat a sample mass of 100g of VX are shown in Table 4-4. The quantities of other materials required to treat (1) agents other than VX and (2) energetics will be different from those listed in the table, but the amounts will be large. (No plan has been presented for handling grit and fine solids from water-jet cutting.)

TABLE 4-4 Materials Required for Processing 100 g of VX Using the ACTODEMIL Process

Material	Mass Required (g)
VX	100
KOH	234
Humic acid	28
HNO_3	225
H_2O_2	1,740[a]
Cement	10,000+[b]

[a]30% solution (estimated).

[b]Estimated assuming a concrete mixture of 50% cement and 50% water.

Waste Streams

The major waste stream produced by the ACTODEMIL treatment process is the product of the stabilization operation, incorporating all of the effluent from hydrolysis/hydrogen peroxide treatment.

Two products result from hydrolysis: precipitated humic acid containing adsorbed salts and organics; and a solution of salts and organics. Both are treated with Fenton's reagent to oxidize the remaining organics completely without oxidizing more than 10 percent of the humic acid. Spent activated carbon from the air purification system is added, as is spent scrubber solution from off-gas treatment systems. Both the liquid and solid streams are then stabilized with cement. Neither the peroxide treatment nor stabilization with cement has been demonstrated.

Other waste streams are: decontaminated DPE suits and other dunnage; metal parts cleaned to a 3X condition and packaged for off-site shipment; and solids from water-jet cutting (if used).

Air and fumes from various parts of the process are collected and treated prior to discharge. Sources include the explosion-containment vestibule, explosion-containment room, toxic cubicle, demilitarization room, and reactors. The gases evolved during the energetic destruction process, including ammonia, amines, and nitrogen oxides, are removed from the gas stream by the acid scrubber. The gas treatment is similar to the treatment planned for the Aberdeen/Newport neutralization processes in which gaseous emissions are first scrubbed with dilute nitric acid solution to remove any amines and then scrubbed by caustic solution to remove any residual agent. Mustard contains volatile low molecular weight chlorinated hydrocarbons that are difficult to hydrolyze or to remove with caustic or acid scrubbers or with activated carbon adsorbers. ARCTECH does not address the management of these compounds in their process.

The gas then passes through a chiller to reduce the water vapor content and a gas heater to elevate the temperature above the dew point. Treatment with activated carbon filters follows. The filtered air then passes through a baseline heating, ventilating, and air-conditioning filtration system for final treatment and is then discharged to the atmosphere.

The mass balances vary with the munition being treated, and the process is still under development. A sample mass balance based on small-scale laboratory tests is shown in Table 4-5. This sample covered only the hydrolysis step and did not include the final oxidation with hydrogen peroxide (still to be demonstrated) or stabilization with cement.

ARCTECH did not estimate the amount of cement required to stabilize the entire product (salts plus water). However, the amount will be large (e.g., 10,000 parts cement per 100 parts original VX).

Start-up and Shutdown

The processes unique to ARCTECH are relatively simple. The hydrolysis reactors are the most complex because the contents are first heated with steam to the operating range (80 to 100°C, depending on the reaction) and then maintained at the desired temperature with chilled water. Shutdown involves only draining the reactors and turning off the heating/cooling system. If a process upset occurs, the reactions in the hydrolysis reactors will continue, so the process cannot be quickly shut down.

EVALUATION OF THE TECHNOLOGY PACKAGE

Process Efficacy

Effectiveness

Laboratory and pilot-scale tests have shown that the agents and energetic materials are completely destroyed. Hydrogen peroxide or Fenton's reagent is used to complete the destruction of agents.

Like the Army's neutralization process selected for the Aberdeen, Maryland, and Newport, Indiana, sites, the ACTODEMIL technology utilizes base hydrolysis. The Army has established the efficacy of the neutralization technology at both laboratory and pilot scales and is now moving forward with implementation of a full-scale system for the destruction of bulk agents at those two sites.

The 1998 GEOMET study (ARCTECH and ICF Kaiser, 1998) showed that agent is effectively destroyed using ARCTECH's hydrolysis approach. Moreover, no Schedule 1 compounds were found in reaction product streams. However, several Schedule 2 compounds were detected in the liquid and solid fractions after humic acid precipitation. For GB, the compounds detected included isopropyl methylphosphonic acid, methylphosphonic acid, diisopropyl methylphosphonate, and dimethyl methylphosphonate. Similar results were obtained for VX, with EMPA, methylphosphonic acid, and diisopropyl methylphosphonic acid detected. Thiodiglycol, 1,4-oxathiane and 1,4-dithiane were found in products from the hydrolysis of HD, HT, and H.

The 1998 GEOMET study also evaluated the use of hydrogen peroxide or Fenton's reagent to destroy Schedule 2 compounds, and the results were inconclusive. Thiodiglycol was destroyed to below detection limits. The Schedule 2 compounds from hydrolysis of VX and GB were reduced in concentration but not to below detection limits. This issue must be addressed.

ARCTECH's research on the treatment of energetics has ranged from laboratory-scale tests on a variety of energetic materials to larger scale testing on bulk propellants and energetics. Large-scale testing on single-, double-, and triple-base propellants at Hawthorne Army Depot had favorable results. With the exception of single-base propellant, the reactions were apparently rapid and resulted in the complete destruction of the energetics. No solid residues remained, and the reaction products were nonenergetic and nontoxic after 24 hours residence time. Ammonia was evolved during the reactions.

To test for treatment of PCBs, a 10,000 ppm sample of Aroclor 1253 oil was placed on a sample of aluminum. The sample was washed with a-HAX solution for 30 seconds. The a-HAX solution was then acidified with nitric acid, which precipitated the humic acid. No

TABLE 4-5 Feed and Product Masses and Concentrations for Hydrolysis of 100 g of VX in the Presence of Humic Acid

		Product Streams					
Total Feed		Liquid			Solid		
VX	100 g	EMPA	46.4 g	3.21%	EMPA	1.20 g	2.51%
KOH	234 g	Thiol	28.8 g	1.99%	Thiol	0.06 g	0.13%
HA	28 g	MPA	4.2 g	0.29%	MPA	0.11 g	0.23%
Water	908 g	"Pentanami de"	15.8 g	1.09%	Chloride	0.34 g	0.70%
HNO_3	225 g	Water + Salts	1,351.8 g	93.42%	HA	28.00 g	58.33%
					Water	18.29 g	38.10%
Total	1,495 g	Total	1,447.0 g	100%	Total	48.00 g	100%

Source: ARCTECH and ICF Kaiser, 1998.

PCBs were found in the remaining liquid. Chloride ion concentration in the reaction mixture increased during the reaction, providing evidence of PCB dechlorination, but the completeness of dechlorination was not measured. The technology provider concedes that adsorption of PCBs on the precipitated humic acid is possible and that these adsorbed PCBs could be thermally desorbed. But this has not been tested.

The unique aspect of this technology is the use of humic acid to separate products of agent hydrolysis from solution. The separation factors[2] for the products (i.e., concentrations on humic acid vs. concentrations remaining in solution) are, therefore, important parameters for determining the effectiveness of humic acid. These concentrations can be obtained from material balances.

One separation discussed in the literature is for atrazine/humic acid in ethyl alcohol (Sanjay and Fataftah, 1997). The separation factor, 375, calculated from the data is very large and represents a very selective separation. The concentration of atrazine was very low, however, only about two ppm (by weight) in the solution. The separation factor would be expected to decline as the concentrations increased.

Tests on agents with preliminary mass balances have been presented in the ARCTECH data-gap resolution technical report (ARCTECH and ICF Kaiser, 1998). Separation factors can be calculated from the reported data. The concentration levels of the materials of interest are not very high (e.g., 2 to 3 percent) (see Table 4-5), although they are much higher than in the "atrazine" case noted above. There were two product streams from the experiment: a wet humic acid (solid) stream from the centrifuge containing hydrolysate product; and a liquid solution. The total feed and two product streams are summarized in Table 4-5. (Some minor components have been omitted).

None of the products normally expected from the hydrolysis reaction appears to have concentrated on the HA-solid. For example, EMPA concentration is 3.2 percent in the liquid, 2.5 percent on the solid. If the EMPA concentration is recalculated assuming that it is simply dissolved in the water associated with the wet solids, its concentration in the liquid is then 2.9 percent (virtually the same as in the bulk solution).

The data for other major components is hard to explain. The thiol (a major product of VX hydrolysis) appears in the product liquid but is reported to be almost nonexistent on the solid. The committee postulated that perhaps this is because thiol was not extracted by the analytical procedure.

It appears that a relatively small amount of humic acid will be used in the process (the a-HAX solution is 3 percent humic acid). As a consequence, the final solid product—solid humic acid with any adsorbed salts and water—is a very small fraction of the total. For example, as Table 4-5 shows, the liquid product is 1,447 g (96.8 percent of the total product), and the amount of EMPA in the liquid is 46.4 g. In contrast, the mass of solid product is only 48 g (3.2 percent of the total product), of which 1.2 g is EMPA.

Two conclusions can be drawn from these data: (1) separation factors appear to be very modest for

[2] The separation factor is defined as $a=[y/(1-y)]/[x/(1-x)]$, where y and x are concentrations of the material of interest in the two phases. For this application, y represents the solid phase, and x represents the liquid phase.

some important components; the separation factor for EMPA, for example, is $\alpha \approx 1$ (essentially no separation); and (2) even if the separation factor were large, there is not enough HA-adsorbent present to make a significant difference (i.e., the solid product is a very small fraction of the effluent stream).

The final process step is oxidation with Fenton's reagent (H_2O_2). Reported data show EMPA is reduced to either the ppm range or below the quantifiable limit. Unfortunately, the amount of H_2O_2 was not given, but, it can be estimated assuming that the organics are oxidized to CO_2, H_2O, etc. The theoretical requirement for VX oxidation is 522 g of H_2O_2 for 100 g of VX (i.e., for the example mass balance in Table 4-5). This corresponds to 1,740 g of 30 percent H_2O_2 solution. Thus, hydrogen peroxide will be a major input to the process.

Sampling and Analysis

ARCTECH has not provided a definitive sampling and analysis plan. However, the proposal states that agent will be monitored at the outlet of the off-gas vent from the overall process, at the air intake for the operation, and in worker activity and observation corridors. The agent monitoring techniques are the same as those in the Army's baseline process and at Aberdeen and Newport. Monitoring of agent in aqueous reaction matrices mimics techniques used at Aberdeen and Newport. The committee does not anticipate that new analytical techniques will be necessary.

ARCTECH's proposal states that the reactors used for hydrolysis of agent and energetics will be operated batch-wise on a 24-hour cycle. The reaction mixture from the previous batch will be sampled at the beginning of this cycle, which provides 6 to 12 hours of reaction time after the last addition of agent or energetics to the reactor. However, the process was changed after the proposal was submitted. Hydrogen peroxide or Fenton's reagent could now be added to the hydrolysis reactors after the hydrolysis reaction is completed. Whether or not this would lengthen the reaction cycle or change the sampling plan is not clear.

ARCTECH's proposal also mentions measurements of pH and oxidation/reduction potential and states that decontaminated dunnage and DPE will be tested for agent before being sent off site for disposal.

Maturity

The ACTODEMIL technology is at an early overall stage of development. The hydrolysis of agent with a-HAX has been done only on a small bench scale. Hydrolysis of certain energetics has been done on a much larger scale (i.e., hundreds of pounds of energetics). Many questions about the effect of using humic acid have not been answered. The proposed oxidation step using H_2O_2 has not been demonstrated. Although the water balance is still uncertain, continued development would most likely show the generation of a large aqueous waste stream, which would have to be stabilized and shipped to a landfill.

Robustness

Robustness—the ability of a process to accommodate wide variations in the composition and quality of feedstock—is difficult to assess at this early stage of development. Because the complete removal of agent and energetics from munitions might be difficult, the technology provider proposes shearing rockets in 10 places instead of the seven places in the baseline technology. Water jets may be used to help remove energetics from munitions. The opened munitions would be immersed in an a-HAX solution to hydrolyze the energetics and any remaining agent. Preliminary tests of this procedure were successful, but whether or not this approach provides adequate and sufficiently rapid destruction of agent and energetics under all expected operating conditions remains to be demonstrated.

Tests indicate that once the agent and energetics have been removed from the munitions, they can be successfully hydrolyzed. The ACTODEMIL process is a batch process, and the reaction time can be extended if agent and energetics are not destroyed in the normal operating period. Therefore, the robustness of the hydrolysis step is expected to be adequate.

The remainder of the process, comprising humic acid precipitation, chemical oxidation, and solidification or stabilization, is not sufficiently developed for a meaningful assessment of robustness. Studies on a larger scale and studies under off-normal conditions should be done.

Monitoring and Control

Because of the mild reaction conditions and the conventional nature of the equipment for this technology, monitoring and control should be straightforward and conventional.

Applicability

The process as currently developed can destroy all types of agent without generating other Schedule 1 compounds. Certain Schedule 2 compounds will be produced, however, and their complete destruction remains to be demonstrated. Complete destruction of several types of explosives and propellants has been demonstrated. The process can also decontaminate agent-contaminated metal parts and DPEs. Dissolution of aluminum parts, but not steel, has also been demonstrated. The ability of humic acid to adsorb reaction residues selectively was not demonstrated. Overall, the concept appears to be applicable to all munition types, but because data are lacking in several key areas (e.g., destruction of Schedule 2 compounds, feasibility of recycling process liquor), it is impossible to draw definitive conclusions about the applicability of the process.

Process Safety

The unique equipment for this process is associated with neutralization, water-jet cutting (if used), oxidation, and solidification. Reverse assembly, water-jet washout and cutting, as well as energetics and agent neutralization, will take place in an explosion-containment room to minimize the potential effects of explosions during the handling and processing of agent and energetics. The hydrolysis processes will operate at low temperature (80 to 100°C) and ambient pressure, and hydrolysis and oxidation will be performed in a batch mode. Thus, the effectiveness of treatment can be ascertained prior to release for solidification. Solidification is a routine industrial process that occurs downstream of the primary and secondary detoxification processes, which include a hold-test-release operation. The solidification system should pose no unique hazards.

Worker Health and Safety

The hazardous materials unique to this process are potassium hydroxide, hydrogen peroxide, and nitric acid, all of which are widely available and widely used industrial chemicals. If a process upset occurs, the incomplete hydrolysis products will be very hazardous. Procedures are expected to be established for safe shutdown and restarting of the process. The air effluent during an upset will continue to be treated by the scrubbers and the activated carbon filters.

The disassembly processes are derivatives of the baseline processes and are not considered to create new or increased risks to workers. Water-jets are being considered for washing out energetics, and water-jet cutting may be used to make mine-initiator energetics more accessible. As discussed in Appendix G, water-jet washout and cutting are commonly used systems and should not pose unique hazards in this application because they will be used in the explosion-containment room.

Public Safety

Release of agent or other regulated substances in plant effluents is expected to be extremely unlikely. The destruction of agent and energetics will be verified after Fenton's reagent or hydrogen peroxide is added to the hydrolysate in a hold-test-release operation. The air effluent from all processes and areas will be continuous and will be cleaned using baseline acid and caustic scrubbers and activated carbon adsorption. No hold-test-release operation is contemplated for the air or gaseous effluents. Instead, these streams will be monitored continuously.

ARCTECH claims that any flammable gases generated in the neutralization process will be adsorbed by the humic acid, but this has not been thoroughly demonstrated. Therefore, the vent system should be designed and operated as though flammable gases will be present.

The technology provider claims that the reactions between the a-HAX solution and the agents are mildly exothermic. The reactors will have water-filled jackets for heating or cooling. Off-gases will pass through reflux condensers cooled with chilled water. The Army's studies on VX have shown that the heat released during the hydrolysis reaction is moderate and that failure of the cooling system to remove the heat of reaction will cause only a 5 to 8°C temperature excursion. Because of the similarity of ACTODEMIL to the Army's hydrolysis technology in terms of the basic reaction during the destruction of chemical agents, ACTODEMIL is not expected to result in any catastrophic thermal excursions. This same analysis also applies to the hydrolysis of other agents. However, reactions involving hydrogen peroxide and organic compounds are known to be violent and exothermic at times. This issue has not yet been addressed by the technology provider.

The maximum amount of agent or energetic that can be treated in a single batch in the reactor will be specified and will comply with the design basis requirements. An upset in the flow rate of a feed stream could cause minor changes in the reaction and less efficient destruction. This condition can be countered by holding a batch until it has been checked and extending the reaction time if necessary.

Based on laboratory-scale, pilot-scale, and demonstration tests conducted with energetics, ARCTECH has extensive data on the amount of heat released during their destruction. The ranges of values for heat released from single-, double-, and triple-base propellants have been established. One of the design features is that enough a-HAX solution is used for the reaction with energetics to ensure that the total heat energy released by destruction of the propellants and explosive compounds will not cause a thermal runaway. This will have to be verified experimentally.

ARCTECH plans to hydrolyze different types of energetic materials simultaneously in the same reactors. As discussed in Appendix E, the committee is concerned that this could lead to the formation of compounds that are both energetic and sensitive. Therefore, energetic materials should be processed in separate reactors unless testing shows that the formation of sensitive compounds does not occur.

Human Health and the Environment

ARCTECH claims that the process is environmentally sound because there are no large process gas streams. Formation of dioxin and furans is very unlikely because processing temperatures are less than 100°C. However, ARCTECH has not completely characterized the products of agent hydrolysis, and development of processes to completely destroy Schedule 2 compounds in agent hydrolysate has not been completed.

Characterization of Effluent and Impact on Human Health and Environment

Metal parts will be decontaminated to a 3X condition and sent to Rock Island Arsenal for further treatment. This will pose no unusual threat to human health and the environment.

The process is likely to produce large solid and aqueous waste streams that will be stabilized and shipped to a hazardous-waste landfill. Although similar waste streams are routinely handled in a manner that poses no threat to human health or the environment, a waste stream this large (about 1,200 tons per day) is likely to be viewed negatively by the public and by regulators.

Completeness of Effluent Characterization

The experimental results show that parts can be cleaned to 3X condition. Other effluents, especially those containing agent hydrolysate, are still poorly characterized.

Effluent Management Strategy

Sending the 3X decontaminated metal parts to Rock Island Arsenal is an acceptable procedure. Stabilizing a large aqueous waste stream and disposing of the stabilized waste in a landfill may raise concerns among regulators and the public.

Resource Requirements

The amount of water and hydrogen peroxide could potentially be large. Nevertheless, the projected amount of hydrogen peroxide is expected to be available from commercial sources.

Environmental Compliance and Permitting

This process could potentially produce large amounts of stabilized aqueous effluent and humic acid solids, which could be problematic when applying for a permit. Also, if the technology provider indicates that unusually large amounts of hydrogen peroxide will be required, the permitting issues surrounding the transportation of this material could be significant.

STEPS REQUIRED FOR IMPLEMENTATION

In order to develop and operate a full-scale ACTODEMIL system, the following steps will have to be taken:

1. Verify the advantages of using humic acid and complete the fundamental definition of the process.
2. Establish the quantity and concentration of hydrogen peroxide necessary to complete the destruction of hydrolysis products.
3. Run repeated hydrolysis reactions using recycled reaction liquor to determine the amount of reaction liquor that can be recycled under steady-state conditions.
4. Establish a water balance based on previous steps. Then, establish procedures and prepare mass balances for the disposal of precipitated humic acid, aqueous effluent, and the stabilized or solidified waste stream.
5. Perform larger scale demonstrations of all steps. To date, only energetics hydrolysis has been demonstrated at a substantial scale. Then, operate an integrated system that includes all steps of the process.

FINDINGS

Finding AR-1. The use of the Army's baseline disassembly method, with minor modification as proposed by ARCTECH, is anticipated to be effective for providing access to agent, energetics, and metal parts for treatment.

Finding AR-2. Hydrolysis is anticipated to be effective for destroying chemical agents.

Finding AR-3. Destruction of Schedule 2 compounds in the product streams from agent hydrolysis using Fenton's reagent or hydrogen peroxide has not been demonstrated.

Finding AR-4. The presence of humic acid during the hydrolysis of agent appears to have little or no effect on agent destruction, with the possible exception of decreasing the amount of gaseous products formed.

Finding AR-5. Hydrolysis is expected to be effective for destroying energetic materials because of the long residence times.

Finding AR-6. ACTODEMIL requires hydrogen peroxide in unknown but potentially large quantities to destroy the products from the hydrolysis reactions. This is expected to lead to large quantities of aqueous wastes, which will be stabilized/solidified and, therefore, will generate a large solid-waste stream (almost certainly RCRA hazardous) that will have to be disposed of in a landfill. This could cause extensive permitting delays.

Finding AR-7. The technology provider claims that humic acid preferentially binds with the by-products of the destruction reactions, allowing recycling of the reaction liquor. This claim was not substantiated by the data provided by the technology provider.

Finding AR-8. The ACTODEMIL technology for chemical weapons demilitarization is immature, and will require a good deal more to be done to define and demonstrate the process.

5

Burns and Roe Technology Package

INTRODUCTION AND OVERVIEW

The Burns and Roe team's technology package is shown schematically in Figure 5-1. This package utilizes the Army's baseline disassembly technology to separate the components of munitions (e.g., chemical agent, fuzes, bursters, metal casings, etc.), with subsequent treatment by high-temperature plasma to decompose the chemical agents, propellants, and wooden, fiberglass, or plastic packing materials. All metals, including the munition casings, are melted by the plasma. An explosion chamber is used to deactivate explosive components by energetic initiation (detonation or deflagration). Debris and gas from the explosion chamber are then also treated using high-temperature plasma. The technology provider's approach for performing the required major demilitarization operations is summarized in Table 5-1.

Because plasma waste treatment, which is integral to the proposed system, is a unique technology, plasmas and their characteristics are discussed first. The Burns and Roe technology package is then described in detail.

Background on Plasma

Electric arcs and discharges have been of interest to scientists and engineers for decades because they involve high-temperature, conductive gases (plasmas). Typically, an arc can be established between two conducting electrodes (e.g., graphite or metal) in a variety of atmospheres. The plasma is comprised of molecules, atoms, ions, and electrons at temperatures of 1,000°C to 20,000°C (1,832°F to 36,032°F) depending on the current and voltage, the gaseous environment, and the pressure of the constricting gas. Either physical or magnetic constriction can be used to increase temperatures.

Because plasma arcs between electrodes generally involve voltage drops of 100 V or more, chemical bonds (whose strengths range from 2 to 10 electron volts [eV]) will be broken, and ionization processes (at 4 eV to 25 eV) will occur. Thus, material exposed to a plasma environment will be transformed into atoms, ions, and electrons, with only a few molecules remaining. This makes the potential use of plasma arcs, torches, melters, and other plasma devices attractive for destroying undesirable molecules (e.g., hazardous wastes). High-temperature plasmas can also produce endothermic neutral species (e.g., C_2H_2, C_2N_2, and NO) or gaseous molecular ions (e.g., SiO^+ and CO^+). When the plasma is cooled to room temperature, most of the molecules are thermodynamically stable, but some metastable species (that are stable at higher temperatures but unstable at lower temperatures) might survive. In addition, metastable species could be formed during cooling, which could also be present at room temperature.

DESCRIPTION OF THE TECHNOLOGY PACKAGE

Disassembly of Munitions and the Removal of Agent/Energetics

The technology provider proposes using the baseline approach to disassemble munitions and segregate the agent, energetic materials, and munition bodies. (See

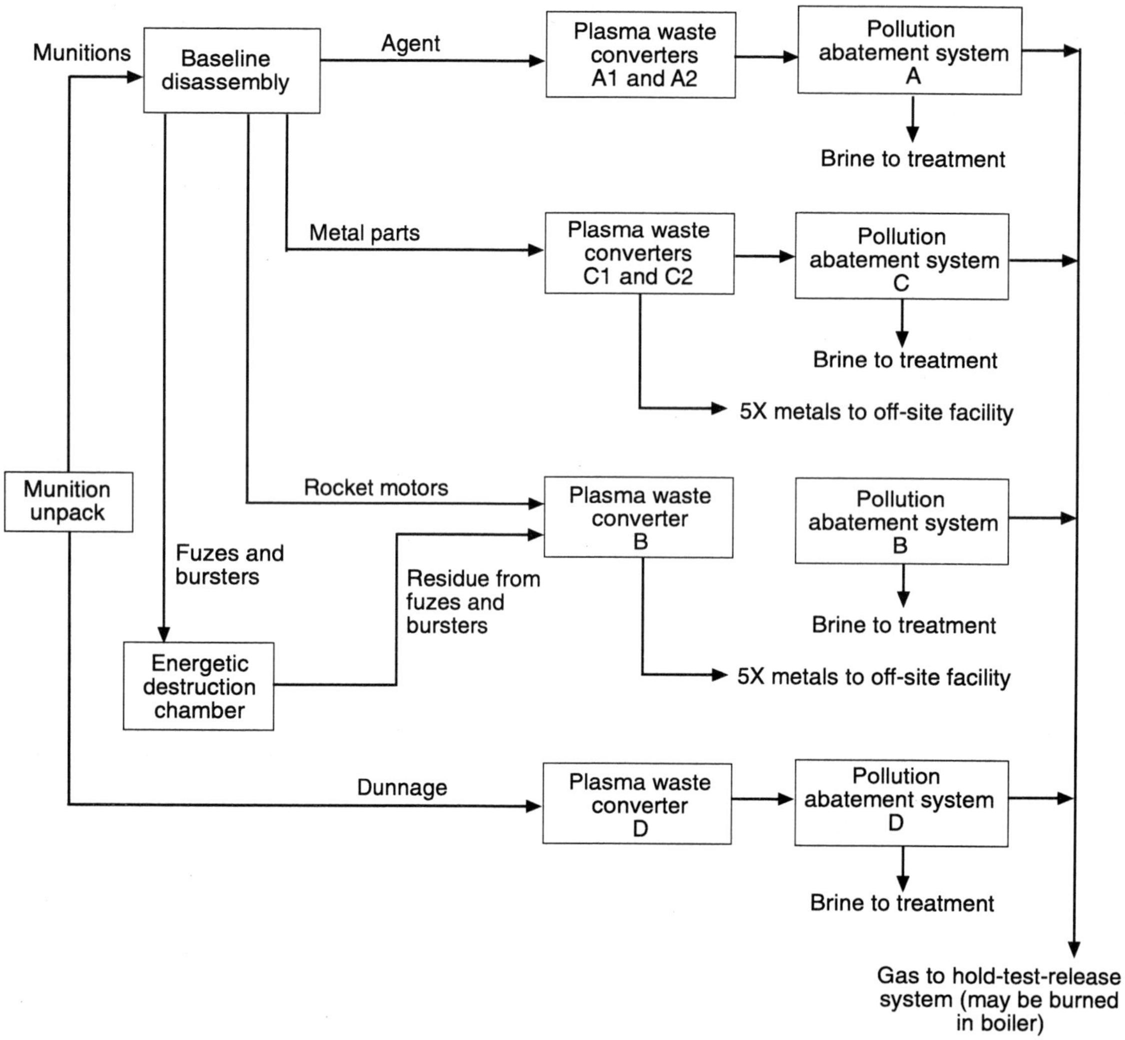

FIGURE 5-1 Schematic diagram of the Burns and Roe technology package.

Appendix C for a description of the baseline disassembly system.) The only modifications to the baseline disassembly process, which occur subsequent to removal of agent and energetics, are: (1) limiting the number of munition bodies per tray to nine, and (2) modifying the conveyors to accept smaller trays.

Description of the Plasma Waste Converter

The Burns and Roe technology package uses specialized plasma waste converters (PWCs) to treat all materials, including chemical agent. Six PWCs of four different types are proposed: two to treat agent and munition bodies (PWC A1 and PWC A2), one to treat pieces of rocket motors containing propellant (PWC B), two to treat metal parts to the 5X standard (PWC C1 and PWC C2), and one to treat dunnage (PWC D). The basic operation of all the PWCs is the same.

A typical PWC (Figure 5-2) is a cylindrical, refractory-lined vessel with an opening in the roof through which a plasma torch is inserted. (For larger PWCs, more than one torch may be used through more than one opening.) There are no airtight seals between the plasma torch and the vessel roof, and the PWC is operated at slightly negative pressure to prevent gas from exiting through the opening. Thus, air is always leaking into the PWC. Each plasma torch is a nontransferred torch[1] consisting of a cylindrical pipe containing water-cooled copper electrodes. The plasma

[1]In a nontransferred torch, both the anode and cathode are contained within the torch.

TABLE 5-1 Summary of the Burns and Roe Approach

Major Demilitarization Operation	Approach(es)
Disassembly of munitions	Army baseline disassembly process.
Treatment of chemical agent	Thermal destruction using plasma waste converter (PWC).
Treatment of energetics	Initiation of explosives in explosion chamber (residual passed through PWC); destruction of propellant in PWC.
Treatment of metal parts	5X treatment of metals (complete melting) in PWC.
Treatment of dunnage	5X treatment of dunnage in PWC.
Disposal of waste	**Solids.** Slag for recycling; salts from scrubbers to appropriately permitted landfill. **Liquids.** None. **Gases.** Hold and test; feed to boiler or thermal oxidizer (combustion) if test results are acceptable.

feed gas passes through the torch, and the plasma is formed in the torch between the anode and cathode. No other materials, such as agent or energetics, are introduced into the plasma torch.

The torch creates a plasma with a temperature, as reported by the technology provider, in the range of 15,000°C (27,032°F).[2] The plasma exits the torch into the PWC chamber and impacts onto solid and liquid material (e.g., metal from weapons) at the bottom of the chamber. In an agent-treatment PWC, agent is introduced into the hot plasma near the bottom of the PWC chamber. Steam is introduced with the agent at a controlled rate to convert elemental carbon or soot (created by dissociation of the feed stream molecules) to CO.

The plasma exiting the torch cools very quickly in the chamber by a combination of the following mechanisms:

- mixing with infiltration air
- sensible heat required to heat the waste feed to the PWC
- the endothermic chemistry of degradation of the agent and other organic waste materials introduced into the PWC
- decomposition of the steam introduced for soot control
- decomposition of the CO_2, if used as a plasma feed gas
- formation of NO_x from nitrogen in the weapons material feed, the plasma feed gas (if N_2), or infiltration air
- heat losses through the PWC shell

For the demonstration system, the technology provider indicates that the air in-leakage rate is approximately 30 standard cubic feet per minute (SCFM) (Burns and Roe, 1999), compared to a plasma feed gas flow rate of approximately 20 SCFM and a PWC total gas outflow on the order of 140 SCFM (Burns and Roe, 1998b). The temperature of the exit gas for the PWC is

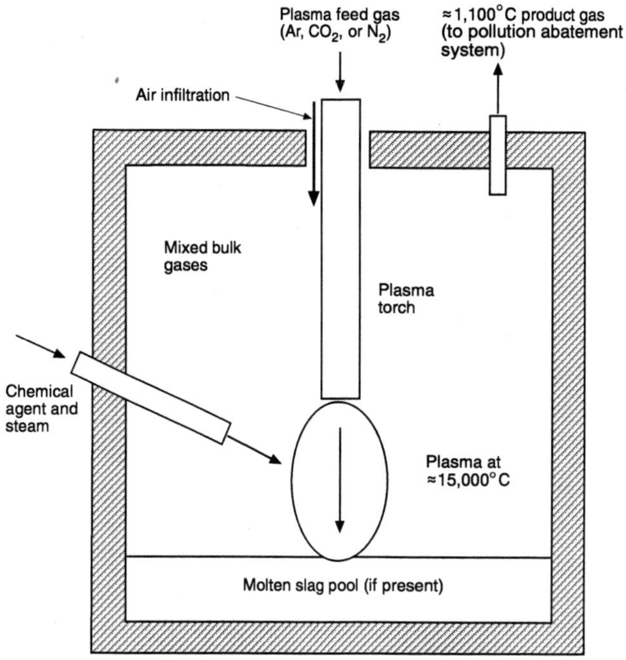

FIGURE 5-2 Schematic diagram of a typical plasma waste converter (PWC) for treating agent.

[2]Because plasmas can contain molecules, atoms, ions, and free electrons, several plasma temperatures can be defined. The committee considers the temperature listed here to be reasonably representative of the very high temperatures of the plasma components.

estimated to be approximately 1,100°C (2,012°F) (Burns and Roe, 1997).

Treatment of Chemical Agent

After being drained from the munitions, chemical agent is pumped to storage tanks and, subsequently, into one of two PWCs designated A1 or A2, identical units approximately 7 ft in diameter and 9 ft high (external dimensions) designed for liquid feed of up to 1,200 lb/hr. Agent and steam are injected into the PWC where they mix with the hot plasma. Product gas then passes through a pollution abatement system (PAS) consisting of four major components:

- a vertical down-flow duct with water/caustic spray to quench the gas
- a countercurrent, multistage acid-gas scrubber with vertical upward gas flow and gravity-driven, downward flow of scrubbing liquid
- a mist eliminator
- a cartridge filter for the removal of fine particulates

Upon exiting the PAS, the product gas enters a hold-test-release system consisting of a compressor, a condenser, and storage tanks. The tank contents are sampled for chemical agent and other components (not yet specified). If no agent is detected, the technology provider plans either to (1) burn the gas in an on-site boiler or oxidizer or (2) sell it as a fuel. If agent is detected, the gas is recycled to PWC A1 or A2 for reprocessing.

Treatment of Energetics

Two general types of energetic materials will be treated: M55 rocket propellant (designated M28), which is configured to burn rather than detonate; and burster and fuze materials, which are intended to detonate upon initiation.

M55 Rocket Propellant

After being sheared, the severed M55 rocket motor pieces containing the M28 propellant are fed into PWC B—7 ft in diameter and approximately 9 ft high (external dimensions)—designed to destroy energetic materials at rates of up to 1,500 lb/hr. The unit is fabricated of 2.5-inch-thick carbon steel to contain any accidental explosions. PWC B also includes a feed chute and hydraulic ram for introducing the waste material into the vessel.

The waste materials mix with the hot plasma, and product gas is discharged to the same PAS and hold-test-release system described previously. Some rocket-component materials (e.g., metal parts and fiberglass shipping and firing-tube pieces) do not remain in the plasma field long enough to vaporize but melt, forming a molten pool at the bottom of the vessel. The metals are tapped and drained to form ingots. Nonmetallic slag is also periodically tapped and drained.

Other Energetics

To lower the risk of detonations in the PWC, the technology provider proposes deactivating explosive material in an explosion chamber. This chamber is commercially available (designed and manufactured by Bofors) and is used by both the military and industry to deactivate small quantities of explosives. It is made of thick high-strength steel and is designed (1) to withstand multiple detonations of a specific mass of TNT, and (2) to contain the product gas from the detonations. Both the quantity and type of energetic material treated per batch must be known in advance to ensure that the unit's explosive rating is not exceeded.

In this application, the bursters and fuzes from rockets, projectiles, mortars, and land mines are fed into the explosion chamber, where they are thermally initiated. The gas from the explosion chamber is then slowly vented to PWC B for further treatment. The solids are removed and are also fed to PWC B. The gaseous and molten products from PWC B are treated in the same way as the M55 rocket products discussed in the preceding section.

Treatment of Metal Parts

Bodies of projectiles and land mines, drained of agent and emptied of energetic materials, are placed in trays and conveyed to PWC C1 or C2—identical PWCs designed to decontaminate and melt metal parts. The proposed units are 7 ft in diameter and 9 ft high (external dimensions), with a peak capacity of 6,000 lb/hr of metal munition bodies. Trays of munition bodies are moved by roller conveyor to a scissor lift that raises the trays to a feed chute. The tray is conveyed into the feed chute, and a ram pushes the tray containing the munition bodies into the PWC where both are consumed. Gaseous products flow through the PAS and hold-test-release system described previously. Molten metal collects in the bottom of the PWC, which is tapped periodically and the metal cast into ingots. This metal will be considered as treated to 5X condition because it will meet the criterion of "heated to at least 1,000°F for at least 15 minutes."

Treatment of Dunnage

Dunnage is gravity-fed to a shredder for size reduction and fed to PWC D (approximately 5 ft in diameter and 5 ft high [external dimensions]) designed to destroy dunnage at rates of up to 1,200 lb/hr. Gaseous products flow though the PAS and hold-test-release system described previously. Molten materials collected in the bottom of the PWC are periodically recovered. The technology provider proposes to control the formation of graphite, soot, and other carbonaceous material by adding steam to form CO.

Process Instrumentation, Monitoring, and Control

Monitoring of "traditional" process variables (e.g., temperatures, liquid and gas flow rates, etc.) is accomplished using standard, off-the-shelf, chemical-process equipment and instrumentation. Monitoring for agent is accomplished using the ACAMS and depot area air monitoring system (DAAMS) developed by the Army.

Agent feed to the PWC is monitored for flow rate and pressure, with real-time signals relayed to the control room. Automatic feed cutoff valves are employed if operational ranges (yet to be established) are violated. Sensors in the energetic feed chute and the munition metal-body conveyor detect blockage of feed material and initiate appropriate action, which could include PWC shutdown. (Because the feeds are not critical to PWC operation, automatic shutdown because of a feed blockage is not included.)

Inside the PWCs, power feed, plasma feed gas flow, vessel pressure and temperature, and steam flow are monitored and controlled to within established operating ranges. Appropriate pressure, temperature, and flow-sensing instrumentation is used to gather and transmit the information to the control room. Any deviation from the established limits of any of the parameters cited above results in automatic PWC shutdown.

The product gas leaving the PWC is cooled, compressed, and collected in tanks downstream of the PAS. This gas is then sampled and analyzed for chemical agent using both ACAMS and DAAMS agent monitors.

Feed Streams

In addition to the munitions and packing materials, the following materials are fed to the system:

- the plasma feed gas, which acts as the plasma medium to the PWCs (argon was specified in the technology provider's proposal; N_2 and CO_2 are being used in the demonstration unit)
- steam to the PWCs to convert elemental carbon to CO
- caustic to the PAS to support the quenching and scrubbing operations
- make-up water to the PAS to support the quenching and scrubbing operations

The technology provider has generated a mass balance for the proposed system using the maximum possible feed rates to each PWC. This results in a "mix-and-match" configuration that is not representative of any particular munition campaign. Nevertheless, the information reflects the sizes of anticipated flows. Process inputs are summarized in Table 5-2.

Waste Streams

The waste streams from the system will be either gaseous or solid. There is no liquid waste stream. (Wet scrubbers are used to absorb and neutralize products like HCl, HF, SO_2, and P_4O_{10} using a caustic solution, but the scrubber liquid is subsequently evaporated, leaving a salt cake.)

Gases

The product gas from the PWCs is quenched in a vertical down-flow duct with water/caustic spray. It

TABLE 5-2 System Inputs for the Burns and Roe Mass Balance

Stream	Flow (lb/hr)
Inputs to PWCs A1 and A2[a]	
HD	1,170
Decontamination solution, oil, laboratory waste	650
Steam	178
Argon	800
Inputs to PWC B[b]	
M28 propellant	386
Decontamination solution	25
Comp B4	64
Fiberglass	200
Metal parts	270
Steam	46
Argon	400
Inputs to PWCs C1 and C2[a]	
HD	585
Metal parts	8,600
Steam	265
Argon	800
Input to PWC D	
Miscellaneous dunnage	1,165
Steam	450
Argon	200
Input to pollution abatement systems	
Make-up water	7,997
Caustic (30 percent NaOH)	5,867
Total mass input to system	30,118

[a]100 HD-filled 155-mm projectiles per hr
[b]20 M55 rockets per hr

Source: Burns and Roe, 1998a.

then passes through a packed tower for acid-gas scrubbing (using caustic) and a mist eliminator for water removal. Fine particulates are removed by a cartridge filter. (A venturi scrubber is included in the demonstration system for particulate removal, just prior to the packed tower [Burns and Roe, 1998b].)

Following treatment with the pollution control equipment just described, the product gas is held in a pressure vessel where it is sampled and analyzed for agent and other components (still to be determined). If the gas is agent-contaminated, it is recycled to the PWCs for further treatment. If no agent is detected, (1) the gas is used as fuel for an on-site boiler, (2) the gas is shipped off site as fuel for other applications, or (3) the gas is burned in an on-site oxidizer. If the gas is burned in a boiler or oxidizer, gaseous effluents from the boiler or thermal oxidizer are scrubbed and released to the environment via a traditional stack.

Solids

Scrubber Brine Wastes. The brine from the PAS scrubber is processed to recover salts using the baseline brine reduction system. This system dries the brine with rotary-drum dryers to yield low water-content salts (5 to 15 percent by weight). The water vapor is discharged to the atmosphere; the water is not recycled. The salts are sent to an appropriately permitted landfill for disposal. (Consideration is also being given to recycling the brine through a PWC.)

Metal Munition Bodies. The 5X-treated molten metal is drained from the PWCs, cast into ingots, and sold for scrap.

Residues from the PWC Bottoms. Nonmetallic materials that collect in the bottom of the PWCs are mixed with sand in the unit and recovered as a vitrified material (slag).

The process outputs from the technology provider's mass balance are summarized in Table 5-3. The predicted compositions of the gaseous effluent streams from the PWCs are shown in Table 5-4, and the predicted compositions after scrubbing are shown in Table 5-5. (The compositions would be different if a plasma

TABLE 5-3 Mass Outputs for the Burns and Roe System

Stream	Flow (lb/hr)
Outputs from PAS for PWCs A1 and A2[a]	
product gas	1,822
Brine	9,907
Outputs from PAS for PWC B[b]	
product gas	1,036
Brine	83
PbO	4
Outputs from PAS for PWCs C1 and C2[a]	
product gas	1,214
Brine	4,864
Outputs from PAS for PWC D[c]	
product gas	1,400
Brine	539
Metals/silicates/bottom materials from PWCs	9,236
Total mass output from system	30,166

[a]100 HD-filled 155-mm projectiles per hr
[b]20 M55 rockets per hr
[c]miscellaneous dunnage waste

Source: Burns and Roe, 1998a.

TABLE 5-4 Predicted Composition of Product Gas from the Plasma Waste Converters (Prior to Scrubbing)[a]

Compound	PWC A1 and A2[b]	PWC B[c]	PWC C1 and C2[b]	PWC D[d]
HCl	552.52	0.56	68.09	40.86
H_2S	237.93		119.80	
N_2		66.27		
NH_3		0.16		
COS	22.08		9.57	
CO	706.01	385.43	286.61	783.42
C_2H_2	89.98		36.56	
H_2CO	19.65		7.99	
PbO		3.79		
CO_2	67.37	146.83	27.38	298.43
HCOOH	20.87		8.48	
Na_2O	77.68	3.88		
CH_4	17.39	7.82	7.07	15.89
H_2O	95.41	36.55	44.06	8.22
H_2	90.79	29.61	34.40	102.64
Ar	800.00	400.00	800.00	200.00
Total	2,797.68	1,080.90	1,650.01	1,449.46

[a]Plasma feed gas is argon; temperature is approximately 1,100°C (2012°F); all quantities in lb/hr.
[b]100 HD-filled 155-mm projectiles per hr
[c]20 M55 rockets per hr
[d]miscellaneous dunnage waste

Source: Burns and Roe, 1998a.

TABLE 5-5 Predicted Composition of Product Gas from the Plasma Waste Converters after Scrubbing[a]

Compound	PWC A1 and A2[b]	PWC B[c]	PWC C1 and C2[b]	PWC D[d]
HCl	5.53	0.01	2.68	0.41
H_2S	2.38		1.20	
N_2		66.27		
NH_3		0.08		
COS	22.08		9.57	
CO	706.01	385.43	286.61	783.42
C_2H_2	89.98		36.56	
H_2CO	15.72		6.39	
PbO		0.02		
CO_2	66.03	143.89	26.83	292.46
HCOOH	0.21		0.08	
Na_2O	0.78	0.04		
CH_4	17.39	7.82	7.07	15.89
H_2O	5.28	2.67	2.72	5.23
H_2	90.79	29.61	34.40	102.64
Ar	800.00	400.00	800.00	200.00
Total	1,822.18	1,035.84	1,214.11	1,400.05

[a]all quantities in lb/hr
[b]100 HD-filled 155-mm projectiles per hr
[c]20 M55 rockets per hr
[d]miscellaneous dunnage waste

Source: Burns and Roe, 1998a.

feed gas other than argon were used; CO_2 and N_2 are being tested in the demonstrations.)

Start-up and Shutdown

PWC start-up procedures involve first initiating the plasma feed gas flow, the torch cooling-water flow, and the plasma-torch power. The vessel is then allowed to reach the prescribed operating temperature, which usually takes approximately two hours from a "cold" start-up. Weapons material feed and steam flow (to convert elemental carbon to CO) are then initiated. The technology provider estimates that once the vessel reaches the desired temperature, steady operation at capacity can be achieved in approximately five minutes.

Upon shutdown, the weapons material feed is stopped, the steam flow is cut off, the power to the torch is turned off, and the plasma feed gas flow is stopped. The torch cooling water continues for approximately 30 minutes to protect the electrodes during cool down of the vessel.

EVALUATION OF THE TECHNOLOGY PACKAGE

Process Efficacy

Effectiveness

Because assembled chemical weapons contain mainly the elements C, O, H, N, S, P, halogens, and various metals (e.g., Al, Fe, Co, Ni), one can predict from standard thermodynamic calculations that CO, H_2, CO_2, H_2O, H_2S, HCl, HF, N_2, NO_x, SO_x, and various metal oxides will be formed at ambient temperatures (e.g., 20°C to 35°C; 68°F to 95°F), depending on the availability of oxygen. Compositions at higher temperatures can also be calculated, and Tables 5-6 and 5-7 show the mole fractions of the equilibrium products at 2,227°C (4,040°F) for various agent and energetic feed materials predicted by the technology provider, assuming argon is the plasma feed gas (Burns and Roe, 1997). These results were not generated via testing but were calculated by a chemical-equilibrium computer program (IVTANTHERMO) that uses standard thermodynamic data (Burns and Roe, 1997). Most

TABLE 5-6 Theoretical Equilibrium Composition of Product Gas from Plasma Treatment of Agents[a]

	Mole Fraction		
Species Produced	GB Feed	VX Feed	HD Feed
H_2	3.6×10^{-1}	5.0×10^{-1}	2.6×10^{-1}
Ar	2.5×10^{-1}	1.1×10^{-1}	2.7×10^{-1}
CO	1.9×10^{-1}	2.4×10^{-1}	
H_2O	7.6×10^{-2}	7.7×10^{-2}	1.4×10^{-1}
HF	5.0×10^{-2}		
PO	3.9×10^{-2}	1.5×10^{-2}	
N_2		1.1×10^{-2}	
SH		6.8×10^{-3}	
H	1.5×10^{-2}	1.8×10^{-2}	1.3×10^{-2}
CO_2	6.4×10^{-3}	5.9×10^{-3}	
PS		4.4×10^{-3}	
S_2		3.4×10^{-3}	
S		2.5×10^{-3}	
SO		1.7×10^{-3}	
P_2	4.4×10^{-3}	1.2×10^{-3}	
PO_2	1.3×10^{-3}		
OH	6.4×10^{-4}	5.6×10^{-4}	1.4×10^{-3}
P	5.5×10^{-4}	2.9×10^{-4}	
PH	2.3×10^{-4}	1.4×10^{-4}	
HPO	9.3×10^{-5}		
COS		1.3×10^{-4}	5.5×10^{-2}
PN		6.5×10^{-5}	
PH_2	4.1×10^{-5}	2.9×10^{-5}	
P_2O_3	3.6×10^{-5}		
O	1.8×10^{-5}	1.3×10^{-5}	4.5×10^{-5}
PF	1.2×10^{-5}		
HCO			1.0×10^{-1}
HCl			1.1×10^{-1}
C_2H_2			2.1×10^{-2}
COOH			9.3×10^{-3}
H_2CO			6.1×10^{-3}
Cl			3.6×10^{-3}
HCOOH			2.7×10^{-3}
CS			3.7×10^{-4}
HClCO			2.2×10^{-4}
C_3H			1.5×10^{-4}
C_2HCl			9.1×10^{-5}
C_2H			3.5×10^{-5}
CH_4			2.9×10^{-5}
CH_3			2.5×10^{-5}

[a]Plasma feed-gas is argon; temperature is 2,230°C (4,040°F).

Source: Burns and Roe, 1997.

of the higher temperature species should revert to CO_2, H_2O, CO, and H_2 when cooled. However, the product distribution and composition might be partially kinetically controlled rather than thermodynamically determined. Thus, species that are stable at higher temperatures but metastable at lower temperatures might be present after cooling.

TABLE 5-7 Theoretical Equilibrium Composition of Product Gas from Plasma Treatment of Energetics[a]

	Mole Fraction		
Species Produced	TNT Feed	Tetryl Feed	NC Feed
H_2	1.7×10^{-1}	1.5×10^{-2}	2.6×10^{-3}
Ar	9.1×10^{-1}	8.9×10^{-1}	8.9×10^{-1}
CO	5.4×10^{-2}	5.9×10^{-2}	2.2×10^{-2}
H_2O		5.3×10^{-3}	2.3×10^{-2}
N_2	1.2×10^{-2}	2.2×10^{-2}	1.8×10^{-2}
SH			
H	3.2×10^{-3}	3.1×10^{-3}	1.3×10^{-3}
CO_2		3.3×10^{-3}	3.1×10^{-2}
PO_2			
OH		2.2×10^{-4}	2.3×10^{-3}
O		2.9×10^{-5}	7.4×10^{-4}
C_2H_2	2.4×10^{-3}		
C_3H	9.0×10^{-5}		
C_2H	1.6×10^{-5}		
HCN	4.0×10^{-3}		
CN	3.9×10^{-5}		
NO		1.7×10^{-5}	3.9×10^{-4}
O_2			2.7×10^{-3}
NO_2			3.9×10^{-8}

[a]Plasma feed-gas is argon; temperature is 2,230°C (4,040°F).

Source: Burns and Roe, 1997.

The plasma temperature in the PWCs is estimated by the technology provider to be in the range of 15,000°C (27,032°F). Although one would expect very high destruction efficiencies at this temperature, much of the agent may not be exposed to such a high temperature in the PWCs. The plasma arc is created in an enclosed torch through which only the plasma feed gas (e.g., argon, N_2, or CO_2) flows. The arc heats the gas, which ionizes, dissociates, and then flows into the chamber surrounding the torch. The chemical agent is injected into the side of the chamber, not through the torch (see Figure 5-2). Inside the chamber, the agent mixes with the plasma. The maximum temperature to which each agent molecule is exposed is unknown, but the temperature gradient within the chamber is very large, as is evidenced by the estimated gas exit temperature of 1,100°C (2,012°F) (Burns and Roe, 1997).

The heterogeneous conditions in the PWC could cause organic intermediates to form. The prevailing view is that organic intermediates[3] are formed during an initial vaporization and pyrolysis phase, prior to

[3]In combustion systems, these organic intermediates are usually referred to as products of incomplete combustion (PICs). Typical examples include benzene, toluene, and naphthalene.

oxidation (Dellinger et al., 1986; Glassman, 1987; Dempsey and Oppelt, 1993). If complete mixing of the intermediates and stoichiometric quantities of oxidant occurs, all of the intermediates would be oxidized. However, in practice, the mixing will not be perfect, and some intermediates may bypass the plasma zone without being complete oxidized, even if sufficient oxidant is present. In addition, larger particles may not have sufficient time to decompose completely in the plasma zone. This imperfect mixing might also allow a small fraction of the original fuel or waste to pass downstream intact, without being pyrolyzed or oxidized.

The oxidant in the PWCs may come from the plasma feed gas (if air or CO_2 is used), or it may come from air in-leakage into the PWC from the surrounding room. For the demonstration system, the technology provider has indicated that the air in-leakage rate is approximately 30 SCFM (Burns and Roe, 1999), compared to a plasma feed gas flow rate of approximately 20 SCFM and a PWC total gas outflow on the order of 140 SCFM (Burns and Roe, 1998b). Thus, the air in-leakage forms a significant fraction of the total gas flow for the demonstration unit. If the available oxygen is less than the amount theoretically required for the complete oxidation of organics (including organic intermediates), then the lack of oxygen, coupled with incomplete mixing, could lead to significant quantities of organic intermediates being formed and passed downstream from the PWCs to the PAS.

For the reasons described above, the committee doubts that all of the chemical agent feed would actually be exposed to the ultra-high temperature plasma and believes that some residual toxic materials would remain or form. Nonhomogeneous temperature distributions, gas turbulence, and incomplete mixing may limit the absolute effectiveness of this process (i.e., the target destruction efficiencies would not be achievable at the required throughput rates). Actual chemical agents and munition components must be processed in the proposed PWCs (or other units of similar size and design) to prove the efficacy of the process and to optimize design parameters, such as flow rates, reduction/ oxidation conditions, and residence times. Design parameters will be very sensitive to equipment configuration, scale-up, plasma feed gas, and the type of chemical weapon and feed rate. The results of the demonstration tests, which are being performed at a reduced scale, could provide some (but not all) of the data to address these concerns.

The technology provider supplied a comparison between the theoretical products and the products measured by GC (gas chromatography) for a mixed feed of polyethylene, cellulose, water, and air to a PWC of unspecified size (Burns and Roe, 1997). This is shown in Table 5-8. The technology provider states that "due to the limitations of gas chromatography, the samples were analyzed only for the primary components of the [product gas]." Thus, approximately 6 percent of the gas was not accounted for by the GC analysis, and there are discrepancies between the theoretical and measured concentrations. Regarding these discrepancies, the technology provider observed that "the computer tends to underestimate hydrocarbon species and oxygen, while overestimating carbon monoxide and hydrogen" (Burns and Roe, 1997). This statement reinforces the committee's concern that hydrocarbon species not predicted by equilibrium calculations (including trace organic species that are of environmental concern) could be present in the product gas from the PWC.

The technology provider has indicated that soot formation in the PWC could be significant. This phenomenon is predicted by the thermodynamics. If CO_2 is

TABLE 5-8 Comparison of Experimental and Predicted Gas Compositions Subsequent to Plasma Treatment[a]

	Mole Fractions	
Species	From Gas Chromatography	Predicted by Equilibrium Calculations
H_2	32.89	40.23
N_2	29.50	29.94
O_2	2.67	0.18
CO	20.70	26.66
CH_4	3.12	0.00
CO_2	3.87	2.12
C_2H_2	0.61	0.00
C_2H_4	0.49	0.00
H	na[b]	0.02
OH	na	2.0×10^{-3}
NO	na	1.0×10^{-3}
O	na	5.5×10^{-5}
NH_3	na	1.0×10^{-6}
N	na	1.0×10^{-7}
C	na	2.0×10^{-12}
Other	na	0.85
Total	93.85	100.00

[a]Plasma feed-gas is air; waste/air feed is approximated by $C_{25}H_{104}O_{44}N_{17}$.

[b]The technology provider did not analyze for these compounds.

Source: Burns and Roe, 1997.

used as the plasma feed gas, it is also possible for soot to form if the resulting CO disproportionates to C(gr) + CO_2. The technology provider has proposed adding steam to the feed to convert elemental carbon to CO. This approach has been demonstrated in various plasma systems used for processing municipal waste. However, careful control of the steam-to-carbon ratio would be required to control the formation of soot. Otherwise, there could be significant emissions of soot. In practice, this control could be difficult to achieve with waste streams of various compositions.

The committee was informed by the technology provider that argon was not used for the demonstrations because it was too expensive (Burns and Roe, 1999). Burns and Roe had planned to use air as the plasma feed gas until it was suggested that members of the Dialogue might consider the use of air to be akin to incineration (Hindman et al., 1999). The technology provider then tested CO_2 as the plasma feed gas but ultimately decided to use N_2. Because different plasma feed gases have different thermodynamic and chemical properties, the choice of the plasma feed gas can have a significant impact on the performance of the system. For example, the power requirements will vary with the plasma feed gas. Electrode wear may also depend on the type of gas, and the composition of the product gas will certainly vary. Therefore, tests performed with one plasma feed gas may not be indicative of performance with a different gas.

The volatile low molecular weight chlorinated hydrocarbons in mustard that can be difficult to destroy are not expected to pose a special difficulty for the plasma treatment units, although this has yet to be demonstrated.

Sampling and Analysis

From the responses to the data-gap questions (Burns and Roe, 1998a), it appears that the technology provider expects to use the same sampling and analytical procedures being used in the baseline incineration system. These are probably adequate.

Maturity

Research on plasma-arc technology dates back to the early 1900s, and many practical industrial applications have been developed, including arc melting of metals; electric arc welding; plasma processing of ores; plasma spraying of metallic or oxide powders; and plasma generation of atomic, ionic, and molecular spectra for analytical systems. Plasma arcs have also been used for treating hazardous wastes. According to the General Accounting Office (GAO), research plasma-arc furnaces have ranged from 2 to 8 ft in diameter, with power levels of 150 kW to more than 1 MW (GAO, 1999). Wastes treated include solvents, paint, batteries, incinerator ash, and radioactive materials. Of the research initiatives by the U.S. Department of Energy and DOD over the past 10 years on plasma treatment of hazardous waste, two have reached the implementation stage: (1) a Navy project to destroy hazardous materials on shore (scheduled for operation in 2000), and (2) an ongoing asbestos destruction project at Port Clinton, Ohio. Other projects are still in the research phase. Although organic wastes have been destroyed using plasma-arc furnaces, much of the research to date has focused on the vitrification of inorganic substances within wastes (e.g., radioisotopes) rather than on the destruction of organic wastes.

A subgroup of the committee visited an Ontario Hydro Technologies site (Toronto, Ontario) on April 5, 1998, to observe a prototype PWC and to be updated on progress in the development of equipment for use with real chemical agents and energetics during the ACWA demonstration phase (see Appendix B). The visiting team was shown a basic version of the technology provider's PWC system. The items observed included (1) a long pipe with water-cooled copper electrodes that operated as a nontransferred DC-plasma torch emitting hot plasma, and (2) a cylindrical furnace system about 6 ft tall and 3 to 4 ft in diameter (external dimensions) with an opening on the top for batch feeding. The electric power requirements (DC) for this unit were on the order of 100 kW to 500 kW, depending on the type of waste and processing rates (e.g., 100 to 500 lb/hr).

During the subcommittee's visit, the small DC-plasma torch that fed the plasma into a refractory-lined furnace was used to demonstrate the treatment of the following materials:

- a simulated, double-base propellant (nitroglycerin and nitrocellulose)

- metals
- plastics
- household materials

All of these materials, which were hand loaded by members of the visiting team, were melted or decomposed in the plasma discharge, as viewed on a TV monitor. No spectroscopic monitoring devices—optical or mass spectrometric—were in operation. A ram feeder and a refractory trough on the device were apparently intended for removing slag. The proposed units for scrubbing SO_X and HCl and for extracting desired products (e.g., CO, H_2, and metals or silicate products), which are shown schematically in the system diagrams, were not present at the Ontario Hydro site. The demonstration system being tested at Edgewood, Maryland, is designed to perform these functions (Burns and Roe, 1998b).

The PWCs proposed by the technology provider have never been tested with actual munitions or chemical agents. According to Burns and Roe (1997), tests conducted by Acurex Environmental Corporation at the EPA's Air Pollution Prevention and Control Division of the National Risk Management Research Laboratory showed that a PWC could destroy simulants of nerve agents, blister agents, and energetics. The PWC tested was a refractory-lined stainless steel vessel sized to process 25 lb/hr of material. It was equipped with a 50 kW to 100 kW, nontransferred, water-cooled DC torch, and the plasma feed gas was argon. The total amount of material destroyed was not reported. In all cases, the destruction efficiencies were stated to be in excess of 99.9999 percent.

Work by MSE, Inc., of Butte, Montana, in 1993 under Department of Energy sponsorship and using a plasma centrifugal furnace made by RETECH, Inc., was cited to validate the effective use of a plasma furnace for the destruction of MK 72 Mod 5 fuzes. A Startech report on PWC processing of pyrotechnic-contaminated materials for Ensign-Bickford in October 1995 gave no details about the PWC used or about the quantities of material processed (Burns and Roe, 1998a). Major products were identified by GC. A unit developed at Drexel University used an inductively-coupled argon-based plasma device to process energetics and agent simulants (Burns and Roe, 1997).

The full-scale units proposed by the technology provider have not been produced yet. A smaller prototype, larger than the units at Ontario Hydro, is being tested between February and May of 1999 at an Army test facility in Edgewood, Maryland.

Robustness

Based on the many practical applications of plasma technology in the industrial sector, it can be considered a robust technology. However, robustness for destroying chemical weapons, and especially large segments of energetic rocket propellants, remains to be demonstrated. Adaptation of plasma devices for the destruction of chemical weapons will require (1) special handling equipment for the safe introduction of shells, rockets, and land mines; and (2) further development of the torch and chamber designs to ensure the destruction of agent and the production of effluents that can be scrubbed and burned or converted to slag and sent to a landfill. Meeting these requirements will entail a much more extended development and testing program than the one being undertaken for the ACWA demonstration phase. The program would have to ensure that the energy released from the processing of rocket propellants can be controlled.

Monitoring and Control

The committee was not given detailed design parameters for the full-scale units. A smaller demonstration unit has been installed and is being operated. The technology provider plans to use the monitoring and control systems currently in use at DOD chemical-disposal facilities and laboratories. Demonstration testing may show that the proposed monitoring and control strategies are effective, but the committee does not have sufficient information to make an evaluation at this time.

Applicability

Conceptually, plasma technology is applicable to all assembled chemical weapons and could be used at any of the chemical weapons storage sites. However, the proposed process would have to overcome the engineering hurdles described above to treat the various components of assembled chemical weapons.

Process Safety

The unique equipment proposed by the Burns and Roe team includes an explosion chamber, the PWCs, a PAS, and a product-gas collection and storage system.

The pressure and temperature in the explosion chamber vary cyclically. When energetic material is being fed, the chamber is at ambient conditions. When the energetic material is initiated, the peak temperature and pressure associated with the deflagration or detonation of up to 2,000 grams of TNT are reached. When the product gas is vented, the pressure and temperature return to ambient conditions.

The PWCs operate at very high temperatures but at slightly negative pressure. Although the plasma temperature is estimated by the technology provider to be in the range of 15,000°C (27,032°F), the interior PWC wall surface is expected to be much lower (1,650°C [3,000°F]; the temperature of exterior PWC surface will be lower still (about 66°C [150°F]). The PWC discharges molten material into ladles for cooling.

The full-scale PAS receives gas from the PWCs at about 1,100°C (2,012°F) and at rates up to 2,000 lb/hr. The gas is quenched, scrubbed, and filtered before being pressurized to 100 psig for storage at ambient temperature. The explosion chamber and the PWCs operate in a batch mode, and the product-gas collection system includes a hold-test-release step before the gas is stored in bulk. Thus, despite the uncertainties about adequate exposure of the various input materials to the high temperature of the plasma described earlier, the presence of any chemical agent can, in principle, be determined prior to the gaseous effluent being released to storage for subsequent use.

Worker Health and Safety

ACWA demonstration tests are planned to confirm that excessive energy will not be released when rocket propellant reacts in PWC B and that the propellant will not detonate. The results of these tests should be thoroughly analyzed.

The explosion chamber and the PWCs, including the feed systems, dunnage shredder, and molten material discharge systems, are operated remotely and are interlocked. Ease of maintenance should be integral to the PWC design, particularly the replacement of the torch electrodes, the repair of the PWC refractory liner, access to the interior of the PWC, and operation of the molten-material discharge valve. Although worker interactions with high-temperature equipment/material or rotating equipment should be minimized by the controls and design features, worker hazards will probably be higher in the presence of high-temperature systems than in the presence of lower temperature systems.

Other worker hazards include the use of a large amount of argon, CO_2, or N_2 (all asphyxiants), the production of pressurized flammable gas, and an electrical power system of 440 to 700 V and 800 amps. None of these hazards is unique, however, and the risks can be minimized with proper precautions. Worker interactions with hazardous chemicals will be limited to caustic for the off-gas scrubber and acid for the neutralization of scrubber brine.

Public Safety

A substantial amount of flammable gas will have to be stored, whether the product gas is burned on site or shipped off site. A large explosion or deflagration involving this gas could cause an on-site hazard and, potentially, an off-site hazard from the direct thermal effects or overpressure forces. A greater concern is the potential damage from explosions to containment structures that could lead to a release of agent. Explosion hazards are common in industry and can be minimized by good design and operation.

Cooling water is circulated through the plasma torch to keep it from melting at the high plasma temperatures. A leak in the cooling system could spray water into the plasma. If the leak is sudden, rapid vaporization could cause a pressure pulse that might overload the downstream gas-handling equipment. Then, untreated agent could be released into the surrounding room through the torch opening in the top of the PWC. Similar "puffing" has been observed in combustion equipment when excessive back pressure occurs. If the leak is gradual, the resulting steam would dissociate in the plasma forming hydrogen and oxygen gas that could recombine and explode if the mixture is in the flammable range above its autoignition temperature. The effect of liquid water introduced into a plasma in the presence of other species present in PWCs must be determined before larger scale experiments are performed. The normal PWC operating conditions appear to be outside the flammability range for hydrogen, but the effect of the additional water (from a leak in the torch cooling system) could create an explosive combination. (For example, the presence of a steam diluent would raise the autoignition temperature; whereas, an argon diluent would decrease the autoignition temperature [Kumar and Koroll, 1995]). If some water is not completely vaporized, it would fall into the molten material at the bottom of the PWC, and a metal-water reaction could create a pressure pulse. These mechanisms should be investigated further, unless the probability of the failure of the torch is determined to be very low.

The technology provider is aware that torch failure is a concern, and the potential for an explosion has been reduced by the torch design and by redundant flow and pressure controls that would actuate fast-closing valves on the water feed as well as the waste feed in the event of a failure.

Testing is planned to validate that agent does not reform and that other hazardous materials (e.g., Schedule 2 compounds and dioxins) do not form as the PWC effluent cools. The potential formation of metastable species (e.g., C_2H_2, HCN, C_2N_2) that could be quenched from the rapid cooling of product gas should be thoroughly investigated.

Human Health and the Environment

Burns and Roe states that there will be no gaseous air emissions and no liquid discharges from the integrated system and that the solid waste will consist only of metal ingots, vitrified material, and possibly scrubber salts. The technology provider has also indicated that the scrubber salts might be recycled to the PWC and vitrified with sand to produce a very stable solid waste. Thus, it is claimed in the proposal that there would be virtually no impact on human health or the environment. However, the committee has identified some issues that must be addressed during the development of the integrated process.

Effluent Characterization and Impact on Human Health and Environment

The primary solid-waste streams include fly ash material caught in filters, scrubber salts from the PAS, and metal ingots and slag from the PWCs. The treatment temperature for metal parts is expected to exceed the required 5X conditions; therefore, metal parts treated in the PWCs should receive a 5X designation.

The technology provider plans to explore the option of recycling liquid scrubber effluent with fluxing agents, such as lime and sand, in a PWC to generate a vitrified solid waste. The treatment of scrubber liquor by vitrification in the plasma unit has not been proven. The committee's concerns relate to the behavior of salts at high temperature and whether the acid components could be incorporated into the melt without being released. For example, NaCl salts could react with SiO_2 at high temperatures to form gaseous $SiCl_4$; also, NaF salts could react with SiO_2 to yield SiF_4.

The PWCs produce the primary gaseous discharge. The technology provider proposes that this gas will be passed through a PAS to a holding tank. The committee is concerned that a PAS designed for fully oxidized gas may not be as effective for gas generated under reducing conditions in the PWCs. Whether reducing conditions exist will depend on the plasma feed gas. The performance of the PAS must be evaluated at the design operating conditions and for the actual product gas.

The committee concluded that some significant design changes may be required to the baseline PAS to optimize its performance for the product gas. For example, the PAS being used by Burns and Roe in the ACWA demonstrations includes a venturi scrubber to control particulates. Because the product-gas flow rate for the PWC being demonstrated is only on the order of 140 SCFM, the gas velocities may not be sufficient for the venturi to remove particulates effectively.

The technology provider presented no data on the effluent characterization if the product gas is burned as boiler fuel. Contrary to the technology provider's claim that there would be no air emissions, the committee concluded that small amounts of acid gases, including NO_x, SO_x, HCl, and HF, may be generated during the burning of plasma-generated product gas. The NO_x and SO_x originate from the high-temperature burning of the gas; HCl and HF are produced with the chlorine and fluorine (originally in the agent) that may not be completely removed by the scrubber (typical scrubber efficiencies are 99.9 percent or greater). In addition, the boiler burner systems must be designed to burn relatively low heating-value gas efficiently to prevent the release of unburned organics and other trace species of environmental concern.

Semivolatile and volatile metals, such as lead, cadmium, and arsenic, are expected to volatilize in the PWC and condense downstream as a fine fume of submicron-sized particles. Thus, the particulate control device will be challenged by a fine metal fume with particulate sizes that are difficult to capture. In addition, this metal fume would be pyrophoric. The technology provider indicated during meetings with the committee that high-efficiency particulate air (HEPA) filters would be added to the PAS to control the metal fume. This addition will have to be evaluated to determine if a nonflammable filter would be required. Also, the pressure drop associated with the additional filter would have to be accounted for in the design of the system.

Completeness of Effluent Characterization

The technology provider has stated that the operating temperature in the plasma zone is about 15,000°C (27,032°F) and that molecules subjected to this temperature will be dissociated into atomic components. However, all components may not be subjected to the temperature of the plasma. The specific characteristics of the product gas depend on the constituents in the waste material being fed to the chamber and the temperatures of the plasma and bulk gas. The technology provider currently has no data on the actual characteristics of the PWC effluent gas for the feeds expected from the disassembly of chemical munitions. The technology provider has calculated the thermodynamic equilibrium constituents of the gas with the assumption that the gas reached the plasma temperature (or at least a very high temperature). These calculations do not address product constituents from material that does not reach the high temperatures of the plasma zone because of bypass or because of kinetic limitations that allow metastable molecules to persist. A more complete chemical analysis of product gas generated at pilot scale will be necessary, including both major constituents and trace species of environmental concern.

After being treated in the PAS, gas from the plasma unit will be passed to a pressurized holding tank. Based on the data in Table 5-5, and assuming compression to 100 psig, cooling to 25°C (77°F), and a holding time of 1 hour, the required tank volume will be approximately 14,300 ft^3.

The gas will be sampled and analyzed for agents by ACAMS and DAAMS. The technology provider also proposes using a continuous emissions-monitoring system consisting of a Fourier transform infrared (FTIR) analyzer and an unspecified particulate-matter monitor. Once the product gas has been certified to be agent free, it may be used as a boiler fuel. The proposed gas analysis will not analyze for trace organic by-products or metals. Although the FTIR system can measure some species to ppb levels, it may not be sensitive enough to characterize fully trace organic and metallic species of environmental concern on a continuous basis.

The committee is also concerned about the deposition of materials on the walls of the hold-up chamber that could be vaporized or resuspended when the chamber is evacuated. A rigorous characterization protocol for the hold-test-release system must be developed and validated prior to implementation, regardless of the final disposition of this gas stream.

Effluent characterization and chemical analysis of the product gas are scheduled as part of the ACWA demonstration phase. A careful, meticulous study of the effluent gas will be critical to the evaluation of this technology. The data should include the identification of any organic intermediates that would be included in an HRA. The committee is concerned that the effluent characterization when CO_2 or N_2 is used as the plasma feed gas will not be valid when argon is used. Testing should be done with the specific plasma feed gas proposed for the full-scale system.

Effluent Management Strategy

The strategy for effluent management proposed by the technology provider is designed to eliminate all liquid discharges and hazardous-waste discharges. Bulk metals are melted in the plasma units and turned into solid ingots, which are expected to meet the 5X decontamination criteria. Dunnage and miscellaneous solid waste will be treated on site in a separate PWC.

Scrubber discharge, which includes aqueous waste containing salts, may be treated on site in a PWC or processed in a brine reduction system prior to disposal in a landfill. If a PWC is used, the scrubber discharge liquid will be mixed with sand or other fluxing agents and then fed to the plasma unit where the water will be evaporated and the solids vitrified. No data were provided to the committee on this process, and the committee questions whether the material can be melted without releasing acid components of the scrubber effluent during vitrification as $SiCl_4$ (gas) or SiF_4 (gas). If the baseline brine reduction system is used instead, the technology provider will have to demonstrate that excessive particulate emissions from the system will not occur. The Tooele Chemical Agent Disposal System brine reduction system failed its environmental testing because of such emissions, and the brine is currently being shipped off site for disposal in a landfill.

Management of trace metals is also a potential concern. The technology provider did not characterize the fly ash that would be collected in the filters of the PAS, which could contain a large amount of the trace metals volatilized in the plasma unit. This effluent stream will be a hazardous waste that will have to be solidified prior to final disposition. If the formation of soot is not prevented by steam injection, a significant amount of finely divided carbon could be present in the effluent as graphite or soot.

In general, the product gas from the PWCs will consist of a variety of organic compounds of uncertain composition. The high temperatures and oxygen deficient (or even reducing) conditions (depending on the plasma feed gas) lead the committee to believe that many of the compounds that can be present in trace quantities in the emissions from combustion systems will probably be present in higher concentrations in the gaseous streams from this process. Although the technology provider proposes capturing and holding this stream for analysis, the committee believes that this will be difficult. The technology provider has presented no data to demonstrate the feasibility of this type of gas capture, containment, and characterization.

The committee also questions the feasibility of burning this gas in a boiler. The high chlorine, sulfur, phosphorus, and nitrogen content of the raw materials will result in a complex mixture of compounds that will have to be removed from the gas stream prior to burning, a difficult, if not daunting, task. The elemental moieties will also create a gas stream with a composition very different from traditional gaseous fuels. The predicted composition after scrubbing (Table 5-5) includes several toxic compounds listed in the Clean Air Act Amendments. The committee, therefore, believes that this technology may encounter significant difficulties in satisfying the risk-assessment and risk-minimization requirements for boilers and industrial furnaces.

Resource Requirements

The major resource requirements for this process are water, power, argon (or other plasma feed gas), and caustic. During operation, 40 gallons per minute of water, 600 SCFM of argon, and 8500 lb/hr of caustic will be required (Burns and Roe, 1997). Although the annual consumption of these materials was not estimated by the technology provider, the committee estimates that 5,000 hours of operation per year would require 12 million gallons of water, 180 million cubic ft. of argon, and 42.5 million lb of caustic. The technology provider estimates that the plasma torches will require 6 MW of power. Each PWC requires 0.5 kWh of electrical energy per pound of material feed. The technology provider has used this value to estimate energy requirements for all of the feed types, including agents, energetics, metal components, DPE suits, and dunnage.

Environmental Compliance and Permitting

Only a few plasma units have received permits for waste processing in the United States to date. The technology provider has not provided a definitive permitting strategy for the unit beyond declaring that the system would not be permitted under RCRA incinerator-permitting procedures. The regulatory definition of an incinerator includes plasma-based treatment systems that burn waste with oxygen in enclosed chambers or uses afterburners. In the Code of Federal Regulations 40 CFR 260.10 Definitions, a plasma or incinerator is defined as "any enclosed device using a high intensity electrical discharge or arc as a source of heat followed by an afterburner using controlled flame combustion and which is not listed as an industrial furnace." Because boilers are industrial furnaces, the proposed configuration would probably *not* be interpreted as an incinerator unless oxygen is used as the plasma feed gas.

The committee identified two alternative permitting routes that might be followed for a plasma-treatment process that generates gases that are subsequently burned in boilers. First, the permitting process could follow 40 CFR 264, Subpart X, Procedures for Miscellaneous Treatment Units:

> A miscellaneous unit must be located, designed, constructed, operated, maintained, and closed in a manner that will ensure protection of human health and the environment. Permits for miscellaneous units are to contain such terms and provisions as necessary to protect human health and the environment, including, but not limited to, as appropriate, design and operating requirements, detection and monitoring requirements, and requirements for responses to releases of hazardous waste or hazardous constituents from the unit. Permit terms and provisions shall include those requirements of other rules that are appropriate for the miscellaneous unit being permitted.

With Subpart X permitting, the permit writer uses the relevant rules as a guide, and permitting authorities are likely to use the most recent incinerator standards as the appropriate rules, as they generally do with thermal-treatment units.

If the plasma-generated product gases are burned in a boiler, regulatory authorities could also opt to impose the boilers and industrial furnace (BIF) permitting procedures. In this case, the authorities could regulate the unit as a boiler burning hazardous waste. BIF rules have been developed, and the permitting procedures have been well defined. The EPA has announced plans to develop Clean Air Act Maximum Achievable Control Technology (MACT) standards for boilers burning hazardous waste over the course of the next few years. These new standards will probably be in place prior to the construction of a full-scale PWC system. Thus, although the PWC system is not likely to be regulated as an incinerator, the permitting procedures would be similar.

STEPS REQUIRED FOR IMPLEMENTATION

The full-scale implementation of this technology will require demonstration with actual chemical agents and weapons. Some of these studies are scheduled for the ACWA demonstration. There is little doubt that the highest plasma-torch temperatures will destroy mustard, GB, and VX, but no testing has been done to demonstrate that the agents remain in the plasma zone long enough to be destroyed. Nor are there detailed analytical data to indicate side reactions or unpredicted products that could result (e.g., dioxins, SO_XF_Y, and OF_2).

A more thorough evaluation of the proposed technology will be possible when a full-scale PWC design is available for modeling gas flow rates and evaluating the exact placement of nozzles and ports through which munition materials would be introduced into the hot plasma zone. The following list includes the most critical steps the technology provider must take before proceeding to implementation:

1. Determine the effect of sudden water injection into the plasma torch in the presence of argon, nitrogen, carbon dioxide, and other species present in the plasma system. Include an evaluation of the effect of gases present in the PWC on the flammability range of hydrogen gas.
2. Determine the likelihood of the release of untreated agent and other hazardous contaminants from the PWC if the gas generation rate is unexpectedly high (e.g., due to a cooling-water leak, the inadvertent introduction of explosive material into the chamber, or a rapid deflagration of propellant).
3. Conduct a thorough analysis of the product gas generated from each PWC using the plasma feed gas proposed for full-scale operation. This analysis should include the identification of organic intermediates that would be of concern in an HRA.
4. Establish the efficacy of pollution-control equipment in removing hazardous compounds (e.g., NO_x, SO_x, HCl, and metals) from the product gas.
5. Perform a larger-scale demonstration of PWC operation, that includes the hold-test-release step.

FINDINGS

Finding BR-1. No tests have been done involving actual chemical-agent or propellant destruction in a PWC. Tests with agent and M28 propellant were planned for the demonstrations being conducted between February and May of 1999, but no data were available to the committee at the time of this writing.

Finding BR-2. Scale-up from the small PWC units in existence to the very large units proposed is likely to present significant scientific and engineering challenges.

Finding BR-3. Tests performed with one plasma feed gas may not be indicative of PWC performance with a different gas. Because different plasma feed gases have different thermodynamic and chemical properties, the choice of the plasma feed gas could have a significant impact on the performance of the system. For example, the electrical power requirements will be determined, in part, by the plasma feed gas. Electrode wear may also depend on the type of gas, and product gas composition will vary.

Finding BR-4. The technology provider's proposal for recycling the liquid-scrubber effluent through the PWC to vitrify the salts may not be practical. If scrubber liquor is fed to a PWC, some of the contaminants may simply revolatilize. In addition, NaCl and NaF salts could react with SiO_2 at high temperatures to form gaseous $SiCl_4$ and SiF_4, respectively (both hazardous materials).

Finding BR-5. The maintenance of negative pressure within the PWC has not been demonstrated under munition-processing conditions. Pressure excursions that produce positive pressure in the PWC vessel could release product gas to the surrounding room. Some upsets that could result in moderate to severe pressure excursions are listed below:

- A leak in the torch-cooling system could release water into the PWC, and rapid steam formation could pressurize the vessel. Water leakage might also lead to more severe pressure excursions or even explosions.
- Energetic material that remained in a mortar or projectile and was introduced into a PWC could detonate upon heating, which would generate a pressure pulse. The severity of this pulse would depend on the type and quantity of explosive.
- An improper cut of the rocket motor could allow a larger-than-design piece of propellant to be introduced into the PWC. If the gas production rate from the propellant exceeds the capacity of the downstream PAS, the vessel could overpressurize.

The technology provider should investigate the likelihood of such events and determine their potential impacts on the operation of the PWCs.

Finding BR-6. Combustion of plasma-converted gas in a boiler faces three major hurdles: (1) to avoid being permitted under RCRA as a boiler burning hazardous waste, the gas may have to be delisted; (2) the gas may require significant scrubbing to remove compounds that are unsuitable as boiler feedstock; and (3) the boiler will have to be configured to burn gas that has a low heating value efficiently in order to avoid generating unacceptable emissions.

Finding BR-7. Although a PWC may not be considered to be an incinerator by permitting authorities, the most likely permitting procedures for a PWC would be similar to those used for incinerators.

6

General Atomics Technology Package

INTRODUCTION AND OVERVIEW

The technology package submitted by General Atomics is summarized in Table 6-1 and the flow charts in Figures 6-1 through 6-4. This package is comprised of four basic technologies.

- The munitions are disassembled using the Army's *baseline disassembly* process, modified to include *cryofracture* of projectiles and mortars once the energetic materials have been removed.
- The chemical agents and energetic materials are decomposed (separately) using *caustic hydrolysis*.
- SCWO (supercritical water oxidation) is used to treat the hydrolysates of agent and energetic destruction.
- *High-temperature heating* is used to decontaminate metal parts to a 5X level.

General Atomics' technology package is designed to treat the following materials:

- projectiles and mortars containing explosives and agent
- rockets containing explosives, propellant, igniters, and agent
- land mines containing explosives and agent
- dunnage, including pallets, metal banding, and DPE, some of which may be contaminated with agent

DESCRIPTION OF THE TECHNOLOGY PACKAGE

Disassembly of Munitions and the Removal of Agent/Energetics

General Atomics proposes using baseline disassembly methods with the modifications described below.

TABLE 6-1 Summary of the General Atomics Approach

Major Demilitarization Operation	Approach(es)
Disassembly of munitions	**Rockets and mines.** Army baseline disassembly process with minor modifications. **Projectiles and mortars.** Army baseline disassembly process to remove energetics with minor modifications, followed by cryofracture of downloaded munitions to provide better access to agent.
Treatment of chemical agent	Caustic hydrolysis; SCWO treatment of hydrolysate.
Treatment of energetics	Water-jet wash-out of energetics from casings; caustic hydrolysis; SCWO treatment of hydrolysate.
Treatment of metal parts	Heat in electrical metal parts furnace to 5X.
Treatment of dunnage	Shred, macerate, slurry; caustic hydrolysis of slurry; SCWO treatment of hydrolysate.
Disposal of waste	**Solids.** Analyze and send dry salts to landfill, possibly after stabilization. **Liquids.** Analyze and discharge condensate from evaporation of salts to wastewater treatment plant. **Gases.** Discharge to atmosphere after HEPA filtration and activated carbon adsorption; continuous monitoring for agent.

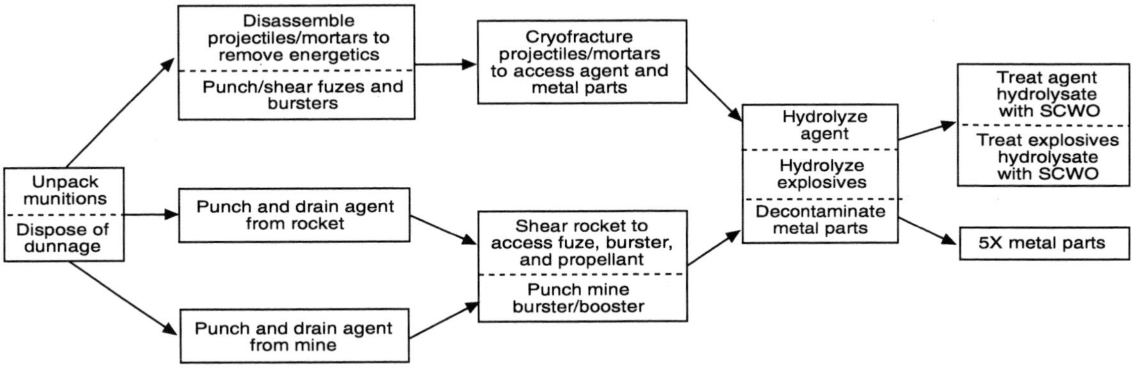

FIGURE 6-1 Schematic drawing of General Atomic's proposed technology package.

Rocket Disassembly

M55 rockets are processed through the baseline disassembly rocket-shear machine, and the drained agent is pumped to a surge tank prior to hydrolysis. The rocket pieces (containing the propellant and other energetic materials) are gravity-fed through the discharge chute of the explosion-containment room to the next treatment step, hydrolysis of energetics. The following modifications to the rocket-shear machine are included in the General Atomics proposal:

- an increase in the number and change in the locations of the cuts to reduce the size of the pieces of energetics fed to the hydrolysis reactor
- an increase in the number and diameter of the holes punched in the agent cavity
- the addition of a flushing step, whereby hydrolysis solution is injected through pressure nozzles inserted into the agent cavity

These modifications are intended to facilitate the draining of agent and minimize cross-contamination and co-processing of agent and energetics in the hydrolysis step.

Land Mine Disassembly

General Atomics proposes modifying the baseline mine-unpacking operations and mine machine to improve draining and flushing of mine bodies and to increase access to the bursters and the booster. The drained liquids and flush solutions are pumped to surge tanks from which they are fed to the hydrolyzers. The mine bodies and energetic materials may, still be contaminated with residual agent and energetics, at this stage. They are gravity fed through the discharge chute of the explosion-containment room to the hydrolysis step.

Projectile and Mortar Disassembly

General Atomics proposes using the baseline reverse-assembly process to remove the energetics from projectiles and mortars. Energetic materials are conveyed through the explosion-containment room discharge chute to the energetics hydrolysis treatment step. General Atomics proposes modifying the baseline process by using cryofracture to provide access to the interior of the munition bodies. General Atomics believes that cryofracture will provide better access to the agent than the baseline process method in which burster wells are pulled and the agent cavity is drained.

In the cryofracture operation, the munitions (with explosive charges removed) are loaded, via a special carrier, onto a cryocooling conveyor. The conveyor lowers the munitions into and slowly moves them through a bath of liquid nitrogen (77 K; –321°F) to embrittle the casing. At the bath exit, the munitions are transferred to a hydraulic press where they are fractured into pieces while they are still cold.

The fractured munition components and chemical agent are then discharged through a chute to the projectile rotary hydrolyzer (described in the next section) for further treatment. The metal parts are washed with alkali solution applied through high-pressure nozzles to clean out any remaining agent and transferred to a chemical reactor for further treatment.

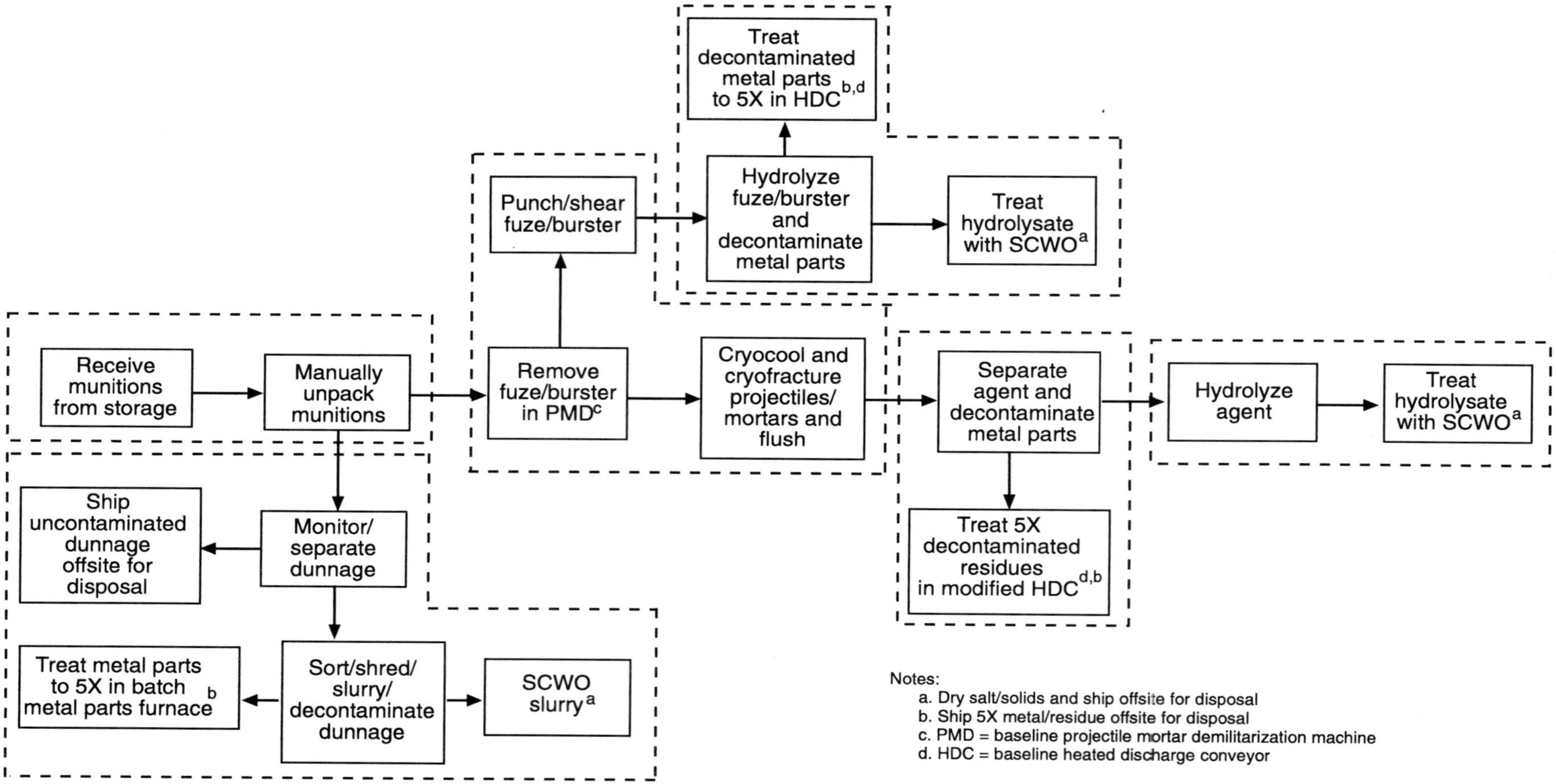

Notes:
- a. Dry salt/solids and ship offsite for disposal
- b. Ship 5X metal/residue offsite for disposal
- c. PMD = baseline projectile mortar demilitarization machine
- d. HDC = baseline heated discharge conveyor

FIGURE 6-2 Block flow diagram for treatment of projectiles/mortars.

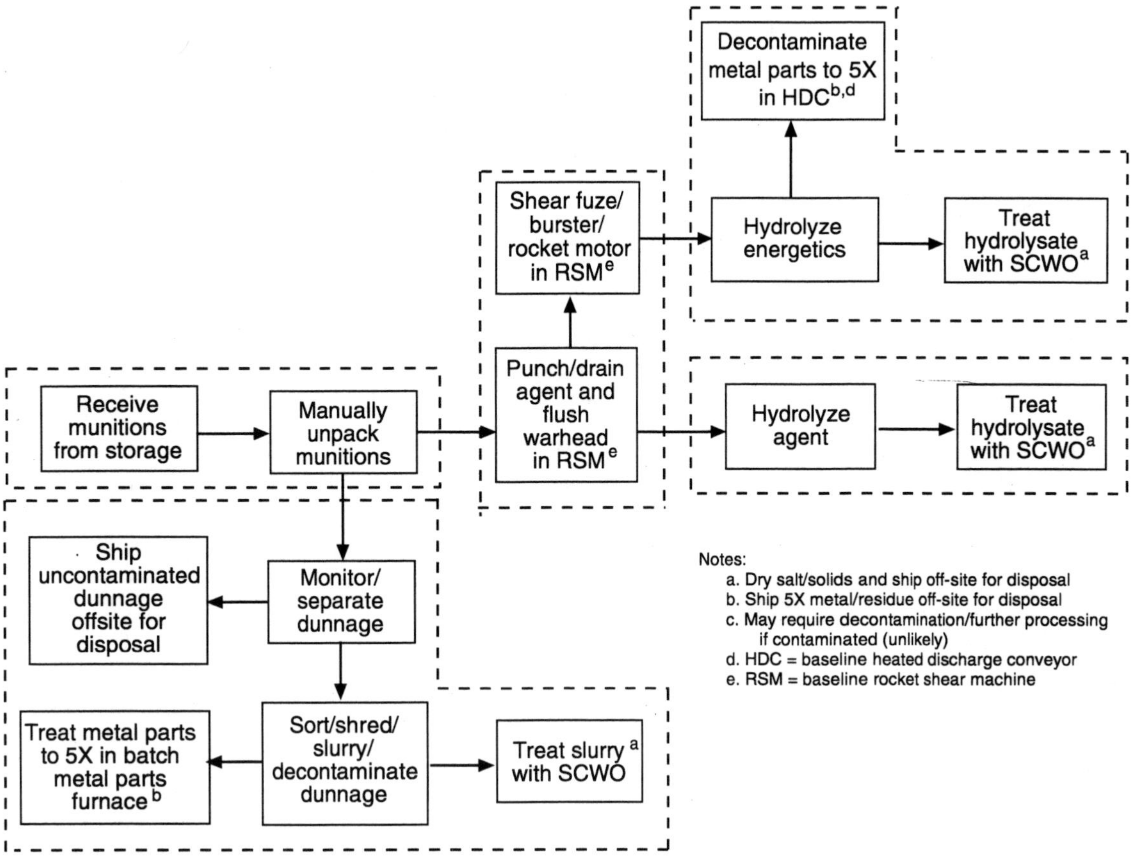

FIGURE 6-3 Block flow diagram for treatment of rockets.

Treatment of Chemical Agent

The agent drained from the rockets and mines is continuously fed from the surge tank to the agent hydrolysis reactor (AHR), a continuously stirred tank reactor (Levenspiel, 1962), where it is hydrolyzed using the Army's agent neutralization process (see Appendix D.). The AHR is operated at 90°C and atmospheric pressure and is blanketed with nitrogen. Although agent hydrolysis should not generate any gases, the nitrogen blanketing is used as a precaution. The AHR effluent gas, mostly nitrogen, is passed through activated carbon filters and released to the plant ventilation system, which is vented through additional activated carbon filters prior to release to the environment.

The hydrolysate from the AHR is pumped to a storage tank where it is sampled and analyzed for agent. If the agent concentration exceeds design specifications, the hydrolysate is returned to the AHR to continue the hydrolysis. Otherwise, the hydrolysate is fed to the agent hydrolysate SCWO system.

The agent from the cryofractured projectiles and mortars may contain shards of solids that cannot be readily treated in a stirred tank reactor. Therefore, this agent is treated in the projectile rotary hydrolyzer (PRH). The PRH is a horizontal cylinder with a high lip at each end and slip seals between each lip and a solid bulkhead. It is operated at atmospheric pressure and equipped with a jacket that can be fed either steam or coolant to maintain a temperature of 90°C (194°F). The PRH rotates slowly, tumbling the contents, mixing them, and moving them down its length.

At the feed end of the PRH, caustic is directed through high-pressure nozzles to flush agent from the solid metal parts, forming a hydrolysate solution. At different points in the PRH, the solids mixed with the hydrolysate are passed over screens that drain the free hydrolysate through a surge tank to the AHR. The PRH cleans residual agent from the munition fragments and acts as the primary reactor for agent hydrolysis. The AHR hydrolyzes any residual agent not hydrolyzed in

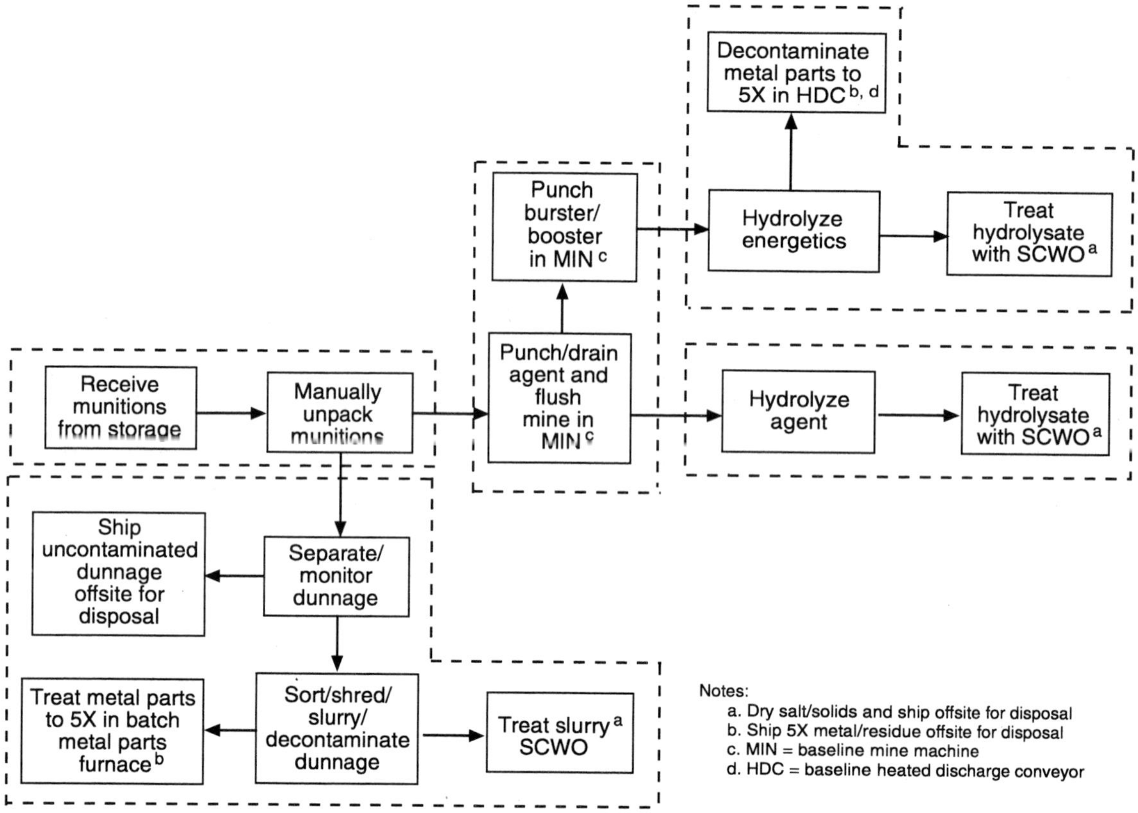

FIGURE 6-4 Block flow diagram for treatment of land mines.

the PRH. The effluent gases from the PRH are scrubbed and passed through carbon filters before being released to the plant ventilation system.

Treatment of Energetics

Energetic materials removed during the disassembly process are conveyed through the explosion-containment room discharge chute into the energetics rotary hydrolyzer (ERH). The ERH, similar in design and operation to the PRH, also operates at atmospheric pressure and approximately 90°C (194°F), which is above the melting point of TNT. The technology provides claims that the combination of rotary mixing, melting, and base hydrolysis converts solid explosives and energetics into a hydrolysate that can be treated by SCWO. (Treatment of the screened metal parts is discussed below.) The results of tests conducted by General Atomics and Los Alamos National Laboratory are included in the proposal to support the use of the ERH (General Atomics, 1998).

Hydrolysis of the energetics is expected to take from one to six hours, depending on temperature, mixing, and concentration of NaOH. At the time the technology package was submitted, neither optimum parameters nor the optimum size of the reactors had been determined. The proposal states that the rotary hydrolyzer for the final process will be oversized to extend residence time, if necessary.

The hydrolysate from the ERH is pumped to a storage tank where it is sampled and analyzed for agent and energetics. If either is found in excess of allowable concentrations, additional time is allowed to continue the hydrolysis. If analysis shows that the hydrolysate is acceptable, it is then fed to the energetics hydrolysate SCWO system.

The effluent gas from the ERH consists of nitrogen, water vapor, a small quantity of hydrogen produced by the chemical reaction of aluminum parts with caustic, and traces of volatile organic compounds. This effluent gas is treated in the same manner as effluent gas from the AHR.

Treatment of Hydrolysate with Supercritical Water Oxidation

The General Atomics technology package includes two of SCWO units—one to treat the hydrolysate from agent neutralization and one to treat the hydrolysate from energetics neutralization. The basic SCWO process is described in Appendix F. Only the specifics of General Atomics' implementation of SCWO are described here.

To treat agent hydrolysate, General Atomics proposes using a vertical, cylindrical SCWO reactor, similar in configuration to systems General Atomics tested for the Army in 1997 (GA, 1997; NRC, 1998). In these tests, mineralization levels for the organic constituents in the hydrolysate were high (i.e., conversion of carbon to carbon dioxide, hydrogen to water, and phosphorus, chlorine, and sulfur to inorganic phosphates, chlorides, and sulfates, respectively). In addition, extremely low levels of light organic compounds were found to be present in the off-gas stream (NRC, 1998). There was, however, significant corrosion and erosion of the titanium from the reactor walls (NRC, 1998). Therefore, General Atomics proposes using a platinum liner in the ACWA application.[1] The reactor operating pressure and temperature are approximately 650°C (1,200°F) and 230 atm (3,400 psi).

Treatment of Metal Parts

The metal parts that collect on the screen at the exits of the ERH and PRH are deposited on electrically heated discharge conveyors, which raise the metal temperature to more than 1,000°F for at least 15 minutes, thereby meeting the 5X decontamination criterion. General Atomics expects that some small fuze-train explosives may escape complete hydrolysis in the ERH and PRH and that these will initiate on the conveyors. General Atomics expects to show in its demonstrations that the initiation of fuzes will not be energetic enough to damage the conveyors.

Treatment of Dunnage

Dunnage consists of materials, like wooden pallets, metal banding, and DPE suits, some fraction of which may be contaminated with agent. General Atomics proposes to treat this material by shredding and separation of the metals, followed by SCWO treatment of the nonmetals and 5X treatment of the metals.

A low-speed shredder breaks up pallets, boxes, and other bulk dunnage or process waste materials. Next, the rough-shredded wood is reduced to small chunks in a hammer mill and then to fine particles in a micronizer. The micronized product is mixed with water and moved to a hydropulper where it is reduced to particles smaller than 0.5 mm in diameter, which is well within the target size (less than 1 mm) for SCWO processing. Dust from the micronizer is collected in a bag-house, drummed, and combined with the reduced process feed for further treatment.

Plastic and rubber materials are treated the same way, except that the hammer mill is replaced with a granulator. The output of the granulator is cryocooled (using liquid nitrogen) prior to being fed to the micronizer.

Carbon steel pieces are separated out magnetically and sent to the metal parts furnace (MPF) for 5X decontamination. The MPF is identical to furnaces used to decontaminate metal, except induction heating is used rather than natural gas. The MPF is vented through activated carbon and HEPA filters.

Process Instrumentation, Monitoring, and Control

The process parameters to be monitored are routine and include pH, temperature, and pressure of all streams. All monitors are commercially available.

Feed Streams

The technology provider generated a mass balance for the processing of 80 VX-filled 155-mm projectiles per hour (GA, 1998; Appendix B). This balance is for the entire plant except for the brine reduction area and the plant ventilation system. Feed streams are shown in Table 6-2.

General Atomics' technology package requires only two reagent feeds: (1) oxygen or air to the SCWO system and (2) sodium hydroxide (caustic) for hydrolysis of agent and energetics and to neutralize the acids produced by the SCWO destruction of the hydrolysate.

The process is a net producer of water (due to

[1]General Atomics experienced problems in fabricating a platinum liner for the ACWA demonstrations, so an Inconel 718 reactor with no liner was used instead.

TABLE 6-2 Process Inflow Streams for the General Atomics Technology Package (80 VX-filled 155-mm projectiles per hour)

Component	Amount (lb/hr)
VX	480
Munition dunnage	420
Nonprocess dunnage	472
Steel	7,102
Aluminum	98
Explosives	220
NaOH	
for munition handling	607
to energetics rotary hydrolyzer	277
Total	884
Water	
with caustic to agent hydrolyzer	911
to agent hydrolyzer	2,370
to energetics rotary hydrolyzer	2,194
with caustic to ERH	416
to dunnage processing	6,548
Total	12,439
Nitrogen	
to cryocooling	6,999
to HDC #1	470
to HDC #2	470
Total	8,407
Kerosene	537
Decontamination fluid	400
Plant ventilation process air	48,956
Total plant inflow	80,415

Source: GA, 1998.

SCWO) and, except for start-up, does not require additional water. Liquid nitrogen is required for the cryocooling processes and for inert blanketing of reactors and tanks.

Waste Streams

The following waste streams leave the plant:

- dried salts from SCWO that probably contain traces of organic materials and potentially hazardous metals
- gases from process vents and from the SCWO systems that are passed through HEPA and activated carbon filters (Table 6-3 shows the air emissions from the process and indicates their general type)
- metal parts that have been cleaned and decontaminated to the 5X condition
- spent carbon (from treating gas and exhaust air)—although this might be treated on site by the SCWO equipment
- spent HEPA filter elements

The mass flows for the waste streams are shown in Table 6-4.

Start-up and Shutdown

Start-up and shutdown are summarized on a process-by-process basis in Table 6-5. The shutdown procedure is the reverse of the start-up procedure–the flow of contaminated material is replaced with a flow of clean hydrolysis solution. No final shutdown (decommissioning) procedures are included in the technology package.

EVALUATION OF THE TECHNOLOGY PACKAGE

Process Efficacy

Effectiveness of Munitions Disassembly

General Atomics introduces two new technologies into the disassembly process—caustic-jet clean-out and cryofracturing of projectiles and mortars. Both technologies have been used in other industrial and munitions disposal processes. Water-jet clean-out of opened shells appears to be a reasonable way of breaking up and flushing away energetic materials in the cut and broken pieces without creating a high shear, which would increase the possibility of accidental detonation or deflagration. The committee, therefore, concluded that this modification was not likely to create operational problems; however, no data are available on the amount of residual material that would remain after water-jet cleaning.

In 1991, an NRC panel evaluated the use of cryofracture for accessing agent and energetic components in chemical munitions prior to incineration (NRC, 1991). At that time, the panel was concerned that the incineration of munition fragments after cryofracture was not adequately understood and concluded that cryofracture was not necessary in conjunction with incineration because the heat of incineration would remove residual agent, including gelled agent.

TABLE 6-3 Potential Air Emission Points for the General Atomics Technology Package

Source	Type	Disposition
SCWO system pressure let-down and liquid-gas separation	particulates, vapors	APCD,[a] environment
Washing of metal parts	droplets, particulates, vapor	room ventilation
Punching and draining of munitions	droplets, particulates, vapor	room ventilation
Cryofracturing	droplets, particulates, vapor	room ventilation
Thermal treatment (5X) furnace	vapor	APCD, environment
Building ventilation system	particulates, vapor	APCD, environment

[a]air pollution control device

In the present application, the ACW Committee has concluded that a key factor in the decontamination of munitions is good access to the residual agent in the munitions to ensure that the washing and hydrolysis are effective. Although improving access to the agent cavity is desirable, the committee found no data to indicate that cryofracturing would provide better access to liquid agent than more conventional approaches, such as shearing.

The ACW Committee then investigated whether the cryofracture process might improve safety or operability over the baseline process for accessing agent in projectiles and mortars. As discussed in Appendix C, the baseline disassembly process has encountered some difficulties in opening the agent cavity (removing the burster well) and draining the agent. Using cryofracture to break open the rounds would eliminate the difficulties with pulling the burster wells, some of which are welded in place. Furthermore, if the agent is gelled or crystallized, cryofracture should represent an improvement over the baseline suction approach. The committee, therefore, concluded that if the robustness and ability of cryofracture to access the agent cavity could be verified, cryofracture might improve performance over the baseline disassembly process.

TABLE 6-4 Process Outflow Streams for the General Atomics Technology Package (80 VX-filled 155-mm projectiles per hour)

Component	Amount (lb/hr)
Salt and other nonmetal solids	
from agent SCWO	1,181
from energetics/dunnage SCWO	949
Total	2,130
Treated steel	7,102
Treated tramp metal and glass	120
Water	
from agent SCWO	3,913
from energetics/dunnage SCWO	10,369
Total	14,282
Total plant outflow	80,414

Source: GA, 1998.

Effectiveness of Agent Detoxification

Agent hydrolysis has been studied extensively, and full-scale plants are being constructed at Aberdeen, Maryland, and Newport, Indiana, to destroy the agent stored there in bulk. These plants are scheduled to be completed before the full-scale implementation of ACWA technologies. Thus, the experience from these facilities should be available to the ACWA program. General Atomics would rely primarily on hydrolysis for achieving a high destruction efficiency for agents (99.9999 percent). (See Appendix E for a detailed discussion of agent hydrolysis.)

Effectiveness of Energetics Destruction

Although this process appears to be capable of destroying the energetic materials, the rate at which these materials will be processed cannot be determined at this time. The rate-limiting step for energetic destruction is the hydrolysis reaction (see Appendix E), which, in this application, is mass-transfer limited (i.e., the chunks of energetic material must be dissolved into the hydrolysis solution, and the rate of dissolution is

TABLE 6-5 Routine Start-up Procedures for the General Atomics Technology Package

Baseline disassembly	Follows the standard procedures developed by the Army.
Cryofracture	Liquid nitrogen bath is filled with nitrogen and munitions introduced.
AHR	Process monitors and controls are activated. Reactor is filled with caustic solution and brought to temperature. Agent or contaminated solids are introduced, as appropriate.
ERH and PRH	Process monitors and controls activated. Reactor is filled with caustic solution and brought to temperature while liquid is recycled. Solid materials are introduced.
SCWO reactors	Process monitors and controls are activated. Reactor is fed a mixture of pure water. Reactor is brought to temperature using auxiliary heaters. Kerosene and oxygen are gradually introduced at operating temperature to begin the chemical reaction. Kerosene is gradually replaced with hydrolysate.
Evaporator/crystallizer	Process monitors and controls and condenser coolant flow are activated. Tested SCWO effluent is introduced. System is brought up to operating temperature. Hydrolysate is introduced. Filter belt is activated.

limited by their surface area). The smaller the pieces, the faster the hydrolysis. At this point, there is little data to confirm that the chunks from the disassembly processes would be small enough to dissolve at the projected rates. If testing shows that the actual rate is different from the projected rate, the capacities or number of reactors will have to be adjusted.

Substituting a rotary reactor for a stirred reactor appears to alleviate the problem of jamming when metal parts are introduced. Tumbling, which will occur in the rotary reactor, is a standard industrial method of mixing solids with liquids.

Effectiveness of Supercritical Water Oxidation

The treatment of the agent hydrolysate by SCWO and the treatment of the SCWO effluent by evaporation and filtration are also being designed into the facility at Newport, Indiana. Experience from operating this facility should be available to the ACWA program for follow-on assessments. However, as is pointed out in Appendix F and in *Using Supercritical Water Oxidation to Treat Hydrolysate from VX Neutralization* (NRC, 1998), although there is a keen interest in using SCWO for treating a variety of wastes, very little production experience is available.

SCWO appears to be capable of decomposing the hydrolysis products into waste streams that can be disposed of in an environmentally sound manner. (Mustard does contain volatile low molecular weight chlorinated hydrocarbons that can be difficult to treat, but they are expected to be oxidized by SCWO. However, this will have to be demonstrated.) The volume of air emissions from the process is small and will be further treated by activated carbon and HEPA filters. With proper monitoring of the adsorbers and filters, these small emissions should easily meet regulatory requirements.

The aqueous stream from the process consists mostly of pure water, the vast majority of which is recycled to the process. In the opinion of the committee, the small amount of excess water appears to be acceptable to a wastewater treatment plant.

The solid waste from the process consists mainly of sodium salts, phosphate, chloride, fluoride, and sulfate. Most likely, these solids will contain some hazardous compounds and elements (how much is not known yet). In the committee's opinion, the nature of this stream and the probable concentration of hazardous constituents should not prevent its being stabilized and safely disposed of in a hazardous-waste landfill. However, this conclusion must be confirmed by further studies.

Sampling and Analysis

The process appears to have no unusual sampling or analytical problems and requires only well developed and generally accepted procedures.

Maturity

The overall process is a combination of several integrated operations, all of which are operated in a batch or semibatch mode. All of the technologies, with the exception of SCWO, have a substantial background, although some require demonstrations for their use with chemical weapons. The least mature operations are described below.

Hydrolysis of Energetics. Energetic materials have been hydrolyzed safely in the laboratory for more than a century, but large-scale hydrolysis has been rare. As noted in Appendix E, the problem is the size of the pieces being hydrolyzed. Because hydrolysis is a solid-liquid reaction, it occurs only at the solid-liquid interface. If a fixed mass of energetic material is broken into many small pieces, rather than a few larger ones, a much larger total surface area is presented to the hydrolyzing solution, and the hydrolysis rate can be predicted with reasonable accuracy. However, the sizes of the pieces of explosives and propellant can vary widely in each batch, which could slow the hydrolysis or even stop it if products are deposited on the surface layers. Avoiding this problem will require additional development for this technology and for other technologies that propose to use hydrolysis to destroy the energetics.

SCWO Operation at the Proposed Scale. The only SCWO system in commercial operation belongs to Huntsman Corporation and is located at their Austin, Texas, facility (Lyon and Ullrich, 1998). This system has been operating for about two years, treating approximately five gallons per minute of wastewater containing approximately 10 percent organic material. This system is about one-fourth the size of the system proposed by General Atomics. A reactor with the same design proposed by General Atomics has been tested on a variety of materials similar to those being treated in this program, but the maximum duration of the tests has been about 40 hours (NRC, 1998). General Atomics has recently shipped a SCWO system for treating shipboard wastes to the U.S. Navy (Hazelbeck et al., 1998), and the Army is planning to use SCWO units for treating VX hydrolysate in Newport, Indiana, with SCWO systems similar in design and size to General Atomics' proposed system.

SCWO Durability. The durability of the components and materials of the SCWO system in the highly corrosive environment generated by the treatment of feed materials containing large amounts of sulfur, phosphorus, and chlorine must be determined (NRC, 1998). General Atomics recognizes this problem and has presented materials-corrosion data in its proposal identifying potential materials of construction that would minimize corrosion. Platinum linings are planned for areas of the reactor that will be particularly vulnerable to corrosion. However, the fabrication of a platinum liner for the ACWA demonstration was not successful. General Atomics proposes developing a scheduled maintenance and replacement program based on anticipated corrosion rates, but these rates have not been established. Therefore, materials of construction remains a critical issue that must be resolved.

The treatment of dunnage using SCWO also raises some concerns regarding durability. First, the slurry stream will be very large. In fact, because of the volume of this stream, General Atomics initially proposed that the dunnage be tested and that uncontaminated dunnage be sent off site for disposal. Second, the ability of existing pumps to pump the slurry up to the high SCWO pressures has not been tested. General Atomics claims that a proprietary pump has been developed, but no data on its reliability was provided to the committee. Third, the behavior of solid materials in the SCWO, but reactor is unknown. Even though metal will be removed during the size-reduction process, some metal shards will probably pass through to the SCWO, but it is not known whether the SCWO system will be capable of handling them.

Process Robustness

The overall process appears to be capable of withstanding the following problems.

Incomplete Drainage of Agent or Gelled Agent in the Munitions. Once the free liquid has been drained from rockets and mines, the whole munition will be flushed with hot hydrolysis solution or water and then immersed in hydrolysis solution for an extended period of time. Any remaining agent will be destroyed by subsequent 5X treatment. Although this process appears promising, it has not been demonstrated that the flushing and immersion steps will fully remove any gelled agent. If gelled agents were to remain, processing rates might have to be reduced to allow for increased residence time in the furnace. Cryofracture, which will be used to access the agent in projectiles and mortars also appears promising but has not yet been demonstrated to improve on the baseline draining method.

Variations in the Composition of Agent or Energetics. Variations in composition of the feed material might affect hydrolysis and SCWO. Because hydrolysis will be performed on a batch basis, any failure in the hydrolysis stage of the process could be corrected by extending the time for hydrolysis or increasing the concentration of reagent. SCWO has already achieved exceptionally high destruction levels for a wide variety of organic compounds. In addition, the semi-batch process allows for the SCWO effluent to be collected and held until it has been thoroughly analyzed. General Atomics indicates that if the effluent does not meet specifications after initial processing, it can be retreated until it does meet specifications.

Difficulty in Removing Nose Closures on Projectiles and Mortars. The use of cryofracture eliminates the need to remove the nose closures on projectiles and mortars. Cryofracture appears to be a robust technology, and the equipment is commercially available at the required scales. One common use of cryofracture is to break up waste tires for use as fuels for boilers or cement kilns or to reduce their volume for disposal in a land-fill.

General Atomics cites research indicating that the cryofracture process has been improved significantly since it was developed in the late 1980s. The cited research is an extensive series of tests General Atomics conducted for the Air Force in which solid-fuel rockets were frozen in liquid nitrogen and fractured successfully. Because the munitions involved in this program are smaller than the Air Force rockets and will not contain explosives, new problems are not expected to arise.

Monitoring and Control

The process does not require any unusual monitoring or control systems. All of the monitoring and control systems are commercially available and appear to be reliable.

Applicability

Before a plant based on this technology package would be ready for operation, many of the technologies will have been used for the destruction of chemical agents at Newport and Aberdeen. To date, some of the technologies (SCWO of propellant hydrolysate, cryofracturing of projectiles and mortars) have only been tested at reduced scale and/or throughput. In spite of these limitations, the process appears to be a reasonable application of the technologies. The General Atomics technology package addresses all of the munitions identified in DOD's RFP; therefore, it should be applicable to any storage site.

Process Safety

The General Atomics technology package would require the following unique equipment:

- a modified (from baseline) rocket-shear machine
- a modified (from baseline) mine machine
- rotary hydrolyzers (based on a modified rotary-drum dryer design) for the hydrolysis of energetics from all munitions and the decontamination of metal parts after the cryofracture of projectiles and mortars
- conveyors to carry projectiles and mortars through the cooling vessels prior to cryofracture
- robots to unpack projectiles and mortar rounds and place them in the cryocoolers (may be similar to baseline pick-and-place robots)
- a modified (from baseline) induction-heated MPF and electrically heated discharge conveyors
- commercial hydropulping equipment modified to shred, mascerate, and form a slurry containing contaminated dunnage
- SCWO reactors for the destruction of the hydrolysates of agent and energetics

Based on the consequences of failures in both the low- and high-pressure systems, the committee concluded that there were no unusual or intractable process safety problems. The cryofracture process (cryocooling and press operation) operates near the atmospheric boiling point (–196°C; –321°F) of liquid nitrogen. The hot water jet and hydrolysis processes operate at relatively low temperatures (90°C; 194°F) and low pressure (near atmospheric), eliminating the problem of significant stored energy. The electrically heated discharge conveyors and MPF operate at 1,000°F (538°C). Hydrolysates will be sampled for the presence of agent and energetics before release to the SCWO reactors. The SCWO reactors operate at 230 atm

(3,400 psi) and a temperature of 650°C (1,200°F). All off-gases will be passed through the facility ventilation system (which includes carbon HEPA filtration).

Worker Health and Safety

The proposed system provides "defense in depth" because agent and energetic destruction are verified after both sequential steps for all munitions. The separation of energetics from agent, followed by the destruction of both materials in caustic solutions minimizes the hazard of explosions. In addition, these processes are operated in structures designed to contain explosive overpressure.

The disassembly processes are derivatives of the baseline processes for disassembling rockets and land mines and are not considered to represent new or increased levels of hazard. These processes will be conducted in vessels or structures designed to withstand explosive overpressure in case an initiation does occur.

The method proposed for the disassembly of projectiles and mortar rounds (cryofracture) is unique. The hazard level from cryofracture appears to be no greater than the level for baseline disassembly. However, the General Atomics hazard analysis does identify scenarios in which fragments containing energetic might be squeezed and initiated during transfer from the cryofracture press to the rotary hydrolyzers. Presumably, these hazards can be accommodated in the equipment design, but the design must also minimize worker exposure to agent during the repair or replacement of parts broken or damaged by explosions.

General Atomics plans to hydrolyze different types of energetic materials simultaneously in the same reactors, which the committee believes could lead to the formation of compounds that are both energetic and sensitive (see Appendix E). Therefore, energetic materials should be processed in separate reactors unless testing shows that sensitive compounds are not formed.

The hydrolysis processes that include dissolution of aluminum parts (e.g., M55 rockets) will generate hydrogen gas, which could conceivably rise to concentration levels that permit ignition in the process areas during upset conditions. Nitrogen purge gas is used to keep hydrogen concentrations below the level of concern.

High pressures, the primary cause of gas leaks into ventilated areas, will be minimized because all parts of the process except the SCWO reactors operate at very low pressures. Process-vessel vapor spaces should always maintain a positive pressure to prevent in-leakage of air and to ensure purge gas flow to process vessel headspaces.

The most significant issue related to worker safety may be during maintenance in DPE suits on specialized equipment (e.g., the rotary hydrolyzers and cryocooling conveyors), which have systems operating in a caustic solution and at very low temperatures. Experience with such systems is limited, and start-up problems may require significant maintenance for the full-scale application, thereby increasing the risk of worker exposure to agent. A second issue related to worker safety is the presence of nitrogen gas in cryofracture equipment, which can lead to very low concentrations of oxygen. As noted by General Atomics, nitrogen is an asphyxiant, so workers will have to be supplied with fresh air, and work areas will have to be monitored.

The SCWO reactors and associated water supply systems will operate at high pressure, representing significant sources of stored energy. Hence, it will be very important that the reactors and associated piping be designed and maintained to minimize ruptures and leaks. Secondary containment should be sized to accommodate a hypothetical "worst case" rupture of the SCWO reactor. SCWO system failures could require extensive repair work that requires either DPE suits or protective gear for hazardous chemicals and, thus, present an opportunity for worker exposure during maintenance.

Fuze bodies and booster pellets that are not dissolved in the caustic solution also represent an explosive hazard in the rotary hydrolyzers and heated discharge conveyors. The technology provider intends to demonstrate a technology that will reduce size mechanically to facilitate full dissolution. In addition, the hydrolyzers and heated discharge conveyors will be designed to withstand initiation of these energetic components.

The primary hazardous materials used during agent and energetic destruction are sodium hydroxide, liquid and gaseous oxygen, and methane (natural gas). Sodium hydroxide will be delivered in solid form and dissolved in water to make a 40 percent caustic solution, which is strongly corrosive to all body tissue. Liquid and gaseous oxygen and methane are handled routinely and safely in many industries and do not represent an unusual hazard to workers.

Public Safety

Accidental releases of agent or other regulated substances to the atmosphere or the groundwater system are extremely unlikely. However, hold-test-release systems for all process effluent gaseous streams are not included in the proposal. Although caustic scrubbing and activated carbon and HEPA filters should be adequate (judging from experience with baseline technology), they do not meet the stakeholder hold-test-release criterion for all gaseous effluents produced during normal operation. The primary cause of a release of material containing agent or other regulated substances would be an explosion or the rupture of a pipe or vessel, but the likelihood of such an event at the conclusion of the design process for the full-scale facility should be extremely small. The design process is assumed to include a QRA (quantitative risk assessment).

Human Health and the Environment

The environmental impact of the proposed process appears to be minimal. All handling of agent and all processing are conducted indoors in sealed rooms that are vented through HEPA and carbon filters. Streams handled are very small and manageable by the standards of almost any industrial-scale process. Process air releases are small enough that they could be collected and stored for analysis to verify their quality prior to release.

Effluent Characterization

The liquid effluents consist of water from the evaporator/crystallizer used to produce the solid filter cake. This material is essentially distilled water and should not pose a significant hazard to human health or to the environment. The solid waste from the process, consisting of dried filter cake, is likely to require stabilization prior to disposal in a hazardous-waste landfill.

Not enough information is available on the process to determine the hazardous constituents (if any) in the gaseous effluent, especially from the pressure let-down of the SCWO reactor. If HEPA and carbon filters are used properly, these discharges should meet regulatory standards. However, this must be confirmed through comprehensive testing.

Completeness of Effluent Characterization

The liquid and solid effluents are well characterized. Only the major components of the gaseous effluents have been characterized. The gaseous emissions will have to be characterized for HRAs and environmental risk assessments as required under current EPA guidelines.

Effluent Management Strategy

The proposed strategy appears to be reasonable and has a number of built-in redundancies that should protect public health and the environment.

Resource Requirements

The power and other resource requirements for the system should pose no difficulties. Because the system is a net producer of water, it does not require a large water inventory. Table 6-3 shows that about 1,800 lb/hr of water is generated in the process. Thus, steam will have to be vented from the dryer or discharge condensate from the brine reduction area.

Environmental Compliance and Permitting

The combination of technologies in the General Atomics technology package is not expected to lead to environmental compliance or permitting problems. All process waste streams except the SCWO off-gas will be evaluated prior to release to confirm that they are either free of regulated substances or that they are at acceptably low concentrations. The SCWO off-gas is scrubbed, monitored, and passed through activated carbon filters.

STEPS REQUIRED FOR IMPLEMENTATION

The following steps would have to be taken to implement the General Atomics technology package.

1. Conduct tests of the cryofracture process to ascertain if it provides better access to the agent cavity in projectiles and mortars then the baseline disassembly process.
2. Sample and analyze air emissions from the demonstration system. The air emissions will have to be measured to a level of detail and accuracy that can be used for HRAs and environmental risk assessments required by EPA (1998a).
3. Verify that energetic materials encased in metal (e.g., rocket or other munitions fragments) will be hydrolyzed.
4. Ascertain how well the SCWO process can handle high-solids materials (shredded dunnage).
5. Ascertain how well the SCWO system can treat hydrolysate containing large amounts of chlorides, sulfur, and phosphates on a continuing basis.
6. Determine erosion and corrosion behavior of the components of the SCWO system.

FINDINGS

Finding GA-1. Cryofracture appears to be an effective method for accessing the agent in projectiles and mortars and might provide an improvement over baseline disassembly in accessing gelled or crystallized agent. This remains to be demonstrated.

Finding GA-2. Hydrolysis of energetics at the scales proposed by the technology provider is a relatively new operation. Chemically, it is possible to hydrolyze all of the energetic materials; however, the rate of hydrolysis is limited by the surface area and, therefore, depends on particle size. (Smaller particles are more desirable because they have a higher surface-to-volume ratio.) The proposed method of removing and hydrolyzing the energetics appears to be reasonable, but further testing is required to determine the hydrolysis rates and to confirm that throughput rates can be achieved.

Finding GA-3. The rotary hydrolyzer appears to be a mature reactor configuration that is well suited for this application.

Finding GA-4. Shredding of dunnage and injection of the slurry directly into a SCWO system is a new and unproven process. While General Atomics claims to have developed a proprietary pump capable of pumping the slurry at high pressures, but it has not been tested under the intense solids loading anticipated. Furthermore, the injection of large amounts of solid material, including wood shreds, cut-up nails, and complex organic materials, such as pentachlorophenol and other wood preservatives, into the SCWO system has not been demonstrated. Considering the difficulty SCWO reactors have encountered with deposition of solids when liquids are treated, the committee believes that this application of SCWO may encounter significant difficulties. (At the time of this writing, processing of solids with SCWO was being performed as part of the ACWA demonstrations.)

Finding GA-5. All of the findings in the NRC report, *Using Supercritical Water Oxidation to Treat Hydrolysate from VX Neutralization*, apply to the General Atomics system.

Finding GA-6. The crystallization and evaporation operations have not been tested for this application. Although these are conventional technologies and are expected to work effectively, testing will be necessary.

Finding GA-7. No hold-test-release facilities are provided for gases from the hydrolysis reactors or the SCWO reactors. These gases will be scrubbed using activated carbon prior to release.

7

Lockheed Martin Integrated Demilitarization System

INTRODUCTION AND OVERVIEW

The Lockheed Martin Integrated Demilitarization System (LMIDS) was designed by a project team of Lockheed Martin and seven other companies (see Table 1-3). The LMIDS includes four primary technologies. First, the chemical agent, the energetic materials, and the metal parts are separated via a modified version of the Army's *baseline disassembly* process. Second, *caustic hydrolysis* is used to decompose the chemical agent, break down the energetic materials, and decompose agent on metal parts and dunnage. Third, the hydrolysates from the hydrolysis processes are further treated using SCWO (supercritical water oxidation). Finally, *gas-phase chemical reduction* (GPCR) is used to decontaminate the metal parts and dunnage to a 5X level and to treat gaseous effluents from the hydrolysis processes. Table 7-1 describes how these four technologies are used to perform the six primary demilitarization operations described in Chapter 1.

The LMIDS segregates the four technologies, assigning a separate process area for each. Figure 7-1 illustrates how the technologies are linked and shows the basic process flow. The four process areas are:

- munitions access and energetic deactivation (Area 100)
- caustic make-up and hydrolysis (Area 200)
- SCWO (Area 300)
- GPRC (Area 400)

Each of these areas is described in detail in the next section.

The technology provider addressed the processing of rockets, projectiles, and mortars but did not explicitly address the processing of land mines. However, the proposal and the data-gap report included statements that LMIDS could accommodate land mines with minor modifications.

DESCRIPTION OF THE TECHNOLOGY PACKAGE

Access to Munitions and the Deactivation of Energetics (Area 100)

In area 100, the munitions are disassembled, and the chemical agent and energetic components are separated using equipment adapted from the baseline process (see Appendix C). Energetic materials are initially deactivated via hydrolysis and metal parts decontaminated with caustic.

Rocket Disassembly

The delivery of the M55 rockets to Area 100 is identical to the baseline process. Once there, the rockets are unpacked from their pallets and, still enclosed in their shipping/firing tubes, are loaded one at a time nose first through an entry airlock into the rocket demilitarization chamber. Inside this chamber, a hole is punched in the firing tube and rocket, and the agent is drained as in the baseline process. The agent is then pumped into an agent weigh tank to verify the amount drained, and the rocket is sheared into pieces. During the shearing operation, the following modification is made to the baseline process: when a shear cut first exposes the propellant, a low-pressure hot-water jet is used to break

TABLE 7-1 Summary of the LMIDS Approach

Major Demilitarization Operation	Approach(es)
Disassembly of munitions	Modified baseline disassembly (multiple lines, modified layout, new drain and wash).
Treatment of chemical agent	Hydrolysis with caustic; SCWO of hydrolysate; GPCR of off-gas.
Treatment of energetics	Hydrolysis with caustic; SCWO of hydrolysate; GPCR of off-gas.
Treatment of metal parts	Wash in caustic; treatment in thermal reactor to 5X; GPCR of volatilized materials.
Treatment of dunnage	Wash in caustic; treatment in thermal reactor to 5X; GPCR of volatilized materials.
Disposal of waste	**Solids.** Decontaminated metal parts to recycling facility; decontaminated solid residue from GPCR to landfill; salts from GPCR to treatment, storage, and disposal facility (TSDF); solids from SCWO to TSDF; uncontaminated packing materials to landfill. **Liquids.** None **Gases.** Gas from GPCR burned in boiler; gas from SCWO released to atmosphere through carbon filters.

up and remove the propellant from the interior of the rocket motor casing.

Initial Deactivation of Rocket Energetics

The sheared metal parts, bursters, fuzes, and fragmented propellant are transported (via gravity feed) into wire baskets in the rocket hydrolysis vessel. The baskets move gradually from the vessel feed point to the discharge point. A 20-percent NaOH solution (caustic) at 90°C enters near the basket discharge point and flows countercurrent to the basket motion. The caustic is circulated to ensure mixing between the caustic solution and the rocket parts. The caustic dissolves the aluminum fuze, exposing the energetic materials. The elevated temperature causes the energetic materials to melt, and these materials are then rendered inert via hydrolysis (see Appendix E for a discussion of the chemistry of energetic hydrolysis). The residence time is set to ensure that when a basket reaches the exit station, the aluminum in the fuzes has dissolved, and the energetic material has been completely removed from the remaining parts. In the exit station, a gas sample from the vapor space over the basket is analyzed to verify that the agent concentration is below an acceptable level. Isotopic neutron spectroscopy is also used to confirm that no significant amounts of energetics are present. If the results are acceptable, the basket is

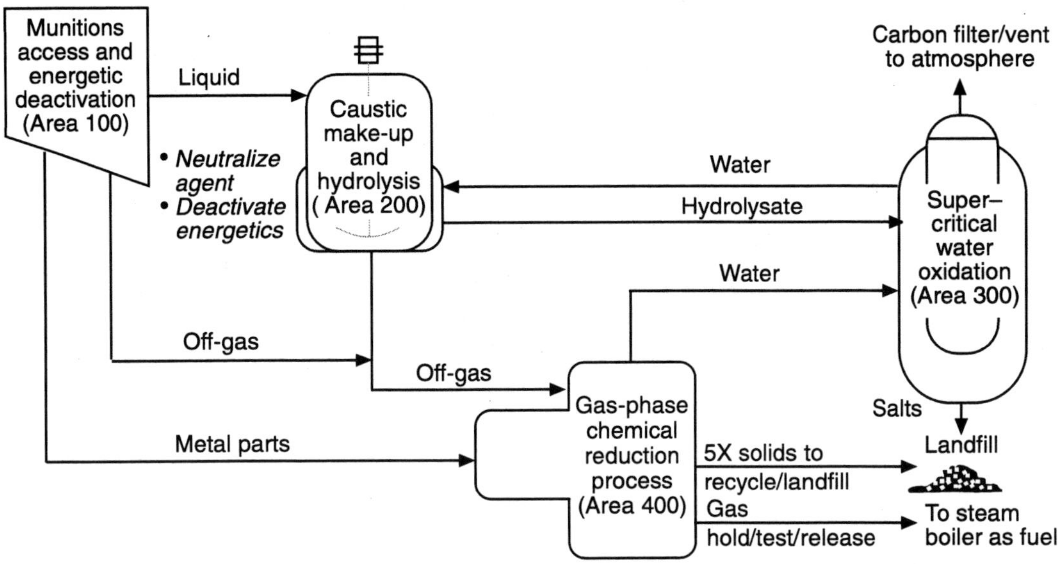

FIGURE 7-1 Process flow for the LMIDS.

moved to the thermal-reduction batch processor in Area 400 for thermal decontamination of the metal parts. Baskets that do not pass this test are returned to the rocket hydrolysis vessel for additional washing.

Projectile/Mortar Disassembly

Mortars and projectiles are delivered to Area 100 in the same manner as they are delivered in the baseline process (see Appendix C). The nose closure, miscellaneous parts, and burster are separated, also according to the baseline process. From this point on, the process differs from the baseline process. Bursters are fed into one of the burster hydrolysis vessels, while fuzes and supplemental charges (if present) are fed into the nose-closure hydrolysis vessel, and nonenergetic nose closures and other miscellaneous parts are collected in baskets for delivery to the thermal-reduction batch processor in Area 400. The treatment of the energetics is described in the next section.

Once the energetic components have been removed, the projectile/mortar is transported out of the explosion-containment room and loaded into a special tray. Once the tray is full, it is conveyed into the projectile hydrolysis chamber where the burster wells are removed using the baseline approach (although the machine has been significantly redesigned). When the burster wells from all munitions have been removed, the tray is conveyed into the projectile hydrolysis vessel.

All operations in the projectile hydrolysis vessel are new. Agent is drained from the projectile bodies by inverting the tray. The drained projectile bodies then undergo initial decontamination by flushing with 90°C sodium hydroxide solution to loosen any heels or crystalline material that may have formed during storage. After flushing, the vapor space is monitored to ensure that the agent concentration is below the level for the 3X standard. If the monitoring produces acceptable results, the tray is sent to the thermal reduction continuous processor in Area 400 for further processing. Otherwise the flush cycle is repeated.

Initial Deactivation of Energetics from Projectiles/Mortars

The energetics from the projectiles/mortars include bursters, fuzes, and supplemental charges. The bursters enter the buster hydrolysis vessel from the explosion-containment room via a gravity feed and are contained in a wire basket. A slightly pressurized caustic solution at 135°C is constantly pumped through the vessel to facilitate melting, dissolution, and hydrolysis of the energetic material. Multiple bursters can be processed simultaneously, and the basket remains in the caustic solution long enough to ensure that all of the energetic material has dissolved completely. The basket is then raised, allowed to drain, and passed into an airlock where the headspace is tested for agent. Isotopic neutron spectroscopy is also used to ensure that no significant amounts of energetics remain. If the agent concentration and neutron spectroscopy results are acceptable, the basket is transported to the thermal-reduction batch processor in Area 400; if not, the basket is returned to the buster hydrolysis vessel.

The fuzes and supplemental charges are treated in the nose-closure hydrolysis vessel in the same manner as bursters. The caustic dissolves the aluminum portions of the fuzes and exposes the energetic materials.

Caustic Make-Up and Hydrolysis (Area 200)

In Area 200, chemical agent and energetics are deactivated, separately, by hydrolysis with hot caustic solution in one of several hydrolysis vessels.

Hydrolysis of Drained Chemical Agent

The chemical agent drained from the munitions in Area 100 is pumped to an agent neutralization reactor in Area 200. The agent neutralization reactor is a 400-gallon, baffled reactor filled with aqueous sodium hydroxide solution (20 percent) heated to 90°C. Agent is introduced at a measured rate until the specified agent loading is reached. Agent feed is then switched to another reactor. Each agent neutralization reactor is operated in a batch mode and is continuously agitated with two high-efficiency impellers to facilitate the complete hydrolysis of agent. The details of the agent hydrolysis reactions are presented in Appendix D.

The hydrolysis reaction mixture is kept at a pH greater than 13 with excess NaOH solution at all times. The batch loading and retention times vary with the type and composition of the materials being neutralized. The technology provider's test results for a GB loading of about 16.3 wt. percent produced a destruction efficiency greater than 99.9999 percent with a retention time of two hours, and a destruction efficiency of 99.97 percent for VX after 30 minutes of reaction time, with reactants consisting of mixtures of VX, Comp B, blended propellant powder, and aluminum alloy 6061 T6. (This feed mixture was generated to simulate the feed from M55 rockets.)

The agent neutralization reactors and all other neutralization (hydrolysis) reactors are blanketed with nitrogen, and an induced draft fan draws a slight vacuum on the reactor headspace. Off-gases evolved during the hydrolysis process are drawn by the fan through a reflux condenser to condense water vapor and other condensable vapors. The condensate flows back into the reactor, and the noncondensible gases flow continuously to the GPCR reactor Area 400 for further treatment.

At the end of the processing period, the reactor's liquid contents are transferred to a holding tank and sampled to ensure that the agent concentration is below the established threshold. If so, the hydrolysate is transferred to a feed tank for SCWO treatment in Area 300. If not, the hydrolysate is recycled to one of the agent neutralization reactors for additional processing.

Because GB has been shown to reform at pH below 13 (see Appendix D), excess caustic is used throughout the hydrolysis process to prevent the reformation of agent. Excess caustic is also necessary for the SCWO treatment of hydrolysate to neutralize the acids formed by heteroatoms (e.g., F and Cl) during the oxidation process. In the LMIDS, all batch neutralization reactors are operated in a staged sequence. While one reactor is receiving agent, the second is in the reaction mode, and a third is either being emptied into a hydrolysate holding tank or being refilled with the caustic decontamination solution. At the end of the neutralization process, the reaction product (i.e., hydrolysate) is discharged to a holding tank where agent analysis is conducted to ensure that agent concentration requirements have been met. Agent destruction efficiencies of 99.99 percent are expected for VX by the technology provider, and 99.9999 percent for GB, HD, H and HT. (Note that neutralization followed by SCWO treatment is expected to result in agent destruction to 99.9999 percent and concentration below detection limits.)

Hydrolysis of Energetics

The caustic solutions from the rocket, burster, and nose closure hydrolysis vessels in Area 100 are continuously fed to an energetics deactivation reactor in Area 200 to ensure that the hydrolysis reactions are driven to completion. (See Appendix E for a detailed discussion of the hydrolysis of energetics.) The energetics deactivation reactor is a 9,200-gallon vessel filled with aqueous sodium hydroxide solution (20 percent) heated to 90°C. The energetics deactivation reactor operates in the same way as the agent neutralization reactor. The off-gases are passed through a reflux condenser before being sent to the GPCR reactor in Area 400 for further treatment.

The liquid solution is held in the energetics deactivation reactor for a specified period of time (to be determined). A sample is then taken and tested for agent and energetic residue. If neither agent nor energetic is detected above the target level, the hydrolysate is fed into the feed tank in Area 300 for mixing with agent hydrolysate and treatment by SCWO. If agent or energetic is detected above the target level, further reaction time is allowed, and another sample is then taken.

Supercritical Water Oxidation of Hydrolysates of Agent and Energetics (Area 300)

The LMIDS uses SCWO for the final destruction of the hydrolysis products of both agent and energetics. (The basics of SCWO are described in Appendix F.) The SCWO process in the LMIDS uses a transpiring platelet wall reactor developed and patented by GenCorp/Aerojet and Foster Wheeler. The inner wall of the reactor is formed of layers of porous platelets that allow the continuous transpiration of deionized water at 315°C (600°F) through the inside wall of the reactor during the SCWO reaction. This inner transpiring wall is contained within a conventional outer wall.

The injection of transpiration water during operation is claimed to separate the SCWO working fluid, which will be at 780°C (1,436°F) and 3,500 psi (238 atm), from the inside surface of the reactor, which is kept at the transpiration water temperature of 315°C (599°F). This reactor technology is purported to have the following advantages over conventional-wall SCWO reactors:

- Contact between the working fluid and the reactor wall is reduced, thereby minimizing or corrosion.
- Deposition of salt on the reactor wall is essentially eliminated.
- The cooled reactor wall allows higher working-fluid reaction temperatures, reducing the residence time necessary for complete oxidation.

The cooler transpiration water (315°C versus 780°C for the working fluid) is intended to dissolve any inorganic salts that reach the reactor wall and carry them to the bottom of the reactor, where, together with the other reactor contents, the reaction mixture is quenched and collected. This design is intended to prevent the deposition of inorganic salts and plugging. A schematic diagram of the transpiring wall platelet liner is shown in Figure 7-2.

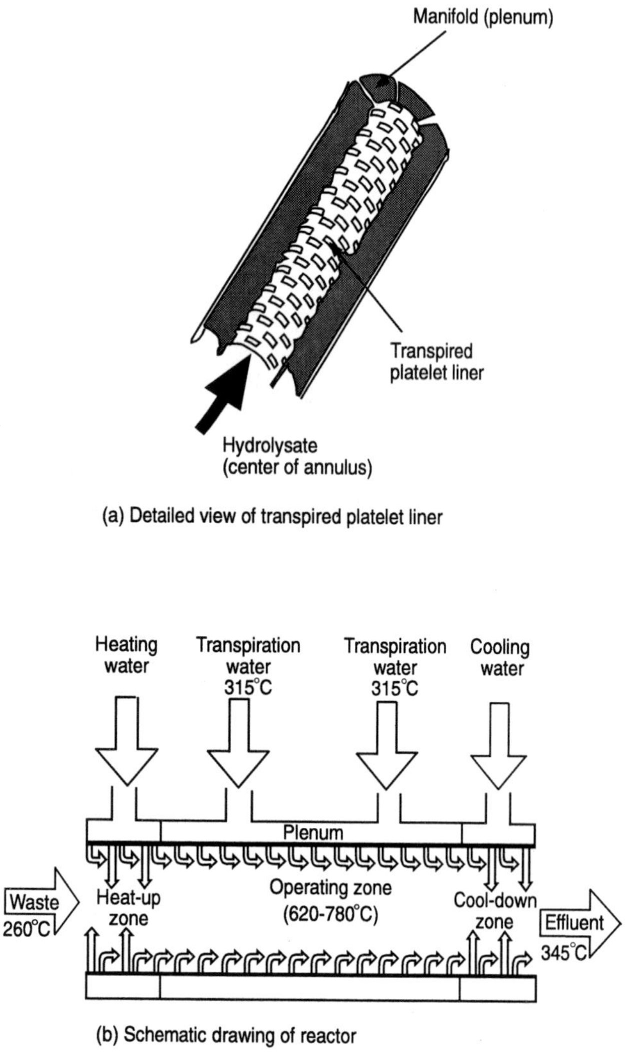

FIGURE 7-2 Transpiring-wall platelet liner.

Prior to injection into the SCWO reactor, the agent and energetic hydrolysates are mixed in the SCWO feed tank, heated, and stirred to maintain a uniform solution at 85°C. The mixed hydrolysate from the feed tank is pumped up to the operating pressure of the transpiring-wall SCWO reactor and mixed with supplemental fuel (kerosene or isopropyl alcohol) to ensure a high temperature. The hydrolysate/fuel mixture is then heated to 260°C by heat exchange with the SCWO reactor effluent in the hydrolysate feed/effluent heat exchanger, and the mixture is fed into the SCWO reactor. Compressed oxygen is preheated to 205°C and fed concurrently to the reactor as the oxidant.

The proposed full-scale SCWO reactor is a 12-ft-long, vertical down-flow, cylindrical reactor that processes about 1,200 lb of hydrolysate per hour. The hydrolysate/fuel oxidation reaction begins at the reactor inlet at 510°C. The oxidation reaction results in temperatures of 780°C at the top of the reactor and about 620°C at the bottom. The reactor is designed to have a total residence time of about 10 seconds. This temperature-residence time combination is believed by the technology provider to be sufficient to oxidize all hydrolysate organics to the desired destruction level. The oxidation products are quenched at the reactor bottom with a water spray to about 345°C. The heteroatoms (Cl, F, N, S, and P) in the hydrolysate react with the excess sodium hydroxide to form sodium salts. Destruction efficiencies of 99.9999 percent are claimed by the technology provider for all agent and energetic hydrolysates.

The effluent from the SCWO reactor, consisting of gases and liquid with dissolved salts, is cooled in heat exchangers and then depressurized through a let-down valve to atmospheric pressure for separation of the gases from the liquid brine. The liquid brine is sent to an evaporator for drying. The evaporator steam is condensed and the water recycled as process water. The dried salts, which are sampled and analyzed for hazardous constituents in accordance with RCRA requirements, are stabilized off site and disposed of in a landfill.

The technology provider claims that the effluent gas stream contains only nonregulated gases, mainly carbon dioxide, excess oxygen, and a small amount of nitrous oxide. Small amounts of carbon monoxide and low molecular weight hydrocarbons may also be present. The gas stream is dried and passed through carbon bed filters to remove traces of volatile organics, and the final gas effluent is monitored and analyzed for regulated constituents as it is vented to the atmosphere.

Gas Phase Chemical Reduction Process (Area 400)

The GPCR process developed and patented by ELI Eco Logic is used to eliminate organic chemicals on decontaminated metal parts and dunnage (from Area 100) or in gaseous process wastes or off-gases from the neutralization reactors in Area 200.

The GPCR (Area 400) process block consists of three parts: a thermal-reduction batch processor; a thermal-reduction continuous processor, where applicable; and a GPCR reactor. The thermal-reduction batch processor is a large vessel that will be loaded with potentially contaminated metal parts (such as pieces of sheared M55 rockets and projectile/mortar casings that were not dissolved in the hydrolysis step), other metal parts, and dunnage. The inlet door is opened, and bins containing the solids are conveyed into the thermal-reduction batch processor chamber. After the inlet door is sealed, nitrogen is introduced to purge oxygen. This nitrogen purge gas is vented through carbon filters. The chamber is then heated until the lowest temperature recorded on the load is 538°C (1,000°F) for at least 15 minutes to ensure 5X treatment. The main process in this chamber is the thermal desorption of organic matter. Gases from the thermal-reduction batch processor are swept to the GPCR reactor. After a batch has been treated, the chamber is purged and cooled with steam, purged with nitrogen, and unloaded.

Dunnage that is (or might be) agent-contaminated is washed and immersed in 20 percent caustic solution, loaded into bins similar to those used for processing metal parts, and placed in the thermal-reduction batch processor, which serves as a pyrolysis reactor. The products are hydrocarbon gases, hydrocarbon liquids, silica residue, and carbon soot. The gases are fed into the GPCR reactor, where they are partially converted into reformer gas (a mixture of hydrogen, methane, carbon monoxide, carbon dioxide, and steam). According to the technology provider, approximately 10 percent of the carbon feed may remain as soot that will require off-site disposal. Other hydrocarbons, such as tars and phenolic compounds, may also be present.

At munition depots where processing results in a large quantity of metal parts of a consistent size and shape (for example, projectile parts), a second type of thermal desorption reactor, the thermal-reduction continuous processor, is used. This unit has three chambers and operates continuously with a residence time of one to two hours. The first (preheat) chamber has a nitrogen purge to remove oxygen, and the second (primary treatment) chamber operates at 750°C with a reducing hydrogen atmosphere. The third (exit) chamber is also purged with nitrogen. All three chambers have airlocks. Metal parts, which have been treated to 5X, are quenched to room temperature and disposed of off site.

The gaseous effluent from the thermal-reduction batch processor and thermal-reduction continuous processor, together with the off-gases from the initial munitions access and energetic deactivation step and from the caustic hydrolysis step, are sent to the GPCR reactor, the third part of Area 400. In the GPCR reactor, a hydrogen-rich atmosphere is maintained, and organic chemicals are reduced to methane and water. Hydrogen chloride, hydrogen fluoride, and hydrogen sulfide are also produced when mustard, GB, and VX/mustard are processed. In addition, the nitrogen from the treatment of VX forms nitrogen gas and perhaps some ammonia, while the phosphorus forms phosphorus acids. The GPCR reactor operates at a temperature of 850°C or above, and the technology provider claims that a residence time of seconds is sufficient for the complete reduction of all organic matter.

Catalytic steam reformers supply hydrogen gas to the GPCR reactor by steam reforming of natural gas. Vertical radiant tube heaters with internal electric heating elements heat the inside of the reactor. The gases enter the top of the GPCR reactor, and their temperature exceeds 870°C when they reach the bottom. When the gases leave the reactor, they pass through primary and secondary caustic scrubbers to remove acid gases, water, and fine particulates. Hydroclones are used to remove solids from the caustic scrubbing fluid. The gas stream exiting the secondary scrubber, which is saturated with water at 38°C, is a mixture of hydrogen, methane, carbon monoxide, carbon dioxide, nitrogen, and trace light hydrocarbons. To ensure that no agent is present, this gas is stored in a series of tanks, where it is sampled and tested. If the agent concentration is below the allowable stack concentration (ASC), the gas is used as an auxiliary fuel for the steam boiler.

Process Instrumentation, Monitoring, and Control

Conventional monitoring and control of the processes are used for the proposed system. The process materials (with the exception of agents and energetics), temperatures, and pressures in this technology package are common in other industrial applications where they are routinely monitored and controlled. The usual collection of equipment for monitoring temperature, pressure, level control, flow, and other parameters normally measured in a chemical plant is used together with the ACAMS and DAAMS equipment developed by the Army. To prevent plugging in the SCWO reactor, the flow of transpiration water, other flow rates, pressure drops, and reactor operating conditions are closely monitored.

Feed Streams

The feed streams entering the LMIDS are listed by area in Table 7-2. With the exception of the chemical munitions, the materials listed are routinely used in large-scale chemical processes. The LMIDS proposal includes the mass balances for five campaigns—a base case and two demilitarization campaigns each for the Blue Grass and Pueblo arsenals. For the purposes of illustration, only the Blue Grass VX base case is considered here. The feed streams will change for the other weapons campaigns at Blue Grass and for the demilitarization campaigns at Pueblo. In Table 7-2, Areas 100 and 200 have been combined because it was impossible from the flow diagrams to determine the split for the use of caustic and decontamination solutions between these two areas.

TABLE 7-2 Process Inflow Streams (lb/hr) from Outside the Process for Blue Grass VX Base Case Campaign (14 M55 rockets/hr and 14 M121A1 projectiles/hr)

Component	Amount
AREA 100[a] + AREA 200[b]	
NaOH	741
Decontamination solution	
NaOH	7
Water	126
NaOCl	7
Total	140
M55 rockets	
VX	140
Steel	172
Aluminum	171
Comp B	45
Nitrocellulose	189
Nitroglycerine	81
Dunnage	308
Total	1,106
M121A1 projectiles	
VX	84
Steel	1,262
Comp B	34
TNT	4
Dunnage	74
Total	1,458
Nitrogen	652
Total areas 100 + 200	4,203
AREA 300[c]	
Kerosene	114
Oxygen	1,074
Waste Oils	14
Total area 300	1,202
AREA 400[d]	
Natural gas	472
Steam	557
Hydrogen	5
Nitrogen	25
Other dunnage	151
Total area 400	1,210
Total plant inflow	6,615

[a]munitions access and energetics deactivation
[b]caustic makeup and hydrolysis
[c]supercritical water oxidation
[d]gas phase chemical reduction

Source: Lockheed Martin, 1998.

Waste Streams

The waste streams for the LMIDS are listed by area in Table 7-3. All solid wastes (other than the metal parts which have been treated to 5X condition) are treated and then disposed of in a hazardous-waste landfill. Ventilation air from contained process areas is passed through carbon filters and monitored for agent before release. There are no liquid effluents other than rainfall runoff and cooling water (which will not be in contact with hazardous materials). The six waste streams produced by the process are listed below:

TABLE 7-3 Process Outflow Streams (lb/hr) to the Environment for the Blue Grass VX Base Case Campaign (14 M55 Rockets/hr and 14 M121A1 projectiles/hr)

Component	Amount
AREA 100[a]	0
AREA 200[b]	0
AREA 300[c]	
Vent gas	1,107
Treated solids to landfill	1,559
Total Area 300	2,666
AREA 400[d]	
Treated metals	
Steel	1,421
Other	11
Clean solids to landfill	313
Product gas to boiler	2,183
Total Area 400	3,928
Total plant outflow	6,594

[a]munitions access and energetics deactivation
[b]caustic makeup and hydrolysis
[c]supercritical water oxidation
[d]gas phase chemical reduction

Source: Lockheed Martin, 1998.

- The SCWO off-gas is continuously monitored, passed through carbon filter beds, and then released to the atmosphere. This gas stream is expected by the technology provider to contain mainly carbon dioxide, oxygen, nitrogen, small amounts of water vapor, and trace amounts of nitrous oxide and light hydrocarbons.
- Treated ventilation air from the process containment areas is passed through carbon filter beds.
- Decontaminated metal parts that have not dissolved in the hydrolysis reactors are processed to a 5X condition in the GPCR reactor.
- Sodium salts produced from elements in agent and energetics hydrolysates (fluoride, chloride, sulfate, nitrate, nitrite, and some phosphate salts) are fed to the SCWO reactor. Tests show that the salts contain up to 10 ppm of organic materials, principally acetone and acetic acid. According to the technology provider, the salts will be free of agent and contain no CWC Schedule 2 compounds. Chemical agents will yield a large amount of salt (VX will yield salt equal to about 150 percent of its original mass). The yield of solid salts from the energetics will vary with the energetic but will be approximately equal to the weight of the energetic. Salts from the SCWO process will be sent off site for stablization and placement in a hazardous waste landfill.
- Residues of carbon (e.g. char and soot) and silica will contain traces of hydrocarbons from GPCR processing of dunnage, fiberglass shipping and firing tubes, DPE suits, etc. are shipped off site for disposal in an approximately permitted landfill.
- GPCR off-gases that contain low molecular weight hydrocarbons (methane and ethylene) and small amounts of hydrogen chloride and hydrogen sulfide are passed through an activated carbon filter and a caustic scrubber and then burned in the facility boiler after passing through a hold-test-release cycle.

Noncontaminated dunnage, such as the wood and wood pallets used to package munitions, will be transported off site for reclamation or disposal.

Start-up and Shutdown

The LMIDS uses both batch and continuous processes operating in series (i.e., feeding one another). Standard chemical-industry practices (operation, instrumentation, and control) can be used to implement normal start-ups and shutdowns, as well as emergency shutdowns. The technology provider has also provided a plan for final shutdown, which includes disassembly of all equipment except the GPCR, treatment of all other process equipment in the GPCR reactor to the 5X standard, and the removal of all equipment from the site.

EVALUATION OF THE TECHNOLOGY PACKAGE

Process Efficacy

Effectiveness of Munitions Disassembly

Rockets. The punching and shearing processes for the rockets are typical of baseline operations and should be capable of achieving the desired processing rates. A unique feature of this technology package is the use of a gravity feed to drop sheared rocket parts into baskets submerged in the rocket hydrolysis vessel through a chute and gate. This interface must be capable of (1) withstanding explosions to protect the rocket hydrolysis vessel or disassembly equipment and (2) operating reliably in the presence of caustic vapors. This operation should be designed to avoid increasing maintenance requirements and, thus, reducing the munition throughput rates.

Projectiles/Mortars. The initial disassembly processes for projectiles are identical to the baseline processes and should meet reliability and production objectives. Modifications to segregate and feed the various projectile/mortar parts to the downstream processes involve multiple, remotely-operated interfaces with the hydrolysis processes. Here, as in the rocket disassembly process, initial reliability may be a problem because of the action of caustic vapors. Another potential problem to achieving the desired throughput without increasing maintenance in DPE suits is the segregation of products from the projectile/mortar disassembly process into four streams. Safety and reliability problems could arise if the parts, similar in size but different in characteristics, are intermixed.

Effectiveness of Agent Decomposition via Hydrolysis

Decomposition with caustic is a proven process for bulk agent (see Appendix D). However, because of the complexity of scaling up mixing processes and the difficulty of removing agent from the crevices in sheared rocket parts, the time required to lower the concentration of agent to the required levels in the various plant-scale hydrolysis vessels may be longer than anticipated. A longer residence time would require equipment modifications to achieve the design throughput. Addressing this concern will require testing with near production-scale equipment. Significant DPE maintenance may be required for the carts used to convey metal parts into the rocket hydrolysis vessel bath and for the projectile baskets that will invert the casings and position them over wash-out wands in a caustic spray environment using remotely operated equipment. However, the committee believes that with a careful design and well chosen materials of construction, the processing objectives for the hydrolysis of agent on projectile/mortar parts should be achievable.

Effectiveness of Energetics Decomposition via Hydrolysis

Significant unknowns remain in the decomposition and deactivation of energetic materials by base hydrolysis (see Appendix E), and the destruction of energetics in the rocket, burster, and nose closure hydrolysis vessels may take longer than expected because of the uncertain reaction rates.

Aluminum rocket parts will also be dissolved in caustic in the rocket hydrolysis vessel. The committee is concerned that the exothermic reaction of aluminum with caustic could produce hot spots and very rapid reactions. The aluminum reaction also produces hydrogen, which could increase the explosion hazard. The technology provider will have to design and operate the rocket hydrolysis vessel purge-gas and off-gas handling systems with these possibilities in mind.

Effectiveness of Supercritical Water Oxidation

The SCWO process appears to be capable of completing the destruction of both agent and energetics. Mustard does contain volatile low molecular weight chlorinated hydrocarbons that can be difficult to treat. These are expected to be oxidized by SCWO but this will have to be demonstrated. A key area of uncertainty in the technology provider's proposed application of SCWO is the proprietary transpiring wall tubular reactor. Although this concept should be capable of achieving the desired processing rates, there has been no long-term experience using a transpiring wall in the severe operating environment of the SCWO reactor. Current experience relates almost entirely to much larger diameter, non-transpiring wall reactor vessels and shows that plugging and corrosion are the main problems encountered. The technology provider proposes using the transpiring wall to overcome these problems, but this technology has not been demonstrated in extended runs, which will be essential to proving the efficacy of this crucial step in the agent/energetics destruction process. The technology provider will also have to verify that the use of the transpiring wall does not allow waste materials to bypass the reaction zone via the cooler transpiration-water region adjacent to the wall.

Effectiveness of Decontamination of Metal and Other Munitions Parts

Although the GPCR process as proposed has not been applied to metal parts from hydrolysis reactions, prior experience in other applications indicates that this technology should achieve the desired process throughput and decontamination levels. The nature of the particulates produced by the hydrogen reduction of organic materials and salts on the surfaces of metal parts is one area of uncertainty that should be addressed in subsequent testing. These particulates may have characteristics (e.g., the formation of sticky soot) that will increase maintenance requirements.

Effectiveness of Decontamination of Other Contaminated Materials

The nature of the particulates generated from the decomposition of fiberglass rocket shipping/firing tubes, DPE suits, and other organic wastes is still uncertain. Approximately 10 percent by weight of the carbon feed is expected to remain as soot, and the technology provider expects that this solid waste stream will be disposed of off site. The large amount of soot generated in the thermal-reduction batch processor could lead to buildups in gas recirculation paths, which could restrict throughput and require additional maintenance to clear the gas path.

Sampling and Analysis

Sampling and analysis requirements appear to be reasonably well known for this integrated process. Easy evaluations of the composition of the hydrolysate can be made from the hydrolysate feed tanks to the SCWO. Similar observations can be made for solid wastes that cannot be released until agent concentrations in adjacent gas spaces are below allowable levels. The technology provider will also have to ensure that agent does not condense, adsorb, or otherwise accumulate on the internal surfaces of the GPCR off-gas hold-test-release tanks, where it would not be detected in the gas analysis but could subsequently revaporize upon depressurization and venting to the boiler fuel system. (The same problem exists for all gaseous hold-test-release systems that are subject to significant pressure variations.)

Maturity

Disassembly Process. The LMIDS uses much of the baseline disassembly process that has been proven at the Johnston Atoll and Tooele, Utah, demilitarization facilities. Modifying the process to include a wash-out step is based loosely on ton-container wash-out tests for the Aberdeen and Newport sites; however, the specific design modifications have not been tested. One of Lockheed Martin's partners, Aerojet, has more than 30 years of experience with hydromining rocket propellants.

Interfaces between the disassembly process and downstream processes may limit the throughput because the reliability of the remotely operated handling equipment used for the interfaces could be difficult to maintain. Some of this equipment is new or has never been used in the harsh environment of caustic hydrolysis processes. Materials selection and design of this equipment will be very important.

Agent Hydrolysis. Neutralization is a proven technology for the deactivation of chemical agents (see Appendix D), and agent hydrolysis processes for HD and VX are being implemented at Aberdeen and Newport. Hydrolysis for GB has been done on a large scale at Rocky Mountain Arsenal. Therefore, the hydrolysis of agent is a mature and well tested technology that requires simple engineering and control.

Energetics Hydrolysis. Several issues remain to be addressed about the technology provider's implementation of hydrolysis for energetics.

1. The caustic hydrolysis step is intended to dissolve the aluminum fuze and expose the energetic materials. The dissolution of aluminum will result in an exothermic generation of hydrogen gas that will bubble out of the aqueous alkaline solution. The production rate of hydrogen and the release rate of thermal energy will have to be monitored and controlled to ensure that there is no possibility of ignition. Also, an autocatalytic redox reaction could occur when the wet aluminum is in the presence of damp energetic materials. Therefore, both must be destroyed in the hydrolysis.
2. Many of the hydrolysis vessels for energetics involve mechanical conveyors operating in a hot 20-percent caustic solution, which is a severe

environment for this type of machinery. Therefore, a great deal of maintenance (by personnel in DPE suits) may be required to keep this equipment in operation. No information has been provided to the committee on the reliability of this equipment in a hot caustic environment.
3. The required residence time in the hydrolysis reactors is directly related to the size of the sheared pieces. The size of these pieces will have to be accurately characterized because they will affect the design of the hydrolysis reactors for energetics.

SCWO. The SCWO process for the treatment of VX hydrolysate has been extensively examined in a previous NRC study (NRC, 1998), and work on the use of SCWO to treat hydrolysates from agent and energetics destruction is ongoing. Other SCWO systems have been under development for more than 20 years; the Huntsman Corporation of Austin, Texas, for example operates the only commercial SCWO unit for the treatment of organic-laden wastewater. Nevertheless, SCWO cannot be considered a mature technology for destroying agent and energetic hydrolysates.

The proposed LMIDS transpiring-wall reactor provides a novel solution to the problem of corrosion and plugging in SCWO reactors. The transpiring-wall SCWO reactor has been tested at bench scale at Sandia National Laboratories in collaboration with GenCorp/Aerojet and Foster Wheeler. A commercial-scale reactor is currently being built for the Army at the Pine Bluff Arsenal by Foster Wheeler/Aerojet to treat smoke and dye wastes with a high salt content (similar in some ways to agent/energetics hydrolysates). This unit is expected to be operational sometime in 1999 but was not operational at the time of this writing. The proposed SCWO unit is essentially identical to a unit being constructed for the U.S. Navy for shipboard waste disposal. No extensive testing of the design has been done to date. The committee has the following concerns about this technology:

1. The design and manufacture of this unique SCWO reactor may be quite difficult. Fabrication of the transpiring wall may present a significant challenge, both in the choice of materials and in the construction of the platelets.
2. Long-term testing will be necessary to establish that the transpiring-wall reactor will not have corrosion, erosion, and plugging problems in the presence of salts. This technology has not been used with salt-containing supercritical water.
3. Control of the transpiring-wall reactor may be difficult, and its operation may be very sensitive to system fluctuations. Fluctuations in pressure can cause a backflow into the transpiring wall and plugging.
4. Because the SCWO reactor is a unique and unproven piece of equipment in the LMIDS, maintenance may be difficult. Furthermore, because of the difficulty of construction, if the reactor becomes nonoperational, there may be a significant delay before it can be repaired or replaced.
5. The SCWO reactor would be used with a mixture of energetics and agent hydrolysates. Because SCWO is capable of treating a variety of materials (probably simultaneously), this may not be a problem. However, the system should be extensively tested with mixed hydrolysates.
6. The composition of the feed to the SCWO reactor will change with time, either because of a change in the mix of weapons or because of an unscheduled shutdown of either the agent hydrolysis reactor or an energetics hydrolysis reactor. It must be established that the SCWO reactor will continue to operate reliably after a sudden change in the composition of the feed stream.
7. The gaseous effluent from the SCWO reactor will be continuously filtered, monitored, and released to the environment. This off-gas will not be passed through a hold-test-and-release process. Continuous monitoring must be demonstrated to ensure that the released off-gas meets safety criteria.

Gas Phase Chemical Reduction. The thermal-reduction batch processor/GPCR process has been used commercially to treat PCB-contaminated electrical equipment. Two full-scale plants have been operating for more than two years in Kwinana in Western Australia, and one plant was operated for a year at a General Motors of Canada facility in St. Catherine's, Ontario. The process will be used here with only minor modifications. Therefore, there is some experience in operating GPCR reactors of the size to be used in the LMIDS. The committee has the following concerns about these units.

1. The GPCR reactor must be operated in a closed, contained environment, where fugitive hydrogen emissions can be a serious hazard. Therefore, the design and control of this system must take this danger into account.
2. The monitoring and control system on the GPCR units must ensure that no oxygen or other oxidants are present before hydrogen is admitted into the system.
3. The proposed method of separating soot from the gaseous effluent from the GPCR reactor may not be very effective and may result in process reliability problems from the accumulation of soot in other parts of the gas-flow path.
4. The GPCR of materials to soot, silica, and a usable fuel gas must be thoroughly demonstrated. The waste streams may be more complex than anticipated.
5. A method must be developed to ensure that agent (or other hazardous materials) does not condense, adsorb, or accumulate on the internal surfaces of the off-gas hold-test-release tanks.

Scale-up. Some aspects of scale-up from demonstration-size equipment should be relatively easy. The SCWO reactors will be the same size as an existing prototype Navy unit. The LMIDS reactors will be designed to operate in parallel, and no problem with scale-up is planned. Other aspects of scale-up may be more difficult because not all parts of the process scale in the same way. For example, many mass-transfer processes scale with length, whereas surface and surface wash-out phenomena scale with area; still others, such as the homogeneous hydrolysis reaction, scale with reactor volume. The hydrolysis vessels will have to be carefully designed to accommodate all three phenomena simultaneously.

Overall Technology Package. The technologies selected by the technology provider have all been implemented with process streams similar to those in the ACWA program. However, they have not been operated as an integrated unit. Furthermore, some of the methods of implementation are new and all but untried at this time (e.g., the transpiring-wall SCWO reactor and the methods for hydrolyzing agent and energetics remaining on metal parts). Thus, although the basic technologies are reasonably mature, certain facets of their implementation and their integration or interfacing are still at early stages of development. To prevent severe operating problems, the integrated system must be demonstrated prior to full-scale operation. The full-scale process will have to be designed to be "forgiving," allowing easy visibility and easy maintenance of remotely-operated and automated equipment.

Robustness

Robustness, or the ability to operate with a wide range of feed stocks and operating conditions, appears to be reasonable in all process areas except for the vessels used to remove energetics and agents remaining on metal parts. If agent or energetic properties are different than anticipated (e.g., more polymerization of the agent than expected), the cleaning method may require significant modifications to achieve the required throughputs.

Monitoring and Control

The monitoring and control approaches for the processes in this system are widely used and should be readily implemented. The process materials (with the exception of agents and energetics), temperatures, and pressures in this technology package have all been successfully monitored and controlled in prior applications.

Applicability

The technology provider included process design information for rockets and projectiles/mortars and stated that land mines could be easily incorporated as an additional feed stream by adding a disassembly capability for these munitions. Thus, the proposed system has broad applicability.

Process Safety

The technology provider proposes using several unique pieces of equipment:

- a modified (from baseline) rocket-shear machine
- a rocket hydrolysis vessel for sheared rocket parts, propellant and energetics

- a nose-closure hydrolysis vessel for projectile nose closures
- a projectile hydrolysis vessel for projectile bodies
- a burster hydrolysis vessel for projectile bursters
- energetics destruction reactors
- agent neutralization reactors
- a GPCR process that includes a thermal reduction continuous processor for treatment of projectiles and a thermal reduction batch processor for other waste streams
- transpiring-wall SCWO reactors for destruction of agent and energetic hydrolysates

In the LMIDS, the energetics and agent are separated early in the process, reducing the possibility of an energetically driven release of agent. Most of the hydrolysis processes operate at low temperatures (90°C or 194°F) and low pressure (near atmospheric), thus avoiding significant stored energy and reducing process hazards. The burster hydrolysis vessel is slightly pressurized to achieve temperatures up to 135°C but does not represent a significant stored-energy hazard. The SCWO reactors operate at 238 atm and 780°C and do represent major reservoirs of stored energy. The GPCR processes operate at temperatures up to 850°C (1,562°F) at low pressure using a hydrogen gas atmosphere for destruction of trace amounts of agents and other hazardous materials. Commercial facilities employing this technology are operating (or have operated) in Canada and Australia. However, all of these facilities are located outdoors. To maintain control of potential airborne agent emissions, the GPCR equipment for the LMIDS will be located inside a containment or confinement building. Therefore, hydrogen leakage from the GPCR is a safety concern. The team partner responsible for this technology, Eco-Logic, is well aware of this and has commissioned a safety study on this issue (Prugh, 1998). Recommended safety measures are being considered or have already been implemented.

Worker Safety

The LMIDS system is basically "forgiving" in that agent and energetic destruction are verified after both of the sequential processes for all ACWs, thus ensuring that all agent and energetic material have been destroyed. The separation of energetics from agent and the subsequent destruction of both materials in a 20-percent caustic solution minimizes the risk of explosions. In addition, both processes are operated in structures designed to contain explosive overpressure.

Mechanical disassembly processes are derivatives of the baseline processes and are not considered to represent new or increased risk levels for rockets, projectiles, or land mines. These processes will be conducted in vessels or structures designed to withstand explosive overpressure.

The GPCR reactor operates in a hydrogen atmosphere and generates methane and other gaseous hydrocarbons that could burn or explode in the presence of air. Explosion hazards for this process are minimized by purging with inert gas during start-up and shutdown. Considerable industrial experience with high-temperature hydrogen atmospheres has established that these processes can be operated safely. GPCR off-gas could contain small amounts of hydrogen sulfide and hydrogen chloride, which will be removed in a caustic scrubber. Worker exposure to these gases is unlikely. Similar gases in much higher concentrations are routinely handled safely in the petroleum refining and petrochemical industry.

The most significant worker safety issue will probably be maintenance of the hydrolysis vessels in DPE suits. These vessels have conveyor systems that operate in hot caustic solutions. Experience with these systems is limited, and full-scale implementation of this technology, especially during start-up, may require significant maintenance, thereby increasing the risk of worker exposure to agent.

Hydrogen and other combustible gases will be generated in the hydrolysis reactions. Therefore, oxygen must be kept out of the vessel vapor spaces and the associated vapor piping, and these gases must not be allowed to collect in air spaces in the contained process areas. Because these processes operate at very low or negative gauge pressures, the driving force that causes gas leaks into ventilated areas is minimal. The hydrolysis vessel vapor spaces (which operate under a slight vacuum) are purged with nitrogen gas to prevent unsafe oxygen levels from building up from the in-leakage of ventilation air. Loss of the nitrogen purge gas would increase the likelihood of an explosion in the vessels or off-gas piping.

The SCWO reactors and water supply system operate at high pressure and are a source of stored energy. Therefore, it is very important that the reactors be designed and maintained in ways that minimize ruptures and leaks. Failures may not result in release to the environment because of secondary containment, but they could require extensive repair work in DPE suits.

Fuze bodies and booster pellets that are not dissolved in the caustic solution also represent an explosive hazard in the thermal-reduction batch processor. The technology provider intends to demonstrate a technology that will reduce the size of these parts to ensure their full dissolution. In addition, the thermal-reduction batch processor will be designed to withstand initiation of these components.

The primary hazardous materials used in agent and energetic destruction are sodium hydroxide, liquid and gaseous oxygen, hydrogen, and methane. Sodium hydroxide will be delivered in solid form and dissolved in water to make a 20-percent caustic solution. All of these chemicals are handled routinely and safely in many industries. The technology provider has conducted a preliminary hazard analysis and has identified reasonable solutions for events that could create unacceptable or undesirable worker safety risks (Lockheed Martin, 1998).

Public Safety

The likelihood of releases of agent or other regulated substances to the atmosphere or to the facility are expected to be extremely small. Hold-test-release systems are applied to all effluent streams except the containment ventilation air and SCWO off-gas. The ventilation system uses tested baseline air cleaning technology. The SCWO off-gas will be cooled continuously, monitored for agent, and passed through a carbon filter before release to the atmosphere. This approach does not meet the hold-test-release criterion for process effluents that has been requested by some stakeholders; but it is not expected to reduce public safety.

The primary cause for a release of material containing agent or other regulated substances would be a disruptive explosion. The likelihood of such an event is expected to be extremely small at the conclusion of the design process for the full-scale facility. (This design process is understood to include the completion of a QRA.) A preliminary hazard analysis conducted by the technology provider revealed no events with unacceptable or undesirable public safety risks.

Human Health and the Environment

A full evaluation of the impact of the effluents on human health and the environment must await the outcome of health and environmental risk assessments, which cannot be prepared at the current stage of development for this system. However, some general observations and comments can be made at this time.

Process gases leaving the plant will have been treated to remove traces of organic materials (including agent) and will be monitored or tested to ensure that they do not contain agent or other regulated substances at concentrations above levels established by the EPA for release to the environment.

The solid waste streams are 5X metal parts, salt from the SCWO reactors, and carbon residue and silica from the GPCR. All nonmetal solid waste streams are expected to be suitable for release to a hazardous-waste landfill. The 5X metal parts are expected to be suitable for reuse as scrap for metallurgical processes.

Effluent Characterization, Management, and Impact on Human Health and the Environment

Gas Streams. These streams originate in the SCWO reactors, the ventilation air exhaust from contained process areas, and exhaust gases from the steam boiler (fueled partly by the GPCR reactor) and the hydrogen generator. Experience indicates that the gas streams will be free of agent and other regulated substances. Demonstration tests would provide reassurance that the design works as planned for the treatment of process gas and that all regulated substances in the effluent gases are identified and measured.

The gas streams from the SCWO reactors and the ventilation air exhaust from contained process areas are monitored, passed through activated-carbon filters, and released to the environment. (The exhaust gas from the SCWO reactors has been shown to consist principally of carbon dioxide and water vapor with trace amounts of low molecular weight hydrocarbons.) The inlet to the SCWO reactor will also be monitored, and detection of unacceptable levels of hazardous materials will trigger an emergency response, including isolation and shutdown. The exhaust gases from the steam boiler and hydrogen generator will not be monitored. The composition of these gas streams should be characterized during demonstrations of the LMIDS.

Acid gases from the GPCR process will be absorbed in a caustic scrubber, and the effluent gas will be passed through a hold-test-release procedure. If shown to have acceptably low concentrations of regulated substances and to be essentially agent free, the gas will be used as fuel in the LMIDS steam boilers. Permits will be required to use this gas stream as boiler fuel based on full characterization of the GPCR process off-gas streams during demonstration.

Metal Parts. These parts will be treated in the GPCR process and will not be released until they have been treated to the 5X standard. As long as the treatment procedure is monitored, the metal parts should need no further testing and can be released. The cleaned parts are not expected to pose any threat to human health or the environment.

Salts. Salts from the SCWO reactors will contain the sodium salts of fluoride, chloride, sulfate, phosphate, nitrate, and nitrite and are expected to contain trace amounts of low molecular weight hydrocarbons. The committee expects that this stream will be classified as hazardous, although this has not yet been determined by testing. The salts will be verified to be agent free before they are released for off-site stabilization and placement in a hazardous-waste landfill. Stabilization of salt wastes is very difficult. The leaching levels of hazardous constituents in the salt will have to be investigated to determine if additives will be necessary to ensure stabilization and whether a formula for a stabilized mixture can be developed. If stabilization and burial in a hazardous-waste landfill are feasible, they would provide adequate protection to human health and the environment. Demonstration tests are necessary to characterize fully the composition of the salts to identify all regulated substances and determine their concentrations.

Dunnage Material. This material is subjected to high-temperature hydrogen reduction in the GPCR process. The solid process effluent is carbon residue and silica and will be sent off site for stabilization and disposal in a hazardous-waste landfill. Demonstration tests will have to characterize fully the composition of the solids to identify all regulated substances and their concentrations.

Resource Requirements

The chemicals required for processing are sodium hydroxide, liquid oxygen, liquid nitrogen, and kerosene (or isopropyl alcohol). The level of usage is not considered to represent an unusual demand on available industrial sources. The utilities required for operation and maintenance are electricity, water, and gas. The amounts required are similar to the amounts for the Army's baseline facilities except for the larger consumption of methane for the production of steam and hydrogen. Manpower required for operation and maintenance will probably be similar to the manpower for the Army's baseline facilities, assuming that maintenance concerns expressed earlier are addressed as the design progresses.

Environmental Compliance and Permitting

Compliance with environmental regulations will require careful, detailed design of the plant, as well as careful operation and environmental management. There are no inherent reasons why the combination of technologies in the LMIDS technology package should lead to unusual problems. The absence of liquid emissions is an important advantage of the process.

The same is true for permitting. All process waste streams except the SCWO off-gas will be evaluated prior to release to confirm that regulated substances are absent or at acceptably low concentrations. The SCWO off-gas will be scrubbed, monitored, and passed through activated carbon filters.

One aspect of the process that may lead to permitting problems is the use of the cleaned GPCR off-gas as a boiler fuel. Extensive testing may be required to characterize this stream to ensure that it can be used safely.

STEPS REQUIRED FOR IMPLEMENTATION

Overall, the LMIDS appears to be capable of operating as proposed by the technology provider, but the process must be developed further, especially the interfaces between and integration of the process units. If the LMIDS were to proceed towards full-scale implementation, the next step should be to design, build, and operate a pilot-scale system that incorporates all of the unit operations into a fully functional, integrated process. Full-scale implementation will involve interfacing and integrating batch processes (the hydrolysis reactors and the thermal reduction batch processor) with continuous processes (the SCWO reactor, the thermal reduction continuous processor, and the GPCR reactor). These interfaces must be tested at the demonstration stage to avoid implementation problems. Also, all problems with materials of construction and waste characterization will have to be solved before implementation. However, no problems have been identified that would prevent eventual full-scale implementation.

In addition to demonstrating that the overall process is capable of long-term operation, specific objectives for three of the pilot-scale unit operations are described below.

Pilot-Scale Evaluation for Hydrolysis of Energetics

1. Establish that the mechanical equipment used in the energetics hydrolysis vessels can tolerate the harsh conditions without excess maintenance.
2. Determine whether the hydrolysis of aluminum together with energetics presents any problems.

Pilot-Scale Evaluation for SCWO

1. Show that the SCWO reactor platelet wall can be constructed.
2. Demonstrate that the SCWO reactor can be operated for sufficient periods of time without excessive clogging or corrosion.
3. Fully characterize the SCWO gaseous effluent from mixed hydrolysates of agent and energetics.
4. Establish that the continuous monitoring of the SCWO gaseous effluent ensures against unacceptable releases of hazardous materials.

Pilot-Scale Evaluation for GPCR

1. Fully characterize the GPCR gaseous effluent and establish whether it can be used as a boiler fuel.
2. Ascertain whether the large quantity of soot generated in the thermal-reduction batch process will create any problems.

FINDINGS

Finding LM-1. The disassembly methods proposed in the LMIDS are based largely on the baseline disassembly methods. The proposed modifications appear to be reasonable, but testing will be necessary to verify that performance, reliability, and production objectives can be met.

Finding LM-2. Primary agent decomposition and detoxification is achieved using a strong caustic hydrolysis of bulk agent—a proven technology. Overall, the implementation of agent hydrolysis in the LMIDS is sound.

Finding LM-3. Primary decomposition and deactivation of energetics is also achieved using a strong caustic hydrolysis. This technology has been tested but is less mature than agent hydrolysis. The implementation of this technology in the LMIDS is reasonable but will require thorough testing at the pilot scale.

Finding LM-4. The method of removing agent from metal parts—caustic solution jet wash-out followed by the movement of the parts in baskets through a caustic bath—is new and unproven. It is expected that this method can be made to work, but the effort and time required to come to acceptable performance goals may be longer than anticipated and may require alternate methods. Thus, it will be desirable to have alternate plans if the desired detoxification efficiencies are not achieved (e.g., increase the capacity of the GPCR unit to allow for more than the planned agent cleanup load).

Finding LM-5. The hot-caustic environments in the initial hydrolysis vessels will pose severe challenges to the reliability and operability of the equipment inside these vessels, especially the basket transport mechanisms.

Finding LM-6. The SCWO process appears to be capable of completing the destruction of both agent and energetics in the hydrolysates. The key area of uncertainty in the technology provider's proposed application of SCWO is the use of its proprietary transpiring-wall tubular reactor. The demonstration of this technology will be essential to proving the efficacy of this crucial step in the agent/energetics destruction process.

Finding LM-7. The crystallization and evaporation operations have not been tested for this application. These conventional technologies, which are expected to work effectively, must still be tested.

Finding LM-8. The use of GPCR in an enclosed environment raises unique safety concerns because of the presence of hot hydrogen gas. Hydrogen is handled routinely (and safely) in the chemical industry, and the technology provider is aware of the hazards. Implementation of this technology will require a design that ensures that these hazards are thoroughly understood and mitigated.

Finding LM-9. The gas stream from SCWO is not subjected to hold-test-release. Instead, the gas is scrubbed, monitored, and passed through activated carbon. This treatment appears to be appropriate for the anticipated composition of the SCWO off-gases.

Finding LM-10. Pilot-scale testing will be necessary to refine the component technologies and demonstrate that these technologies can be operated as an integrated system.

Finding LM-11. The proposed use of the cleaned GPCR off-gas as a boiler fuel poses unique permitting challenges. Any demonstration must characterize this stream to ensure that permitting as a boiler fuel is possible. If this off-gas cannot be used as a boiler fuel, significant process modifications may be necessary.

Finding LM-12. All of the findings in the NRC report, *Using Supercritical Water to Treat Hydrolysate from VX Neutralization* (NRC, 1998), apply to the LMIDS SCWO system (see Appendix F).

8

Parsons-AlliedSignal Technology Package

INTRODUCTION AND OVERVIEW

The Parsons-AlliedSignal team has submitted a proposal under the acronym WHEAT (water hydrolysis of explosives and agent technology). This technology package is comprised of five basic technologies:

- The Army's *baseline disassembly* process, with modifications including water-jet cutting for rockets, is used to separate agent, energetics, and metal parts.
- *Hydrolysis* is the primary process for detoxifying the agent and deactivating energetics.
- *Biological processing*, supplemented by ultraviolet/hydrogen peroxide treatment, is used to convert the hydrolysis products to materials acceptable for discharge to the environment.
- Metal parts and dunnage are decontaminated to 5X by *heating in high-temperature steam.*
- Gas discharges from the plant go through a *catalytic oxidation* unit for treatment.

Table 8-1 lists how these technologies are used to perform the six major demilitarization operations listed in Chapter 2. A process flow diagram for the Parsons-AlliedSignal package is presented in Figure 8-1, and a detailed description of the package is given in the next section. The technology provider addressed the processing of rockets, projectiles, and mortars but did not consider the processing of land mines.

DESCRIPTION OF THE TECHNOLOGY PACKAGE

Disassembly of Munitions and the Removal of Agent/Energetics

The baseline disassembly process (see Appendix C) is used to a large extent in this technology package. However, some modifications to the baseline process are proposed as described below.

TABLE 8-1 Summary of the Parsons-AlliedSignal Technology Package

Major Demilitarization Operation	Approach(es)
Disassembly of munitions	Army baseline disassembly, augmented with water jet cutting
Treatment of chemical agent	Base hydrolysis; biotreatment of hydrolysate
Treatment of energetics	Water-jet wash-out; base hydrolysis; biotreatment of hydrolysate
Treatment of metal parts	Heat in steam to 5X conditions in metal parts treater; catalytic oxidation of gas
Treatment of dunnage	Treatment in metal parts treater;catalytic oxidation of gas
Disposal of waste	**Solids.** Dry salts and biotreatment product to appropriately permitted landfill; calcined grit to landfill **Liquids.** None **Gases.** Discharge to atmosphere after catalytic oxidation, caustic scrubbing (and possibly carbon filtration)

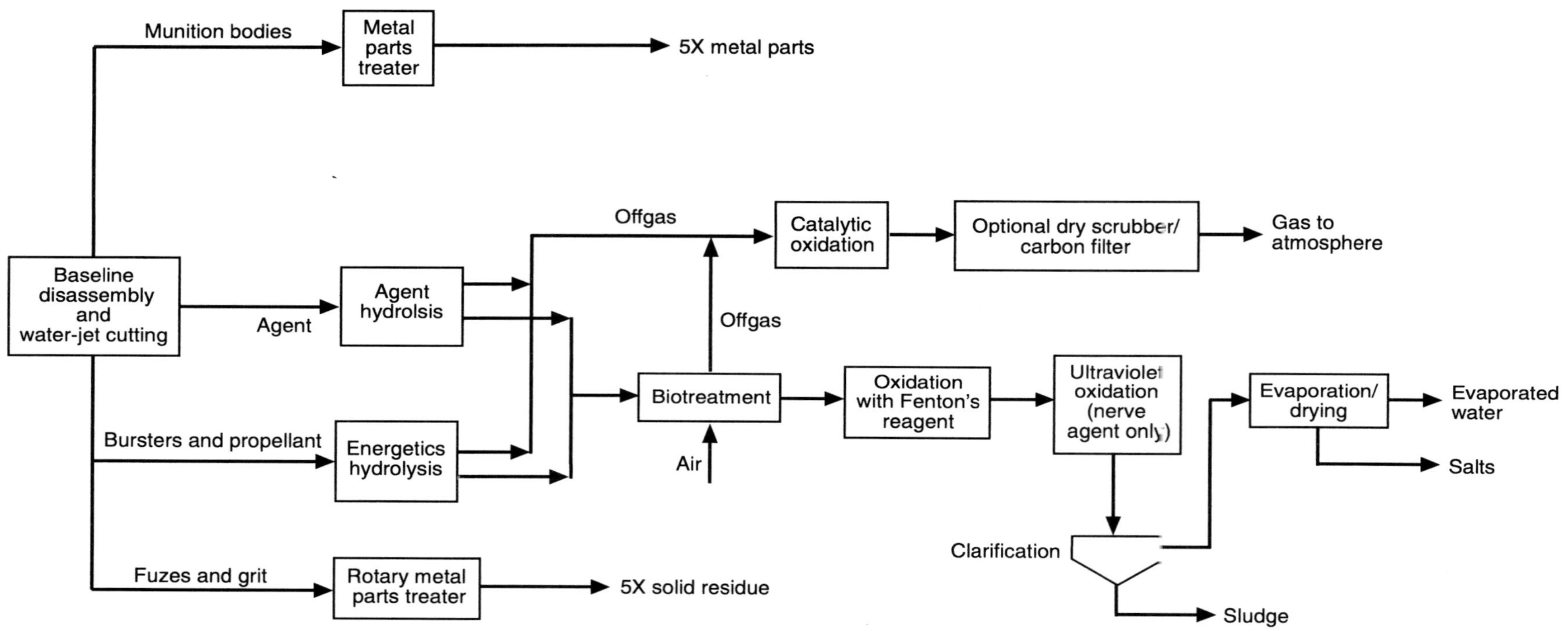

FIGURE 8-1 Process flow diagram for the treatment of assembled chemical weapons by WHEAT.

Projectile Disassembly

The fuze, booster (if present), and burster are removed using the baseline process. The burster tubes and supplementary charges are then placed in a special fixture, and high-pressure hot water is used to dissolve and/or physically remove energetics from the tubing. The wash-out solution is fed to the energetics hydrolysis reactor as a batch. Boosters are washed out in the same manner as bursters. The fuzes are sent to a rotary metal parts treater, which will be described shortly.

The agent is drained from the projectile as in the baseline process. The projectile body is then sent to the metal parts treater, which is described later in this chapter.

Mortar Disassembly

The fuze and burster are removed by the baseline process; subsequent fuze and burster processing is the same as for the projectile. The agent is drained from the mortar as in the baseline process.

Rocket Disassembly

Rockets are disassembled with the rockets still inside their shipping and firing tube. Chemical agent is removed using the baseline punch and drain approach and is pumped to an agent storage tank prior to hydrolysis. Water-jet cutting with abrasive (garnet) is then used to sever the fuze from the rocket, and the fuze is sent to the rotary metal parts treater.

The rocket (with the fuze removed) advances to a wash-out station, where high-pressure (4,500 psi), hot water is used to remove the burster energetics (Parsons-AlliedSignals, 1998). Abrasive water-jet cutting is subsequently used to separate the warhead from the motor section, and a high-pressure (15,000 psi) hot-water jet washes the propellant from the motor section. The burster energetics, propellant, and washout solution are fed to the energetics hydrolysis reactor. The grit is separated from the cutting water; the cutting water is sent to the agent hydrolysis reactor; and the grit is sent to the rotary metal parts treater. Following the removal of energetics, the warhead and motor sections are inspected and sent to the metal parts treater.

Treatment of Chemical Agents and Energetics

The chemical agents and the aqueous dispersion of energetic materials derived from wash-out of the munitions are considered together in this section because they follow the same processing sequence.

Hydrolysis

The hydrolysis reaction conditions for agents are the same as those outlined in Appendix D. Reaction times will of this be specified to ensure very high conversion, (e.g., 99.9999 percent destruction efficiency). Schedule 2 compounds formed from the hydrolysis of each agent will require further treatment by bioreaction.

Energetic material is fed to the hydrolysis reactor as an aqueous slurry, having been reduced to a fairly small particle size (e.g., less than a quarter inch). (An explosives shredder may be used to reduce particles to this size.) The reaction rate is expected to be controlled by diffusion to the solid surface of the particle and is, therefore, dependent on particle size. Reaction conditions are the same as those outlined in Appendix E.

All hydrolysis products are transferred to storage tanks before they are fed to the bioreactors. The hydrolysis reactors are operated as batch reactors; bioreactors are operated continuously. Tanks are sized to accommodate the change between the batch and continuous operating modes.

Both VX and GB produce hydrolysates that contain a small amount of an organic phase. The reactors and storage tanks are constantly stirred to prevent the organic phases from separating from the bulk aqueous phase. GB hydrolysate is more dilute than VX hydrolysate (i.e., 8 percent of the reaction products are from GB compared to 30 percent from VX in their respective hydrolysates). The dilution prevents the formation of a solid precipitate from GB hydrolysis, which is probably sodium fluoride and some iron salt (see Appendix D).

Biological Treatment

Aerobic bioreactors oxidize the hydrolysates (from chemical agents and energetics) to the following products:

- carbon dioxide, water, and biomass (solid products of the biological cell mass produced in the reactions) (the technology provider estimates that about 80 percent of the carbon in the process feed is oxidized to CO_2; the balance is in the organic biomass (sludge), as well as a small amount of organic matter remaining in solution)
- other products, such as fluoride, sulfate, nitrite, nitrate, phosphate salts in solution, and ammonia
- some low molecular weight, partially oxidized species (e.g., acetic acid), as well as some organic compounds that color the aqueous solution (color bodies)

The biological reaction is relatively slow. The liquid residence times in bioreactors (the so-called hydraulic residence time) are typically five to 15 days, although the technology provider believes that five days will be sufficient. The average residence time of biomass can be as long as two months.

The bioreactor design will include an AlliedSignal development called an immobilized cell bioreactor, which holds the biomass that develops in the reactor on a porous screen (sponge). In other bioreactor designs, the biomass floats freely in the liquid. Advantages claimed by the technology provider for the immobilized cell bioreacter are (1) more rapid reaction because the biomass that accumulates on the screen is more concentrated than in the free floating alternative; and (2) lower production of biomass overall—usually an advantage because disposing of the biomass is a problem. The lower biomass production may not be an advantage in nerve agent disposal, however, because the biomaterials use phosphorus primarily to produce biomass, rather than for metabolism, and a bioreactor that produces more biomass will more effectively eliminate the phosphorus from the nerve agent hydrolysate.

Since the original proposal was made, the technology provider has suggested using a combination of an immobilized cell bioreactor and a "conventional" bioreactor. The combined bioreactor would consist of a long box, with its long side horizontal. The reactor would be aerated to supply the oxygen for the bioreactions. The liquid in the reactor would be stirred by the air, which would have the undesirable consequence of keeping the entire process at the lowest reactant concentration (i.e., the exit concentration). To avoid this, the reactor would be compartmentalized, with liquid flowing horizontally from stage to stage within the reactor. The reactor would have two to four stages. The first stages would use the immobilized cell bioreactor design for rapid reaction. The last stage would have free-floating biomass to promote the removal of phosphorus. Bioreactors used only for mustard and energetics (that contain no phosphorus) are expected to use the immobilized-cell technology for all stages.

The combination of a very large volume of feed (low concentration of organics and salt in water), and a hydraulic residence time of five days, dictates the reactor volume. The basic reactor module will be a 40,000-gallon "box"—the largest size that can be transported by highway. Much larger reactor volumes will be needed, however, and this requirement will be met by adding more 40,000-gallon tanks.

The technology provider has suggested a basic module of four 40,000-gallon tanks grouped around a central "facilities" corridor (pumps, blowers, metering equipment and controls, etc.). The number of modules will be determined in the final plant design by the required rate of munitions destruction, as well as on the particular munitions. The technology provider has suggested a configuration with three modules for the Pueblo, Colorado, site and four modules for the Richmond, Kentucky, site. Thus, there would be 12 or 16 40,000-gallon reactors at these sites (total bioreactor volumes of 480,000 and 640,000 gallons).

The reactors are operated with very dilute solutions. For example, experimental work has been done with feed concentrations of less than 0.7 wt. percent; the tentative plant design (for VX) calls for 1.2 wt percent. This means the agent and energetics feed would represent 1.2 wt. percent of the material fed to the bioreactor. The technology provider has suggested that salt contents of up to 4 percent could be tolerated. Because the hydrolysis products, which are the feed streams, are typically in the range of 5 to 10 wt. percent, the hydrolysate feed stream must be diluted by a large factor, as much as 10-fold, before entering the bioreactor.

Effective bioprocessing requires pH control. The product from the hydrolysis reactors has a pH of 10 to 12, which is generally too high for the bioorganisms to tolerate. The pH is adjusted by adding acid to lower the pH to 8.5.

The microbial population responsible for the

bioreactions imposes additional requirements on the feed to the reactor to keep the population alive. Three important food elements are carbon, nitrogen, and phosphorus. The optimum ratio for carbon to nitrogen to phosphorus is 100:5:1 (Lupton, 1998). Because none of the hydrolysates meets this criterion, other materials are added, such as dextrose. The amounts of these extra materials vary depending on the particular hydrolysate mixture being fed, but for hydrolysate from nerve agents, the extra nutrients are a very large addition to the feed to the plant. Mixing chemical agent hydrolysate, which is rich in phosphorus, with energetics hydrolysate, which is rich in nitrogen, can also help realize the optimum feed.

The bioreactor cannot eliminate organic material completely. Enough must remain in solution to sustain microbial life. The solution leaving the bioreactor is expected to have a dark color, due to organic color bodies. The depth of color depends on the feed material. The effluent from hydrolysate from munitions is particularly dark.

One product of the bioreaction is sludge, which is flocculated, dewatered, dried on a drum evaporator, and packaged for disposal in a landfill. The resultant solid will probably contain most of the heavy metals from the original feed and may be classified as toxic.

There are two other product streams from the bioreactor: (1) the aeration air, mixed with product gases from the bio-organisms and volatile materials (including low molecular weight chlorinated hydrocarbons from mustard) and some liquid "spray;" and (2) a large stream of water containing dissolved salts and the remaining organic material.

Catalytic Oxidation

A large volume of gas leaves the bioreactor. This gas is then heated to 425 to 450°C (797 to 842°F) and passed to a catalytic oxidation unit for the removal of trace organics, oxidizable nitrogen, and chlorine compounds. The gas is then cooled and scrubbed using either a liquid or a solid soda scrubber. It may then be passed through a carbon filter before release to the air. A carbon filter has been installed on the Parsons-AlliedSignal ACWA demonstration system. If analytical data show that the filter is necessary, it will be included in the final plant design.

Salt Recovery and Water Management

Most of the water in the liquid bioproduct is recycled. Salts are recovered by evaporation, and most of the steam is condensed and recycled. A reverse osmosis unit is included in the technology provider's ACWA demonstration system to reduce the volume of water that must be evaporated. There is no plan to include the reverse osmosis unit in a final plant design, however.

The salt content of the reactor liquid is affected by the fraction of product liquid withdrawn and evaporated. If the fraction is large (e.g., approaching 1), the salt content can be maintained as low as 1 percent, but at the expense of a large evaporation requirement. The technology provider has suggested a salt content of 2 percent as a reasonable level, although the microbial population can tolerate levels as high as 3 or 4 percent.

The organic matter remaining in the bioreactor liquid from VX or GB disposal may contain low levels (ppm) of Schedule 2 phosphonates. An ultraviolet /hydrogen peroxide treatment is used to reduce these materials to levels below detection limits.

Brine from the bioreactor unit is first concentrated in an evaporator and then evaporated to dryness. The salt is recovered with a rotary-drum dryer. The steam from the evaporator is condensed and the water recycled. The steam from the drum dryer is released to the air. The salt, with a small residual organic component, is packaged in drums for disposal.

Treatment of Metal Parts

Metal parts are heat treated to produce metal cleaned to a 5X condition, which can be released from Army control. The important difference from the baseline approach is that the heating medium is superheated steam instead of combustion gas.

A batch process is used to treat the parts to a 5X condition. The metal parts treater is an autoclave with induction heating. The metal parts are assembled in a basket and placed in the metal parts treater, which is then purged with nitrogen followed by low-pressure steam. The metal parts treater and its contents are quickly heated to about 650°C (1,202°F), and the system is held at that temperature for 15 minutes. At the end of the process, the steam is swept with air to a condenser, and the gases are passed on to a reheater and a catalytic oxidation unit. Organic materials driven off the metal parts and broken down by the heat are oxidized to CO_2, H_2O, and possible acids, such as P_2O_5. The gas from the catalytic oxidation unit is scrubbed with lime or caustic solution, possibly passed through a carbon filter, and exhausted to the environment.

Fuzes and associated small metal parts are also heated in steam in a rotary metal parts treater where the fuzes ignite or explode. The small metal parts are treated to a 5X condition for disposal.

Treatment of Dunnage

Some contaminated dunnage is treated in the metal parts treater. For example, DPE suits are shredded and then vaporized in the metal parts treater, leaving behind a small ash residue. Some dunnage may be held until the plant is closed. Some may be carefully examined for contamination and reused (e.g., pentachlorophenol-treated wooden pallets). Grit from water-jet cutting is heated to a 5X condition in the rotary metal parts treater.

Process Instrumentation, Monitoring, and Control

On-line chemical analysis during baseline disassembly and during operations unique to the Parsons-AlliedSignal process is limited to (1) a determination of pH and specific gravity at critical control points; (2) chemical oxygen demand and levels of all nutrients for the immobilized cell bioreactor; (3) ACAMS monitoring of exhaust gases and of ventilated spaces and critical operations, such as munition overpacks; and (4) monitoring of sulfur and phosphorus content of the catalytic oxidizer inlet stream. The bioreactor is instrumented to monitor pH, temperature, and inlet and outlet chemical oxygen demand.

Feed Streams

The following materials will be required by the plant:

- grit for water-jet cutting
- caustic for agent and energetics hydrolysis and for gas scrubber solutions
- nitric acid for pH control
- nutrients for the bioreactor, with dextrose in the largest amount
- a flocculating agent, probably a polymeric amine
- hydrogen peroxide for final polishing of the bioliquid product
- carbon adsorbent for gas and ventilation air cleanup
- fresh water (no wastewater leaves the plant, but some water is lost as water of crystallization of salts, water with the biosolids, humidity added to the air, etc.)
- decontamination solution (sodium hypochlorite)

Waste Streams

There are three gaseous and four solid waste streams. There is no liquid effluent from this process.

Gas Streams

The largest gas stream is estimated to be 30,000 actual cubic feet per minute (ACFM) of slightly depleted air, with oxygen content reduced from 21 percent to about 19 percent. This stream is treated in a catalytic oxidation unit. A second, much smaller stream, also treated in a catalytic oxidation unit, is a product of the metal parts treaters. A third gas stream is steam from the drum dryers used to dry salts and biomass. This stream is released to the atmosphere.

Solid Streams

Metal Parts. All of the metal parts from the original munitions are cleaned to a 5X condition.

Biosludge. The dried biosludge, together with a small amount of a flocculating agent, probably a polymeric amine, is expected to amount to 10 to 20 percent of the mass of material fed to the reactor (agent plus energetics plus added nutrients).

Sodium Salts of Various Acids. Sodium salts are produced from the heteroatoms (i.e., fluoride, chloride, sulfate, nitrate and nitrite, and phosphate) and are a solid waste. Chemical agents yield a large amount of salt (e.g., VX yields salt equal to about 150 percent of the mass of the original VX). The yield of solid salts from the munitions varies with the munition but will probably be roughly double the weight of the agent and energetics in the munition. The salts may have a small organic residue left from the biotreatment and the treatment with ultraviolet light/peroxide. This remains to be demonstrated.

Grit (Garnet). This material, used in water-jet cutting, is treated at 1,000°F for at least 15 minutes in steam.

Mass Balance

Mass balances for two munitions were provided by Parsons-AlliedSignal: an 8-inch GB projectile and a 4.2-inch mortar with mustard (Tables 8-2 and 8-3). (Both tables show Fenton's solution being used, although that is now considered unlikely.) These mass balances cannot be considered exact because the products from the bioreactor are uncertain. The fraction of the feed (C, H, and N) that will be oxidized to gaseous products and the fraction oxidized to solid biomass are not known with certainty.

Experience suggests that about 10 to 20 percent of the feed (C, H, N) to the bioreactor, including added nutrients, will end up as biomass; the rest will be oxidized to gas products. The values shown in Table 8-3 demonstrate the large amounts of nutrients and dextrose required for bioprocessing nerve agent (GB) hydrolysate.

TABLE 8-2 Mass Balance for Processing HD 4.2-inch Mortars (lb/lb HD)

Component	Amount (lb)
Input Streams	
Flocculent + Fenton's reagent	0.2
Air	278.0
Nutrients	0.3
HD	1.0
Energetic materials	0.1
Sulfuric acid	0.0
NaOH	1.0
Water	9.4
Total input	290.0
Output Streams	
Air	280.0
Sludge (wet)	0.6
Water (evaporated)	7.8
Salt	1.6
Total Output	290.0

Source: Parsons-AlliedSignal, 1999a.

TABLE 8-3 Mass Balance for Processing GB 8-inch Projectiles (lb/lb GB)

Component	Amount (lb)
Input Streams	
Flocculent + Fenton's reagent	7.8
Air	2,897.0
Nutrients	6.3
Dextrose	44.4
GB	1.0
Energetic materials	0.5
Sulfuric acid	2.8
NaOH	2.7
Water	50.1
Total input	3,012.6
Output Streams	
Air	2,930.0
Sludge (wet)	53.0
Water (evaporated)	24.3
Salt	4.9
Total Output	3,012.2

Source: Parsons-Allied Signal, 1999a.

Start-up and Shutdown

The bioreactor is the largest volume unit in the Parsons-AlliedSignal technology package. This unit will be run continuously but with some possible changes in the feed. Experience suggests that the time required for acclimation of the bio-organism to a mustard hydrolysate feed is only a few hours. Acclimation to a nerve agent hydrolysate that contains phosphorus in the form of a phosphonate compound may take several weeks. Disposal campaigns are planned accordingly.

In the event of a nonroutine shutdown, most processing units, which are batch or semibatch operations, can be shut down and held on stand-by status. The bioreactor can withstand a shutdown of the air supply for only a few hours, however. For a longer shutdown, the bioreactor will require an auxiliary feed and air to maintain the microbial population.

EVALUATION OF THE TECHNOLOGY PACKAGE

Process Efficacy

Effectiveness of Munitions Disassembly

The technology provider claims that munitions handling, disassembly, and plant safety design and practices will be modeled closely after the baseline system for which significant operating experience is available. These technologies have all been demonstrated and can be considered to be mature.

Two new technologies will be introduced in the disassembly process: water-jet cutting and water-jet clean-out. Both have been used in the demilitarization of conventional munitions. A brief overview of the state of the art for demilitarizing ordnance using high-pressure jet cutting and clean-out of the energetic materials is given in Appendix G. In view of the significant developments in water-jet cutting technology and its tested use for cutting high-explosive casings, the technique can be considered suitable for application in disassembly operations.

Explosive and propellant will be recovered from the munitions by high- pressure water-jet clean-out, which has been used on a substantial scale in conventional demilitarization operations for many years. However, it has only been used for removing energetic materials from ordnance and not for minimizing the particle size of the energetic material. The committee believes that a small particle size (e.g., 0.25 inch) will be necessary for the hydrolysis reaction. Therefore, the simultaneous removal and size reduction of energetic materials will have to be demonstrated.

Effectiveness of Hydrolysis

In the Parsons AlliedSignal Technology package, hydrolysis will be the main technology for achieving a 99.9999 percent agent destruction. Subsequent treatment in the bioreactor is expected to destroy the hydrolyzed materials, but the bioreactor should handle little if any agent. The hydrolysis reactions of chemical agents have been studied extensively (see Appendix D), and hydrolysis for VX and HD should have accumulated many hours of demonstration (at Aberdeen and Newport) before an ACWA-based plant starts up. The Army has already hydrolyzed several hundred pounds of GB and VX to prepare hydrolysate for ACWA technology demonstrations of SCWO and biotreatment.

The hydrolysis of energetic materials is considered a less mature technology than the hydrolysis of agent (see Appendix E). The design will have to allow for various reaction times and for various quantities, depending on the type of munition being processed.

Effectiveness of Biotreatment

The use of natural microbial consortia for the degradation and mineralization in a biotreatment system depends mainly on providing organic food sources and nutrients to the microorganisms. A sustainable food–to–microorganism ratio must be maintained in the bioreactor to ensure microbial viability. In theory, microorganisms can be made to mineralize almost any organic contaminant, but in practice the toxicity of organic and inorganic constituents in the feed can be a major problem that requires close monitoring and control. Furthermore, biotreatment alone cannot remove all of the organics in the hydrolysate. A final polishing step may be required to meet regulatory levels.

Biotreatment of the hydrolysate from mustard will be used at the Aberdeen Proving Ground facility. Tests using a "sequencing batch reactor" system have been quite successful. The liquid product from this process will go to the sewage treatment plant of the Aberdeen Army base before final release to the environment.

Similar testing of a batch reactor was conducted for biotreatment of VX hydrolysate for possible use at the Newport, Indiana, bulk storage site. However, difficulties were encountered that were not completely resolved (DeFrank et al., 1996). VX hydrolysate contains phosphorus in the form of phosphonate (i.e., with a C-P bond), which appears to be difficult to metabolize. Recent tests by the technology provider, however, have been successful in removing the phosphorus to very low levels (more than 95 percent removal) (Parsons-AlliedSignal, 1999b). Removal of phosphonate appears to depend on the following factors:

- The bacteria will use phosphonate as long as no other form of phosphorus is available. Therefore, other materials, such as phosphates must be rigorously excluded.
- The optimum ratio of major nutrients for the biomass is in the approximate ratio C:N:P = 100:5:1. Therefore, relatively large amounts of nutrients containing C and N have to be added to make use of the phosphorus in the feed. For example, dextrose was added in a ratio of 44 lb per lb of original agent in the hydrolysate.
- Phosphorus is used to produce biomass, rather than for metabolism. Therefore, a combination of reactor types must be used for optimum results

(e.g., an immobilized-cell reactor followed by a free-floating biomass reactor).

In contrast to the difficulties experienced with the bioreaction of nerve agent hydrolysates, the bioreaction of mustard hydrolysate has worked well, although food supplements are required. Mustard does contain low molecular weight chlorinated hydrocarbons that are difficult to biodegrade, and they will be sent to the catalytic oxidizer, either in the effluent air from the bioplant or in gases leaving the evaporator. The catalytic oxidation unit is expected to be effective in destroying them, but this will have to be demonstrated.

The committee anticipates that the ACWA demonstration tests being conducted during the preparation of this report will address a number of questions concerning the bioprocess. First, the pH of GB hydrolysate will be adjusted to 8.5 before it goes to the bioreactor. This pH is low enough for GB to re-form (see Appendix E). This possibility will be monitored during the demonstration. Second, the biosludge, amounting to 10 to 20 percent of the carbon in the feed, will have to be tested for toxicity.

Third, the technology provider has indicated that the final effluent from the bioreactor after post-treatment, should have a biological oxygen demand of < 100 mg/L, a chemical oxygen demand of < 1,000 mg/L, and a total organic carbon of < 100 mg/L. This chemical oxygen demand is somewhat lower than reported in the technology provider's proposal and will have to be demonstrated. (The biological oxygen demand and total organic carbon are usually much lower than the chemical oxygen demand, though exceptions can occur.) A "polishing" step (hydrogen peroxide with ultraviolet light) to reduce them further is provided for the nerve agent hydrolysates.

Fourth, the airflow rate through the bioreactors is far above the stoichiometric requirement for mineralizing the feed. In early laboratory tests reported by the technology provider, the air flow was 25 to 50 times the stoichiometric requirement. In more recent larger scale work, however, the technology provider has demonstrated satisfactory operation with 12-fold stoichiometric requirements (i.e., about 8 percent of the oxygen in the inlet air was used). Operations should require as small an air supply as feasible for the bioreactors because the product air must be treated further (i.e., heated to 425°C [797°F] for the catalytic oxidation unit, cooled for acid gas scrubbing, and possibly reduced in relative humidity for activated carbon adsorption).

Fifth, material vaporized or entrained in the air from the bioreactor may affect the performance of the catalytic oxidation unit. For example, entrained phosphorus could deactivate the catalyst. The gas stream from the bioreactors will have to be characterized. Also, tests will have to show whether entrained material will be a problem.

Sixth, the salt content of the feed to the bioreactor must be kept low for the microbial population. The technology provider has stated that it will be maintained at less than 3 percent. It appears that a salt content of about 1 percent is being used in the ACWA demonstrations. The level will affect the requirements for the salt-recovery evaporator. A low salt content of about 1 percent (which will be demonstrated) appears to be a conservative choice for operation.

Effectiveness of Evaporation for the Production of Salts and Biosolids

The evaporation processes for recovering biosolids and salts are well established. The products will probably be considered toxic, however, because most heavy metals in the feed will end up in the biosolids. Polysaccharides formed in the bio-operation are known to be good sequestering agents for heavy metals. (See Tables D-3 and D-9 in Appendix D for a list of heavy metals found in some samples of VX and HD.) Some metals will also appear in the salts from the evaporation process.

Heavy metals in the biosludge will probably prove to be nonleachable as defined by EPA's TCLP test because they are usually tightly bound to the polysaccharides. Heavy metals in the salts from nerve agent processing will probably also be relatively nonleachable because phosphates are present. However, heavy metals in the salt product from mustard may be more soluble.

Effectiveness of Catalytic Oxidation

Two separate catalytic oxidation units, with subsequent alkaline scrubbing and carbon filtration, are proposed. One treats gas from the metal parts treaters, the other treats gas from the bioreactors. The catalytic oxidation units resemble automobile catalytic converters and use proprietary AlliedSignal technology. They must handle a large gas flow, estimated at 30,000 ACFM, from the bioreactors, and the oxidizable material will be very dilute. The proposed catalyst has been shown to be effective at residence times as low as 0.1 seconds and should oxidize most of the chlorinated materials in the gas stream (arising from mustard). Small molecules are difficult to oxidize, however, and the efficiency of the catalyst in oxidizing materials such as methylene chloride (which might be in the gas) must be demonstrated.

The possible presence of chlorinated dioxins and furans in the product gas should be checked. If they are present, the proposed carbon filters could effectively remove them.

A small amount of liquid mist in the gas leaving the bioreactors should be expected. Phosphorus and sulfur (derived from nerve agents and mustard, respectively) in this mist may negatively affect the catalyst. The catalyst has been shown to destroy nerve agents and mustard in short-term tests, but the technology provider recognizes that long-term data are limited, and durability has yet to be demonstrated. This is one area being investigated during the ACWA demonstration tests. A caustic scrubber and an activated carbon filter to treat the effluent from the catalytic oxidizer are included as an extra safeguard. (The presence of agent in the gas stream from the bioreactors appears to be highly unlikely. Much smaller gas streams [from the metal parts treaters or venting of feed tanks] might contain some agent.)

The technology provider's proposal states that sulfur and phosphorus concentrations in the gases flowing to the catalytic oxidation units will be monitored. Continuous monitoring of the exhaust-gas stream is used in place of a hold-test-release monitoring process.

Effectiveness of Peroxide/Ultraviolet Oxidation

The hydrogen peroxide/ultraviolet light treatment process must be tested to demonstrate its effectiveness. The solution to be treated is colored (brown) and undoubtedly contains finely divided solids in suspension. Ultraviolet radiation is directed into the solution through quartz windows, and it may be difficult to keep the window clean during VX processing. The technology provider's goal is to reduce the total organic carbon to less than 50 ppm and to reduce residual phosphonate to 3 ppm.

Effectiveness of the Metal Parts Treater

The metal parts treater and rotary metal parts treater will heat the metal parts (as in the baseline system) but will use steam (at 650°C [1,202°F]), rather than combustion gas. The process will qualify as a 5X treatment.

Sampling and Analysis

The proposed sampling and analysis methods are generally well established and straightforward and should not pose significant difficulties.

Maturity

The overall process is a combination of many (at least 10) different technologies, all of which have substantial operational backgrounds, although some will require demonstration for their application to chemical weapons destruction. Hydrolysis of mustard and VX are planned at Aberdeen, Maryland, and Newport, Indiana, respectively. Extensive testing and development of the hydrolysis processes has been done for the design of these two facilities. Biotreatment of mustard hydrolysate is planned for the Aberdeen facility, and a similar level of testing and development has taken place, albeit on a different type of bioreactor than the immobilized cell bioreactor. Tests of the biological treatment test work on mustard hydrolysate have been quite successful. Biotreatment of energetic hydrolysates has also been successful, though the full range of materials has not been demonstrated. Ultraviolet/peroxide treatment of the mustard hydrolysate to remove low molecular weight chlorinated hydrocarbons is also being investigated at Aberdeen, but this process differs from the treatment proposed by Parsons-AlliedSignal for ACWA. The concentration of chlorinated hydrocarbons at Aberdeen should be much lower than the concentration of organics to be treated in this application. The bioreaction of a feed with high phosphorus content, particularly with the phosphorus present as a "phosphonate" (i.e., C-P bond), has not been proven.

Individually, all unit operations can be considered mature technologies, and some have been applied to the treatment of assembled chemical weapons materials. Nevertheless, past industrial experience has shown that starting up a process with many steps in series cannot be accomplished without significant difficulties.

Robustness

All of the technologies appear to be reasonably operable and robust in that they can be readily controlled to desired set points and can accommodate modest fluctuations in feed composition. Munitions disassembly, energetics wash-out, hydrolysis, and biotreatment will already have been applied to chemical weapons disposal at Aberdeen and Newport. The sensitivity of the bioreactor to fluctuations in feed or contamination is being addressed in the demonstration. Biological conversions bring their own special problems, however, particularly in dealing with living materials. Problems with feed toxicity or predators for the microorganisms may develop. The problems previously mentioned all appear to be solvable. However, one aspect of robustness that cannot be determined at this time is the ability of the integrated process to handle feed variations through the entire set of technologies.

Monitoring and Control

Overall, the proposed monitoring, control, and instrumentation system appears to be modern in design and well thought out. Parsons-AlliedSignal proposes using as much of the design as possible of the Army's current baseline technology, taking advantage of the design and lessons learned from that experience. The following monitoring and control technologies are new to chemical weapons destruction:

- the design of monitoring, control, and instrumentation of water-jet cutting and wash-out systems
- the demonstration of a biomass accumulation in the reactor that can achieve the necessary conversions
- control of foaming in the evaporator (a small amount of organic material from the bioreactor in the brine would affect the liquid surface property and could lead to foaming in the evaporator, a common problem that can probably be handled by additives)
- demonstration of the steady-state ultraviolet/peroxide oxidation step

Applicability

Many of the technologies in this technology package will have been applied to chemical weapons before the start up of an ACWA plant based on this overall process. Catalytic oxidation (on a large scale), metal decontamination (to 5X) with steam, and water-jet cutting applied to rockets will be new. In the committee's opinion, the technology package is conceptually applicable to the treatment of all assembled chemical weapons. However, mine processing was not addressed by the technology provider, and successful biotreatment of nerve agent hydrolysate must still be demonstrated.

Process Safety

The unique equipment proposed by the Parsons-AlliedSignal team is associated with the following technologies:

- water-jet cutting
- wash-out of agent and energetics
- shredding of energetics
- base hydrolysis
- biological treatment
- ultraviolet/peroxide oxidation
- catalytic oxidation
- decontamination of metal parts with high-temperature steam in the metal parts treaters
- decontamination of grit and munition fuzes in high-temperature steam in the rotary metal parts treater

Water-jet cutting, wash-out, and energetics hydrolysis will be done in explosion-containment areas. The hydrolysis process operates at temperatures up to 90°C (194°F); the bioreactor processes operate at ambient temperature. Both processes operate at ambient pressure. The catalytic oxidizer and the metal parts treaters operate at elevated temperatures, 425°C (797°F) and 649°C (1,200°F), respectively. The hydrolysis reactors and the 5X treaters, which represent the primary detoxifying processes, operate in a batch mode; thus, the effectiveness of treatment can be ascertained prior to release of the processed material to the next step.

The remaining systems are routine chemical processes, and in this application, they occur downstream of the primary (hydrolysis) and secondary (biotreatment) detoxifying processes. These systems should pose no unique hazards. The equipment consists of caustic scrubbers, carbon filters, evaporators, and dryers.

Worker Health and Safety

If a process upset occurs, the primary destruction components (the hydrolysis reactors and bioreactors) cannot be shut down quickly because incomplete hydrolysis products, are extremely hazardous. Procedures are expected to be established for safe shutdown and restarting of the system. The air effluent during an upset will continue to be treated, first in the catalytic oxidizer, then in the caustic scrubbers, and, potentially, in activated carbon filters.

The low-speed shredder for energetics poses a potential worker safety issue. Friction, shear, or heat may result from the inadvertent introduction of metal, an excessive feed rate, or some other cause and could initiate the energetic material. Workers are not expected to be present, however, during normal operations.

The rotary metal parts treater can be designed to accommodate detonations of fuzes (fuze detonation chambers are not unique). Workers are not expected to be present during normal operations.

Only trace amounts of energetics will be present in the metal parts treater under expected operating conditions. Scenarios for the introduction of energetics beyond design conditions will be evaluated to ensure that they are extremely unlikely before the design is completed. Potentially flammable dunnage pyrolysis products are being characterized during ACWA demonstration testing, and the impact of these and other effluents should be considered as the design develops.

The technology provider plans to hydrolyze different types of energetic materials simultaneously in the same reactors. As discussed in Appendix E, the committee is concerned that this could lead to the formation of compounds that are both energetic and sensitive. Therefore, different energetic materials should be processed in separate reactors unless tests shows that the formation of sensitive compounds does not occur.

Small amounts of aluminum particulates created during water-jet cutting of the rocket warhead sections will generate hydrogen during the hydrolysis step. The hydrogen and hydrolysis gases will be vented to the bioreactor off-gas stream and then heated to 425°C (797°F) for treatment in the catalytic oxidizer. The large bioreactor off-gas flow rate will dilute the hydrogen to well below the flammability limit before heating in the oxidizer. Standard operational procedures and designs for flammable gas systems, (e.g., maintaining a negative pressure to avoid release to air spaces) should be adequate to minimize explosion hazards.

The biosludge produced in the bioreactor could contain some pathogenic microorganisms. The potential for worker exposure to these microorganisms is expected to be minimized by appropriate protective gear.

Water-jet technology is commonly used in the demilitarization of conventional munitions and should not pose unique safety issues. The ACWA demonstration includes tests of the capability of this technology to separate the fuze and the rocket motor from the warhead. Even if an ignition occurs, there will be little risk to workers because the cutting is performed remotely in an explosion-containment area.

The energetics hydrolyzer incorporates an external circulation and cooling loop. Pumping an aqueous slurry of energetic materials can be done safely under the proper conditions. If an accident occurs during normal operations, there would be little risk to workers because the loop is in an explosion-containment area. The loop will have to be designed to ensure that energetic material does not precipitate and accumulate in the piping, which could result in an accident during maintenance procedures.

The primary hazardous materials used are sodium hydroxide, nitric acid, sodium hypochlorite, and hydrogen peroxide. These chemicals are used routinely at many industrial facilities and are not unique to the Parsons-AlliedSignal process for demilitarization.

Public Safety

The release of agent and other regulated substances in plant effluents is judged to be extremely unlikely. The destruction of agent and energetics will be verified by hold-test-release operations before the transfer of hydrolysate from the hydrolysis reactors to the bioreactors and before the transfer of bioreactor sludge to the sludge containerization step. The gaseous effluent from the bioreactors will be continuously released through activated carbon filters, if the demonstration test results indicate that this is desirable or necessary. No hold-test-release operation is provided for the gaseous effluent stream from the bioreactors. Because this stream will be continuously monitored for hazardous materials, the release of hazardous materials is considered extremely unlikely.

There is a low probability of agent release in case of a failure of the rotary metal parts treater. This hazard can be mitigated by good design and operational procedures that are confirmed by risk analyses.

GB reformation in the bioreactor because of low pH (see Appendix D) is being investigated during demonstration testing. Any GB vented from the feed tank or the bioreactor is expected to be destroyed in the catalytic oxidizer or captured in the backup activated carbon filters.

Human Health and the Environment

Effluent Characterization and Impact on Human Health and Environment

In the absence of health risk and environmental assessments, a precise statement on the impact of the effluents on human health and the environment cannot be made at this time. However, the gas flow leaving the plant should be free of hazardous material. It will have been exposed to a very high temperature (about 425°C [797°F] in the catalytic oxidation unit), and it will have been through extensive cleanup processes to remove traces of organic materials (including any agent).

Two of the solid materials leaving the plant will be treated as hazardous waste: salts with traces of organic material; and biomass with small amounts of salts.

Completeness of Effluent Characterization

The very large gas flow, primarily from the bioreactor, will have gone through catalytic oxidation and acid gas scrubbing. The composition will have to be determined in detail during initial trials. The gas should then be tested routinely for chemical agent, oxygen, carbon dioxide, and carbon monoxide on a real-time basis. It should also be characterized for low concentrations of hazardous materials, such as dioxins.

Biomass will be tested periodically for leachability and for toxicity. Salt residue will also be tested for leachability. Other effluents that have been treated to a 5X condition will not require further characterization.

Effluent Management Strategy

Salts. Dried salt, probably containing some organic materials, will contain sodium salts of fluoride, chloride, sulfate, nitrate, and nitrite. The technology provider expects this waste stream to be hazardous. Stabilization, either at the plant or at a commercial hazardous-waste treatment facility, may be required. Experimental studies will be necessary to determine the leaching levels of hazardous constituents in the salt and then, if required, determine the additives needed to stabilize the salt. Stabilization and burial in a hazardous-waste landfill should provide adequate protection to human health and the environment.

Biosludge. The composition of the biosludge produced in the bioreactor is unknown at this point, although the technology provider expects it to be hazardous because it will be the sink for most heavy metals from the process. The sludge will also contain microorganisms, some of which might be pathogenic. Testing will be necessary to determine whether or not the waste is hazardous as defined by the EPA. If so, disposal in a hazardous-waste landfill may not be possible because of the biological activity. Incineration of the waste is an alternative. If it is not hazardous, it can most likely be sent to a municipal solid-waste landfill without threat to human health or the environment.

Gas. Exhaust gas from the catalytic oxidizer will pass through an acid gas scrubber (and possibly through an activated carbon adsorber). Whether or not any constituents of concern will be present in this stream is not known at this point. Analyses will be necessary to confirm the presence or absence of low molecular weight hydrocarbons and chlorinated hydrocarbons, oxides of nitrogen, and chlorinated dioxins and furans.

Metal Parts and Garnet Grit. Metal parts and grit are cleaned and deactivated to the 5X condition in the metal parts treaters. The cleaned parts are not expected to pose any threat to human health or the environment.

Resource Requirements

The Parsons-AlliedSignal technology package has three steps that will consume large amounts of energy: (1) producing 650°C (1,202°F) steam for the metal parts treater; (2) heating the vent gas from the bioreactor to about 425°C (797°F) for the catalytic oxidizer; and (3) evaporating the salt solution from the bioreactor. The power requirement has been estimated to be a few megawatts. None of the resource requirements appears to be excessive.

Environmental Compliance and Permitting

There are no apparent reasons that the combination of technologies selected by the technology provider should lead to unusual permitting or compliance problems. The absence of liquid emissions is an important advantage of the process. However, the catalytic oxidation operations are close enough to incineration in concept that regulatory (and public) concerns could be raised.

STEPS REQUIRED FOR IMPLEMENTATION

The following steps would have to be taken to implement this technology package:

1. demonstration of the effectiveness of the biotreatment of various combinations of agent and energetics hydrolysates of sufficient length to give reasonable assurance of long-term performance
2. operation of the bioreactor at the planned salt-content
3. characterization of the off-gas from the bioreactor to evaluate the extent of air-stripping from the reactor and the possible poisoning of the catalyst in the catalytic oxidation unit
4. demonstration of the effectiveness and long-term performance of the catalytic oxidation system in destroying organic constituents in the bioreactor off-gas
5. quantification and characterization of the sludge from the biological process to ascertain if Schedule 2 compounds or other hazardous constituents are present
6. demonstration of unproven steps in the proposed process, including ultraviolet/peroxide oxidation and evaporation operations
7. quantification and characterization of the salts from the evaporation operations to ascertain what organic compounds are present

FINDINGS

Finding PA-1. The biological treatment operation will require further demonstration to prove its ability (1) to handle a variety of feed stocks with reasonable acclimation times between changes, and (2) to achieve high levels of conversion of the Schedule 2 compounds in the hydrolysate. The demonstration will have to last long enough to give confidence in the long-term operation ability of the process.

Finding PA-2. The relative effects of biological treatment and air-stripping on the destruction of organic materials in the bioreactor have not been established. This will affect the composition of the off-gas from the bioreactor.

Finding PA-3. The effectiveness of ultraviolet/hydrogen peroxide oxidation in reducing Schedule 2 compounds to an acceptably low level has not been demonstrated.

Finding PA-4. The bioreactor has been operated only at very low salt concentrations. Operation at design concentrations has not been demonstrated.

Finding PA-5. Additional data should be gathered on the effectiveness of the catalytic oxidation system in destroying organic materials in the biotreatment off-gas.

Finding PA-6. The sludge from the biological process has not been completely characterized.

Finding PA-7. Even though the evaporation operations involve conventional technologies, they have not been tested for this application.

Finding PA-8. The dried salts from the evaporation operations have not been characterized for leachability and toxicity.

9

Teledyne-Commodore Solvated Electron Technology Package

INTRODUCTION AND OVERVIEW

The Teledyne-Commodore team's technology package for the destruction of assembled chemical weapons (summarized in Table 9-1) involves four fundamental technologies:

- The chemical agent, energetic materials, and metal parts are separated via *ammonia-jet cutting and wash-out.*
- *Solvated electron technology (SET)* is used to destroy chemical agent, deactivate energetic materials, and decontaminate metal parts and dunnage. SET solutions, metallic sodium in anhydrous liquid ammonia, are highly reducing and are characterized by an intense blue color from the presence of partially solvated (i.e., ammoniated) electrons. The blue color provides a visual indicator of the reactivity of the solution for destroying agents and energetics. The solid and liquid residuals of the SET process are treated by water *hydrolysis* to destroy the excess sodium.
- The hydrolysate from agent and energetics destruction is further treated by *oxidation* with sodium persulfate or hydrogen peroxide.

Figure 9-1 shows how the four fundamental technologies are linked and identifies the basic process flow. Teledyne-Commodore assigns a separate area for application of each technology to the materials from the munitions. The following section describes each area in detail.

DESCRIPTION OF THE TECHNOLOGY PACKAGE

Munitions Access and Energetics Deactivation (Area 100)

Teledyne-Commodore designed Area 100 to handle three types of munitions: M55 rockets containing explosives, propellant, igniters, and agent; projectiles and mortars containing explosives and agent; and land mines containing explosives and agent. Figures 9-2 through 9-4 outline the approach for each. The proposed disassembly processes differ significantly from the baseline processes developed by the Army, principally in the use of ammonia-jet cutting and ammonia wash-out.

Fluid jets are used routinely in industry for cutting metal and have been used in the demilitarization of conventional munitions. For background information on jet cutting, see Appendix G. Only the particular application proposed by Teledyne-Commodore is described below.

In both the cutting and wash-out operations, ammonia is pressurized to 2,720 atm (40,000 psi) by an intensifier pump and delivered to the work area through a 0.01-inch diameter orifice at velocities of about 1,000 m/s. Pre-intensifier booster pumps are used to ensure adequate pressure and volume of ammonia and to prevent flashing of the liquid into gas during the suction stroke of the intensifier. For the cutting operation, 180-micron abrasive particles, normally garnet, are added to the pressurized ammonia stream through a stainless steel venturi mixing section. A pressure

TABLE 9-1 Summary of the Teledyne-Commodore SET Technology Package

Major Demilitarization Operation	Approach(es)
Disassembly of munitions	High-pressure ammonia fluidized jet cutting and ammonia-jet wash-out.
Treatment of chemical agent	SET reduction; hydrolysis of the condensed-phase products; oxidation of the hydrolysate with sodium persulfate or hydrogen peroxide.
Treatment of energetics	SET reduction; hydrolysis of the condensed-phase products; oxidation of the hydrolysate with sodium persulfate or hydrogen peroxide.
Treatment of metal parts	Agitation of shredded parts in SET solution to a 3X condition.
Treatment of dunnage	Grinding or shredding; mixing with SET solution to destroy agent.
Disposal of waste	**Solids.** product from process and decontaminated dunnage sent to a suitable permitted landfill; metal parts shipped to Rock Island Arsenal for 5X treatment. **Liquids.** stabilized with cement and shipped to appropriately permitted landfill; oil and hydraulic fluids sent to TSDF. **Gases.** off-gases from process and vaporized hydrocarbon residuals burned in boiler.

vessel surrounds the cutting and wash-out processing equipment, providing an intermediate chamber that is maintained at 10.5 atm (140 psig) inside an Army baseline explosion-containment room.

The pressurized fluid from the intensifier pump is passed through a chiller to reduce the temperature of the liquid "significantly" below room temperature because prechilling is reported by the technology provider to enhance the jet cutting properties of the liquid ammonia. The proposal does not include details on the chiller or the optimum temperature of the fluid jet stream. The following sections summarize Teledyne-Commodore's approach to the disassembly of each munition type.

Rocket Disassembly

M55 rockets are loaded from the receiving area through an airlock onto a rotary index table located in a pressurized chamber. The index table automatically rotates the rockets to six successive cutting and wash-out stations for separation of agent, energetics, and metal parts as outlined in Figure 9-2.

Projectile/Mortar Disassembly

Projectiles and mortars are processed similarly to rockets by ammonia-jet cutting and wash-out. Figure 9-3 outlines the sequence of cuts.

Land Mine Disassembly

Figure 9-4 is a diagram of the process proposed for disassembling M23 land mines by ammonia-jet cutting and wash-out. Fuzes are separated from the munitions via jet cutting and are initiated in a detonation chamber using a high-voltage electrical charge. The fuze parts are then washed with SET solution in a dedicated reactor.

Treatment of Chemical Agent (Areas 200, 400, and 500)

SET Reduction followed by Hydrolysis (Area 200)

In Area 200, a mixture of agent and anhydrous ammonia, transferred from Area 100, is collected in a carbon steel vessel, 3.5 feet in internal diameter and 10.5 feet in height. A SET solution is generated by mixing liquid sodium and ammonia to form a 4-percent solution of sodium in liquid ammonia. The sodium is transferred at its melting point of 97.5°C (207.5°F); the liquid ammonia is transferred at room temperature and 10.5 atm (140 psig) (saturated conditions). The reagents are combined in an in-line static mixer that Teledyne-Commodore describes as "flow splitting devices, which provide uniform droplet sizes of both ... streams" (Teledyne-Commodore, 1997). When the sodium comes in contact with the ammonia, it dissolves rapidly, releasing 1,400 calories per mole of sodium introduced. Since a 4 wt. percent solution of sodium in

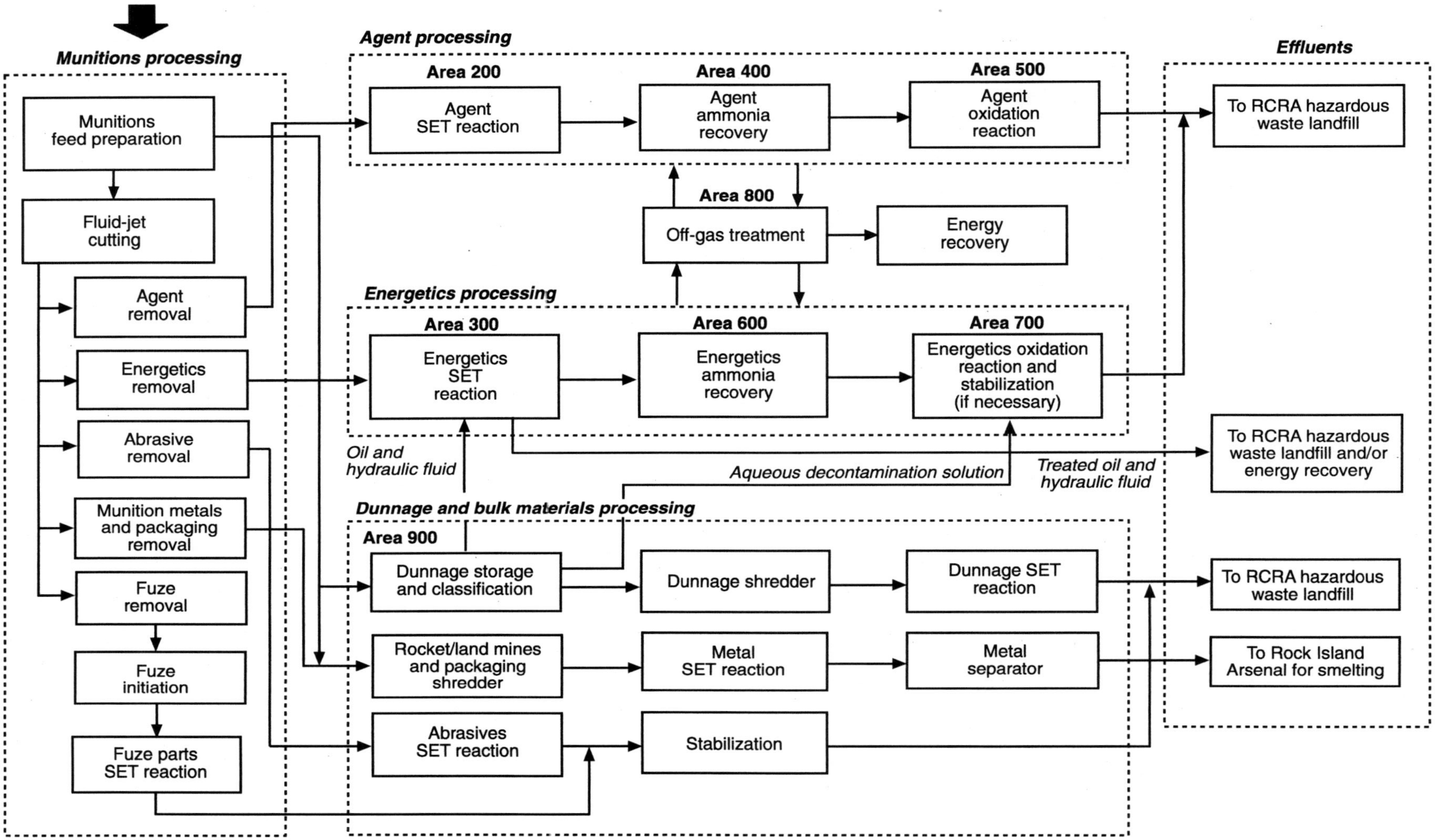

FIGURE 9-1 Block flow diagram of the technology package proposed by Teledyne-Commodore. Source: Teledyne-Commodore, 1997.

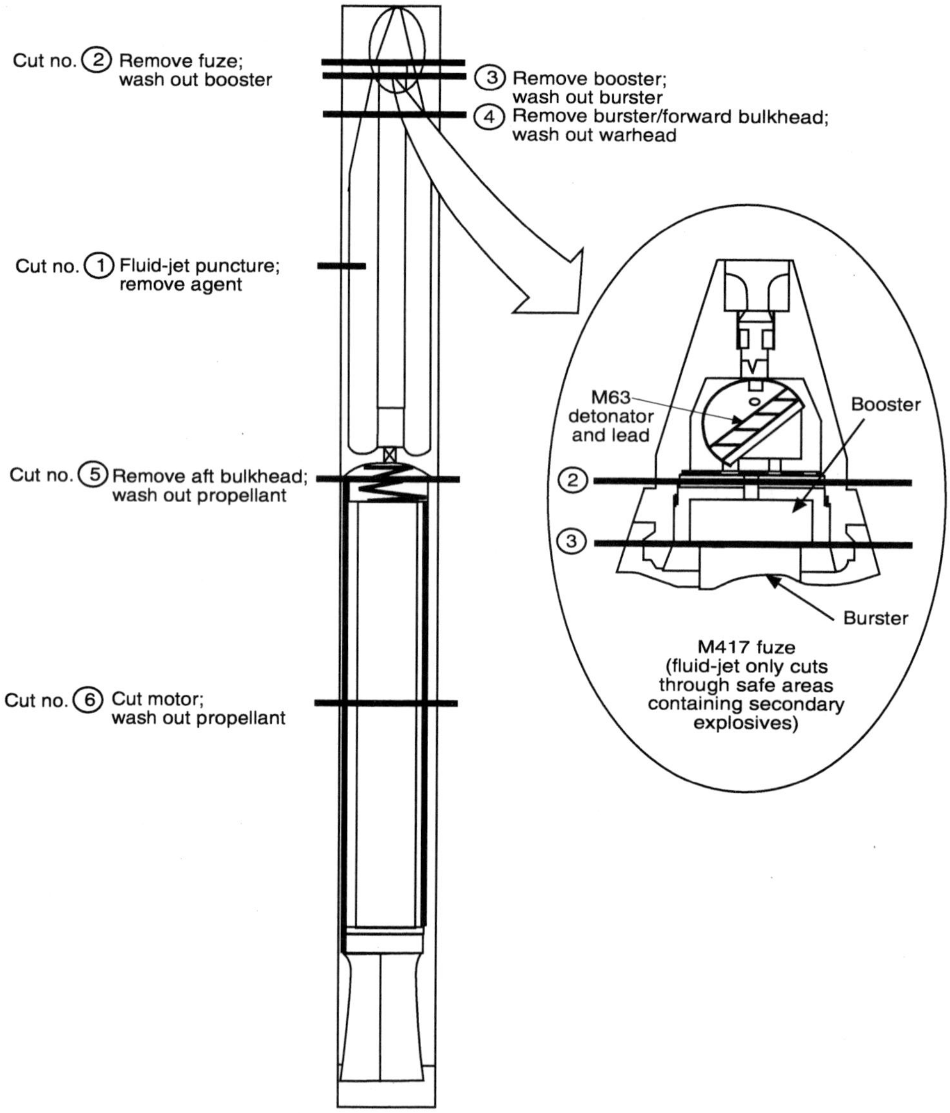

FIGURE 9-2 The sequence of cuts for ammonia-jet cutting and wash-out of M55 rockets. Source: Teledyne-Commodore, 1997.

liquid ammonia contains 0.17 moles of sodium per hundred grams, the heat release on mixing would be 238 calories per hundred grams of solution. The heat of vaporization of ammonia at room temperature and 10.5 atm (140 psig) is 279 cal/g. Therefore, approximately 1 percent of the ammonia would be expected to vaporize in the process of forming the SET solution.

The prepared SET solution is fed to a static reactor mixer, and agent from the feed vessel is introduced below the liquid surface. The reaction is carried out at temperatures of 19 to 23°C (66 to 73°F) and pressures of 8.5 to 12.4 atm (110 to 167 psig). The agent destruction reactions are exothermic, and temperature is controlled by the evaporation of ammonia. The evaporated ammonia and off-gases from the SET reaction are collected in holding tanks for testing prior to venting to the gas-treatment train (Area 800). The slurry from the SET reaction, still at elevated pressure, is treated with water to destroy the excess sodium. The resultant suspension is transferred at elevated pressure to a centrifuge where solids and liquids are separated. The solids are transferred to a carbon steel vessel, 3 feet in internal diameter and 8 feet in height, and mixed with water. The solution is then pumped to Area 500 for oxidation. The liquids remaining in the centrifuge are mixed with water and fed to the bottom of the ammonia-recovery tower in Area 400.

Although reactions of organics with solutions of

metallic sodium in liquid ammonia have been studied since 1865, the reaction products cannot be predicted. In general, the solvated electrons are attracted to the polar bond between carbon and a more electronegative species, such as chlorine, fluorine, phosphorus, sulfur, or oxygen. The result is cleavage of the covalent bond. The strongly electronegative species leaves as an anion. The complementary site on the less electronegative atom (usually carbon) may capture a second electron, thereby becoming negative; or the bonds in the remaining carbon skeleton may rearrange themselves and release a gaseous alkane or alkene. Further reaction occurs when the condensed-phase products of the SET reaction are hydrolyzed.

Teledyne-Commodore's experimental results from treating specific agents with SET in the laboratory are described below.

Agent HD (Mustard). The expected initial reaction of mustard in the SET process is cleavage of the carbon-chlorine bond by a solvated electron to form sodium chloride (NaCl). Measured results, scaled up to 100 pounds of HD, are shown in Table 9-2. The gaseous ammonia released in the SET reaction results from evaporation, which is used to control the temperature of the liquid mixture. In the proposed full-scale system, the gaseous ammonia from the hydrolysis reaction is produced in the agent-ammonia recovery tower (see Area 400).

Teledyne-Commodore was unable to determine the molecular composition of the slurry from the SET reaction prior to hydrolysis. The analytical problem, as described to the committee, was two-fold. One was the tendency for foaming and sudden gas evolution when the pressure was dropped on the SET product slurry. The other was the lack of standard analytical protocols for ammonia systems. The solid and aqueous solution from the hydrolysis step was combined for analysis and was found to contain NaCl, sodium sulfide (Na_2S), and a variety of polysulfides. Agent concentrations were below detectable limits in the solid and liquid products,

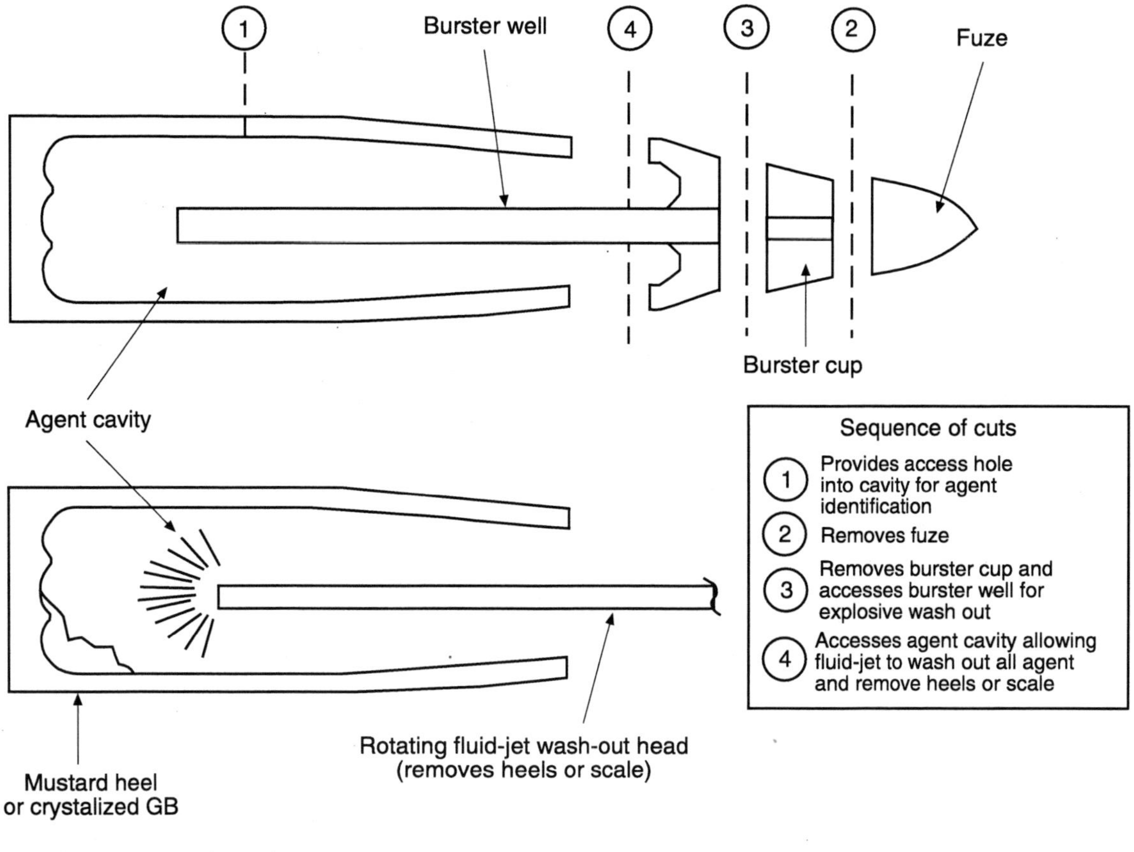

FIGURE 9-3 The sequence of cuts for ammonia-jet cutting and wash-out of projectiles and mortars. Source: Teledyne-Commodore, 1997.

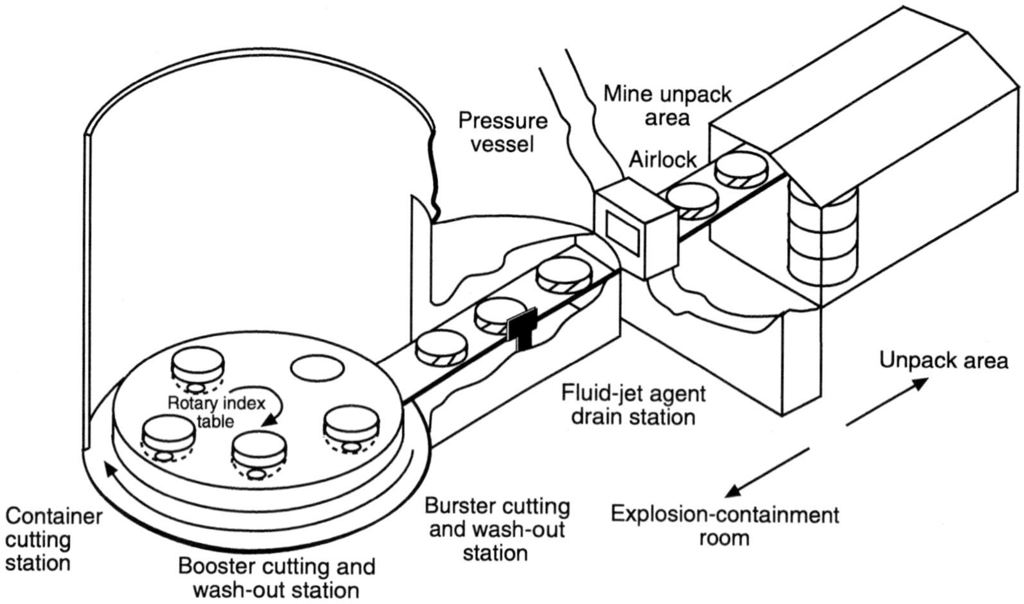

FIGURE 9-4 Schematic diagram of the process proposed for disassembly of M23 land mines. Source: Teledyne-Commodore, 1997.

and no agent was detected in the ethylene off-gas, provided that agent had been introduced below the surface of the SET solution.

Agent GB. The expected initial reaction of GB in the SET process is cleavage of the phosphorus-fluorine bond by a solvated electron to form sodium fluoride. Measured results, scaled up to 100 pounds of GB, are shown in Table 9-3. Release of ammonia gas in the SET reaction again results from evaporation used for temperature control. Release of both ammonia and isopropyl alcohol to the gas phase in the hydrolysis reaction occurs in the agent-ammonia recovery tower in the proposed full-scale system (see Area 400) (Figure 9-5).

TABLE 9-2 Measured Results for the SET/Hydrolysis Reaction of HD based on Laboratory Data and Scaled Up to 100 lb of Agent

	SET Reaction (lb)		Hydrolysis (lb)	
Material	Feed	Product	Feed	Product
NH_3 (liquid)	1,656			
Na (liquid)	76			
HD	100			
C_2H_4 (gas)		22		0.05
H_2 (gas)		0.4		0.04
NH_3 (gas)		84		1,495
Slurry		1,676	1,676	
Water			282	
Solid				67
Aqueous solution				397
Loss		50		
TOTAL	1,832	1,832	1,958	1,959
Heat Release	352,800 BTU		119,363 BTU	

Source: Adapted from Teledyne-Commodore, 1998a.

TABLE 9-3 Measured Results for the SET/Hydrolysis Reaction of GB based on Laboratory Data and Scaled Up to 100 lb of Agent

	SET Reaction (lb)		Hydrolysis (lb)	
Material	Feed	Product	Feed	Product
NH_3 (liquid)	958			
Na (liquid)	41			
GB	100			
C_3H_8 (gas)		8		8
H_2 (gas)		0.1		0.03
CH_4 (gas)		0.1		
C_2H_4 (gas)		0.001		
NH_3 (gas)		44		897
Slurry		1,047	1,047	
Water			112	
Solid				122
Aqueous solution				122
Isopropanol (aqueous)				9
TOTAL	1,099	1,099	1,159	1,158
Heat Release	29,730 BTU		47,207 BTU	

Source: Adapted from Teledyne-Commodore, 1998a.

The composition of the slurry from the SET reaction was not analyzed prior to hydrolysis. Table 9-4 shows the solid and aqueous products predicted based on nuclear magnetic resonance (NMR) analysis using ^{31}P spectra. In small-scale tests, GB was not detectable in any of the reaction products.

Agent VX. Unlike mustard and GB, the VX molecule does not have a strongly polar bond for the solvated electrons to attack. Teledyne-Commodore postulates that cleavage occurs initially at the phosphorus-sulfur bond, the sulfur-carbon bond, or at both simultaneously. Measured results, scaled up to 100 pounds of VX, are shown in Table 9-5. Ammonia is vaporized during the SET reaction, and ammonia and a mixture of alcohols and amines are generated as part of the hydrolysis process in the agent ammonia-recovery tower in the proposed full-scale system (See Area 400) (Figure 9-6). The composition of the slurry from the SET reaction was not analyzed prior to hydrolysis. Table 9-6 shows the solid and aqueous products predicted based on NMR analysis using ^{31}P spectra.

Concentration of Hydrolysate and Recovery of Ammonia (Area 400)

In Area 400, the hydrolyzed liquid from the SET process is pumped to the bottom of the agent ammonia-recovery tower. The recovery tower is a packed carbon steel column with a hot oil reboiler and a chilled overhead condenser. The gases that rise to the top of the water are condensed with chilled propylene glycol and transferred to an ammonia-recycle drum. Noncondensable gases are sent to Area 800 for treatment. The material that collects at the bottom of the recovery tower, which contain the residues from the hydrolysis of the liquid products of SET, is sent to Area 500 for oxidation.

Oxidation (Area 500)

In Area 500, the hydrolysates from Areas 200 and 400 are oxidized with sodium persulfate. The purpose of this oxidation step is to eliminate CWC Schedule 2 compounds by converting all organic phosphorus to

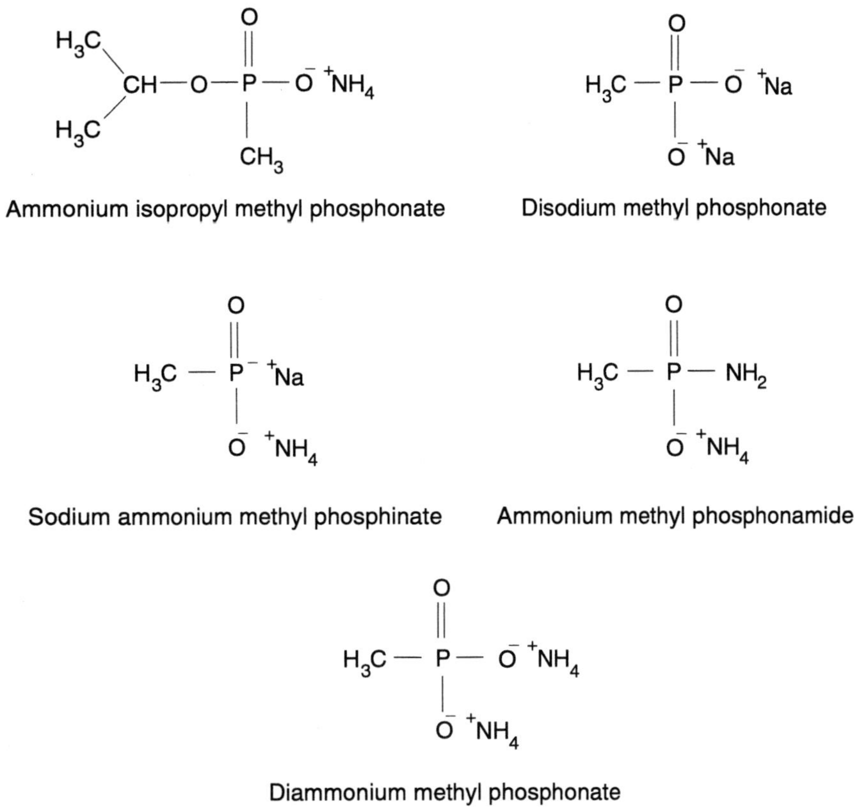

FIGURE 9-5 Formulas for the more complex reaction products from SET/hydrolysis of GB.
[a]These compounds are more commonly called phosphonites.

TABLE 9-4 Predicted Solid and Aqueous Reaction Products of SET/Hydrolysis of GB[a]

Reaction Products	Formula	Moles/Mole of GB	lb/100 lb of GB
Sodium fluoride	NaF	1.00	30.0
Sodium hydroxide	NaOH	0.84	24.0
Isopropanol	C_3H_7OH	0.39	16.7
Ammonium isopropyl methyl phosphonate	See Figure 9-5	0.36	39.8
Propane	C_3H_8	0.25	7.85
Disodium methyl phosphonate	See Figure 9-5	0.25	25.0
Sodium ammonium methyl phosphinate	See Figure 9-5	0.16	13.6
Ammonium methyl phosphonamide	See Figure 9-5	0.13	10.4
Diammonium methyl phosphonate	See Figure 9-5	0.10	7.3
Hydrogen	H_2	0.84	1.2

[a]Prediction based partly on quantitative analysis and partly on compounds identified but not quantified by NMR analysis.

Source: Adapted from Getman, 1998.

inorganic orthophosphate and converting the thiol from VX to sulfonic acid. Teledyne-Commodore is still in the very early stages of investigating persulfate oxidation for final treatment of residuals from the SET/hydrolysis reactions.

The pH of the solution is adjusted to around 10 by addition of 10 N sodium hydroxide, and the temperature of the solution is raised to 80 to 85°C (176 to 185°F). Sodium persulfate is then fed to the reactor as a 32.5-percent solution in water. Gaseous products of oxidation are sent to the gas-treatment train (Area 800). When oxidation is completed, the contents of the reactor are evaporated. Water is condensed for reuse, and the solids are packaged for delivery to a landfill.

Teledyne-Commodore has not yet identified optimum conditions for the oxidation of hydrolyzed residues from SET treatment of GB and VX. The addition of sodium persulfate to the hydrolysates in small-scale tests resulted in highly exothermic reactions, more vigorous with VX than with GB. The persulfate had to be added very slowly to keep the temperature in the range of 95 to 100°C (203 to 212°F). Experiments characterized by Teledyne-Commodore as full-scale reactions of VX and GB residues from SET/hydrolysis with sodium persulfate are described below.

In an experiment on VX, 25 grams of residuals were mixed with 367 grams (287 cc) of 10 N sodium hydroxide and heated to 85°C (185°F) with stirring. The first increment of sodium persulfate caused a vigorous exothermic reaction and a rapid increase in temperature to 100°C (212°F). The solution was allowed to cool, and 468 grams (360 cc) of 32.5-percent sodium persulfate was added over a period of two hours. In the process, 1,020 cc of gas was evolved, but analysis of the gas is suspect because of air leakage into the sample. Based on a review of published literature provided by Teledyne-Commodore, the committee expects that CO_2 and oxygen would be the gaseous reaction products. However, Teledyne-Commodore did not analyze for CO_2, and the air leak precluded a determination of the amount of oxygen, if any, generated in the oxidation reaction. An analysis of the condensed-phase product showed only 69 percent conversion of phos-

TABLE 9-5 Measured Results for the SET/Hydrolysis Reactions of VX based on Laboratory Data and Scaled Up to 100 lb of Agent

	SET Reaction (lb)		Hydrolysis (lb)	
Material	Feed	Product	Feed	Product
NH_3 (liquid)	486			
Na (liquid)	21			
VX	100			
C_2H_6 (gas)		2		0.003
H_2 (gas)		0.05		0.07
C_2H_4 (gas)		0.05		
NH_3 (gas)		35		443
Slurry		554	555	
Water			102	
Loss		16		
Solid				17
Aqueous solution				188
Alcohols and amines (gas)				9
TOTAL	607	607	657	657
Heat Release	52,200 BTU		43,265 BTU	

Source: Adapted from Teledyne-Commodore, 1998a.

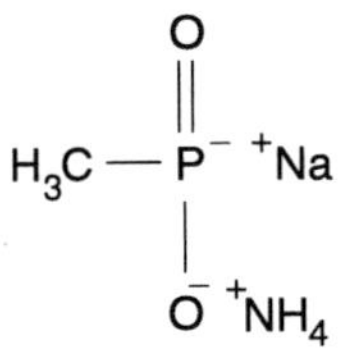

Sodium ammonium methyl phosphinate

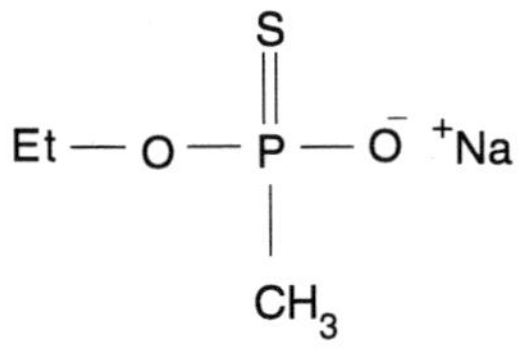

Sodium ethyl methyl phosphorothiolate

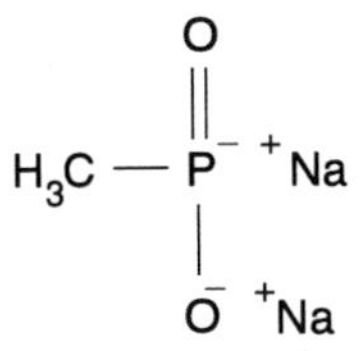

Disodium methyl phosphinate

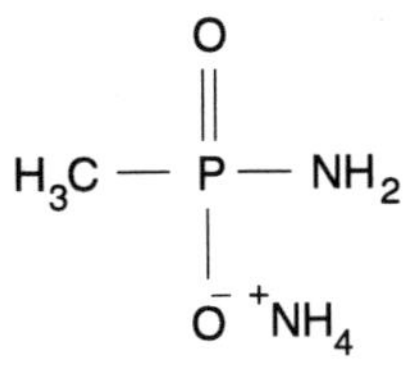

Ammonium methyl phosphonamide

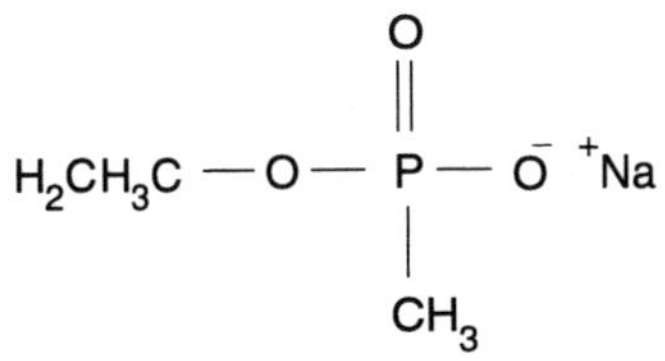

Sodium ethly methyl phosphonate

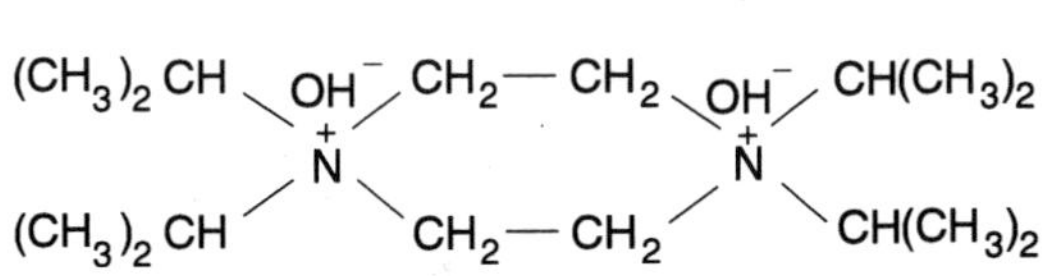

Dihydroxy N,·N, N', N', – tetraisopropyl piperazine

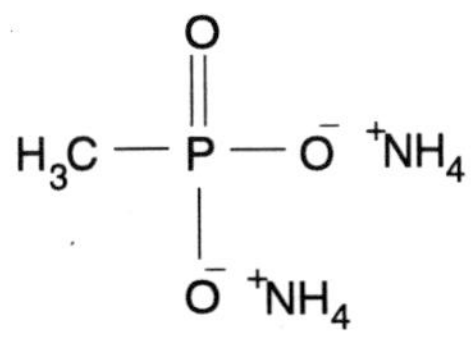

Diammonium methyl phosphonate

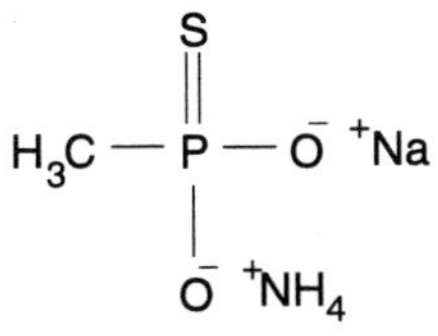

Sodium ammonium methyl phosphorothiolate

FIGURE 9-6 Formulas for the more complex reaction products from SET/hydrolysis of VX.
[a]These compounds are more commonly called phosphonites.

phorus to orthophosphate. Some methylphosphonic acid was found to be present in the oxidized product by NMR analysis, but the amount was not quantified.

In an experiment on GB, 21.7 grams of residuals were mixed with 200 cc (250 grams) of 10 N sodium hydroxide and heated to 90°C (194°F); 375 cc of gas was released during the heating process. The Teledyne-Commodore report does not explain why the temperature increase alone, prior to the addition of persulfate, resulted in a gaseous release. Moreover, the gas was not identified. Over the next hour, 487.5 grams (375 cc) of 32.5-percent sodium persulfate was added in drops to maintain the temperature in the range of 95 to 100°C (203 to 212°F). An additional 925 cc of gas was evolved during the oxidation process. This analysis of the gas is also suspect because of apparent air leakage

TABLE 9-6 Predicted Solid and Aqueous Reaction Products of SET/Hydrolysis of VX[a]

Reaction Products	Formula	Moles/Mole of VX	lb/100 lb of VX
Sodium bis-diisopropylamino ethyl mercaptide	See Figure 9-6	0.73	50.0
Sodium hydroxide	NaOH	0.59	8.8
Ethyl alcohol	C_2H_5OH	0.55	9.5
Hydrogen	H_2 (g)	0.48	0.4
Sodium ammonium methyl phosphinate	See Figure 9-6	0.36	16.0
Sodium ethyl methyl phosphorothiolate	See Figure 9-6	0.24	14.5
Disodium methyl phosphinate	See Figure 9-6	0.16	7.4
Ethane	C_2H_6 (g)	0.16	1.8
Dihydroxy N,N,N',N'- tetraisopropyl piperazine	See Figure 9-6	0.135	14.7
Ammonium methyl phosphonamide	See Figure 9-6	0.12	5.0
Sodium ethyl methyl phosphonate	See Figure 9-6	0.05	2.7
Diammonium methyl phosphonate	See Figure 9-6	0.04	1.5
Sodium ammonium methyl phosphorothiolate	See Figure 9-6	0.03	1.3

[a]Prediction based partly on quantitative analysis and partly on compounds identified but not quantified by NMR analysis.

Source: Adapted from Getman, 1998.

into the sample. For the condensed phase, Teledyne-Commodore reports, "Within experimental error, all phosphorus in the product was present as inorganic phosphate and NMR showed no C-P bonds."

Treatment of Energetics (Areas 300, 600, and 700)

SET Reduction followed by Hydrolysis (Area 300)

The slurry produced from the wash-out of energetic material in Area 100 is collected in a carbon steel vessel, approximately 2.5 feet in internal diameter and 7 feet in height, and diluted with ammonia to a "standard composition." A solution of sodium in liquid ammonia is formed, pumped to a reactor vessel, and mixed with the diluted energetic slurry. The reaction time for destruction of energetics is reported by Teledyne-Commodore to be as long as 30 minutes. In contrast, decontamination of agent by the SET process is reported to be almost instantaneous. For M28 propellant and RDX, the longer reaction time may be attributable to their relatively low solubility in SET solutions. Also, when any energetic material is reacted, side reactions may consume solvated electrons to form polymers.

The products formed in the SET reaction are transferred to a holding vessel, approximately 3 feet in internal diameter and 8 feet in height. Water is added, and the hydrolysate is pumped to the bottom of the energetics-ammonia recovery tower in Area 600. The products of SET/hydrolysis are generally complex polyaromatic hydrocarbons that Teledyne-Commodore believes are formed from the polymerization of the aromatic or heterocyclic ring structures common to most energetics and from small quantities of nitrate, nitrite, and cyanide salts.

Teledyne-Commodore conducted research on SET destruction of explosives, fuzes, and propellants over a nine-month period in the laboratories of the Southwest Research Institute. Most were bench-scale tests conducted at –33.4°C (–28°F), the boiling point of ammonia at atmospheric pressure. One series of tests was conducted at ambient temperature and at a pressure of about 10.5 atm (140 psig), the conditions proposed for a full-scale plant. No major differences were observed in the reactions carried out under these two very different sets of conditions. Energetics tested were TNT, RDX, tetryl, PETN (pentaeryhritol tetranitra), Comp B, picric acid, nitrocellulose, and M28 propellant.

In the laboratory tests, a weighed sample of the energetic was dissolved in liquid ammonia, and sodium was added incrementally until the characteristic blue color of the SET solution was observed. At the end of the reaction, isopropyl alcohol or water was added to destroy excess sodium, and the ammonia was evaporated prior to collecting a sample for analysis. Details of tests on different types of energetics are provided below.

TNT. Sodium was added incrementally to a solution of one gram of TNT in liquid ammonia. After 1.957 grams of sodium had been added, the blue color characteristic of SET remained, and the test was terminated. The residue remaining after the addition of water and evaporation of the ammonia was a paste, consisting of "nondescript polymeric organic materials." The only reaction products identified were those listed in Table 9-7, which total less than 1 wt. percent of the TNT treated. No products of SET/hydrolysis could be identified by either NMR or liquid chromatography/mass spectrometry (LC/MS)analysis. No TNT or other nitrated toluenes were found in the residues at Na/TNT ratios greater than 0.5.

The residue from SET reactions with TNT were sensitive to electrostatic energy as evidenced by the presence of smoke and the presence of combustion products during testing. The residue can be considered electrostatically sensitive because the minimum ignition energy was measured to be as low as 90 mJ in at least one of five trials. The tests were conducted at constant arc gap of 0.25 inches (the distance from the end of the electrode to the inside bottom surface of the anode sample-holder cup).

Teledyne-Commodore cites Batz et al. (1997) for a discussion of the reaction of TNT with SET solutions. Batz observed a white precipitate upon reaction of TNT with a SET solution and noted that, "Evaporation of the ammonia leaves an off-white solid which explodes upon agitation." In the Batz experiments, a polymeric fraction was also formed that was insoluble in water and very sensitive to explosion when agitated.

RDX. One gram of RDX added to 100 ml of liquid ammonia did not dissolve completely after stirring for 1 hour and 16 minutes prior to the addition of sodium. The reaction was judged to be complete after sodium additions totaled 1.3 grams, when the blue color of the SET solution persisted for more than six minutes. The residue after hydrolysis and ammonia evaporation was an off-white dry flaky material. The only reaction products identified after the addition of sodium and subsequent hydrolysis are shown in Table 9-8. The products identified account for less than 1 percent of the RDX treated. No reaction products could be identified by either NMR or LC/MS. Semivolatile analysis identified small quantities of hexamethylenetetramine, a possible impurity in the RDX that survived SET treatment. NMR analysis detected no RDX in the residue. However, the residue was sensitive to electrostatic energy, as indicated by the presence of smoke and combustion products under test conditions. The minimum measured ignition energy was 50 mJ.

Tetryl. Tetryl reacted similarly to TNT. Both energetics are very soluble in liquid ammonia, and the residue in both cases was a pasty, nondescript polymeric species. Total gases released from the SET reaction with tetryl ranged from 32 to 214 ml/g and averaged 3 percent hydrogen, 9 percent oxygen, 10 percent nitrogen, 901 ppm methane, and 117 ppm C_2 hydrocarbons. No tetryl was identified in the residue by high performance liquid chromatography analysis.

In one test, after the water quench to destroy excess sodium, a strong exothermic reaction occurred upon drying of the residue from SET treatment. Teledyne-Commodore notes, however, that if the residue is mixed during the drying process, the temperature does not increase sufficiently to cause carbonization.

Comp B. The addition of Comp B to liquid ammonia turned the solution a deep cranberry-red color. Sodium was added incrementally, and at a Na/Comp B ratio of

TABLE 9-7 Identified SET Reaction Products of Treatment of TNT

Products	Quantity
Nitrite	18 mg/g
Nitrate	98 mg/g
Ionic cyanide	0.595–2.8 mg/g
Gases released[a]	13 ml/g

[a]Mainly hydrogen, with ppm quantities of C_1-C_3 hydrocarbons.

Source: Adapted from Teledyne-Commodore, 1998a.

TABLE 9-8 Identified SET Reaction Products of Treatment of RDX

Products	Quantity
Nitrite	3.1–26 mg/g
Ionic cyanide	0.016–15.8 mg/g
Gases released[a]	55–140 ml/g

[a]0.21 percent hydrogen; 11–455 ppm methane; and 2–115 ppm ethane.

Source: Adapted from Teledyne-Commodore, 1998a.

TABLE 9-9 Identified SET Reaction Products of Treatment of Comp B

Products	Quantity
Nitrite	33 mg/g
Nitrate	2 mg/g
Inorganic cyanide	3.4 mg/g

Source: Adapted from Teledyne-Commodore, 1998a.

1.15:1.0, the SET blue color persisted. The residue after hydrolysis and evaporation of ammonia was a paste that was soluble in both water and methanol. The only reaction products identified (summarized in Table 9-9) account for less than 1 percent of the weight of the Comp B. No reaction products could be identified by either NMR or LC/MS.

Neither RDX nor TNT was identified in the residue. However, the residues were sensitive to electrostatic energy, as evidenced by the presence of smoke and combustion products during testing. The minimum measured ignition energy was 100 mJ.

M28 Propellant. M28 propellant dissolves very slowly in liquid ammonia. To accelerate dissolution, the propellant was ground to a sawdust-like consistency with a wood file. The residue after SET treatment, hydrolysis, and evaporation of ammonia was a soft sticky paste. The only identified reaction products (summarized in Table 9-10) account for less than 1 percent of the weight of the M28. No reaction products could be identified by either NMR or LC/MS analysis.

None of the target explosive compounds was identified in the residue, as determined by high pressure liquid chromatography analysis. A TCLP analysis of the residue showed lead at a concentration of 75.6 mg/L. The nonhydrolyzed, dried, ammoniated residue was sensitive to electrostatic energy, as indicated by flaming. The minimum measured ignition energy was 175 mJ.

In a larger-scale test conducted at ambient temperature and elevated pressure, 97.9 grams of ground M28 and 97.9 grams of sodium were added in increments to 6 liters of liquid ammonia. The test took close to seven hours to complete. The reaction mixture was allowed to evaporate in the reactor, covered with a sheet of plastic, and a slow flow of helium was introduced to exclude ambient air. After 17 hours, the residue had the appearance of drying mud. The next morning, approximately 15 hours later, the plastic had been destroyed and part of it had hardened and turned dark. The residue was a dry, black, carbonaceous material with a charcoal-like odor. In a second similar test, 20 ml of isopropyl alcohol were added to destroy the unreacted sodium prior to evaporating the ammonia in an argon stream. This test was completed without incident.

In a test conducted to determine the effect of operating at 2.5 wt. percent ground M28 at 9.2 atm (120 psig), Teledyne-Commodore observed a spontaneous reaction in the reactor while pressure checking the system at 6.8 atm (85 psig). The undissolved M28 propellant appeared to have melted.

Hydrolysate Concentration and the Recovery of Ammonia (Area 600)

The energetics-ammonia recovery tower operates identically to the agent-ammonia recovery tower. The aqueous bottoms are collected in a carbon steel vessel, 4.5 feet in internal diameter and 12.6 feet in height, and transferred in batches to Area 700 for oxidation.

Oxidation (Area 700)

According to the technology provider, the hydrolyzed residues of SET treatment of energetics contain cyanides at levels that exceed regulatory limits. Alkaline chlorination, the cyanide destruction method in most common use, is not applicable to the hydrolysates because hypochlorite reacts with ammonia to form nitrogen trichloride, a strong irritant. The problem arises from incomplete removal of ammonia from the SET/hydrolysis residual. After conducting several screening tests to find an oxidizer that would destroy cyanide

TABLE 9-10 Identified SET Reaction Products of Treatment of M28 Propellant

Products	Quantity
Nitrite	9.1 mg/g
Nitrate	5.9 mg/g
Ionic cyanide	0.6 mg/g

Source: Adapted from Teledyne-Commodore, 1998a.

in the presence of significant quantities of other reduced species, the technology provider selected alkaline, copper-catalyzed persulfate/hydrogen peroxide oxidation. However, this system has not been tested sufficiently to identify optimum operating conditions.

Gas Treatment (Area 800)

Gases and ammonia vapors from areas 200 and 600 pass through reflux condensers to remove most of the ammonia. Residual ammonia vapors and noncondensable gases are collected in parallel holding tanks for testing. After verifying that no agent is present, the gases are vented through a pressure let-down valve to a water scrubber to remove ammonia as ammonium hydroxide, which is sent to the ammonia recovery tower (Area 400). Noncondensable gases are chilled, reheated to a relative humidity of about 50 percent, and passed through a deep-bed activated carbon adsorber to remove any trace agent; low-boiling hydrocarbons and hydrogen are passed through the bed and recycled as boiler fuel.

SET Reduction and the Hydrolysis of Metal Parts and Dunnage (Area 900)

Treatment of Metal Parts

Munitions metals separated in Area 100 are shredded in Area 900 and subsequently transferred to a metal SET reactor for treatment to 3X condition. Treatment involves agitation with a SET solution in a tumbler reactor operated in batch mode.

The metal SET tumbler reactor is a double cone mixer without baffles. The shredded metal parts are introduced through the top of the tumbler, which is then sealed, vented, and filled with ammonia. Liquid sodium is injected to form a SET solution, and the sealed vessel is rotated slowly to ensure that all surfaces of the metal parts are wetted by the solution. Upon completion of the SET reaction, water is added to the reactor to destroy the remaining sodium. The 3X metal parts are shipped to Rock Island Arsenal for 5X treatment.

The proposed system has not been tested, but in tests on small metal coupons, agent was destroyed to below detectable limits, and none of the metals was found to be reactive with SET solution.

Treatment of Dunnage and Abrasives

Many of the materials that constitute dunnage are porous to agent and, therefore, are decontaminated in SET solution. Teledyne-Commodore proposes reducing the particle size of dunnage and nonmetal wastes other than activated carbon to three-eighths inch or smaller by means of a size-classifying fine shredder with internal recycling. The shredded particles are transferred to a rotary-plow mixer for treatment with SET solution. Carbon, which is already smaller than three-eighths of an inch, is fed directly to the mixer.

The rotary-plow mixer is a horizontal cylindrical vessel with a central rotating shaft. Plow-shaped heads rotate through the vessel to agitate the mixture. After the finely divided solids have been introduced, the vessel is sealed, ammonia and liquid sodium are added, and the mixer is started. Following decontamination, water is added to destroy excess sodium.

The proposed system has not been tested, but small-scale tests have shown that agent was destroyed to below detection limits. In addition activated carbon, DPE, waste oils, and silicone rubber were found to be reactive with the SET solution.

A slurry of ammonia and abrasives from jet cutting is also treated in SET solution. This occurs in one of two abrasive SET reactors, which are tumblers designed to treat slurries. Following decontamination, water is added to destroy excess sodium.

Process Instrumentation, Monitoring, and Control

The ammonia-jet cutting system incorporates sensors for pressure, temperature, and ammonia leakage; these sensors are integrated with control systems. When deviations from normal pressures or temperatures occur, the pressurized fluid is automatically diverted to the supply tank and its recirculation system until the problem is corrected. A control-system failure triggers redundant pressure-relief valves. On-line electric sensors detect ammonia leakage into secondary-containment piping; excessive leakage automatically activates a shutdown and maintenance request.

The rotary index tables are equipped with sensors that monitor the position of munition parts and the operation of the fluid-jet cutting components. Additional sensors monitor clearance and torque on the table to prevent jamming or overload. Controls embedded in the index tables only allow operations to proceed if control algorithm requirements are satisfied. A mishandled munition causes the index table to stop until the workstation is cleared. If a workstation is empty, the control system prevents operation of the fluid-jet components until munitions are properly loaded into the work nests.

SET solutions are prepared by mixing predetermined quantities of liquid sodium and ammonia in an in-line static mixer. Mass-flow controls and interlocks control the process precisely. Two independent mass-flow meters control the flow of agent and energetics into a static-mixer reactor containing the prepared SET solution. Interlocks stop agent flow if SET solution is unavailable. Measurements of conductivity are used to indicate destruction of agent or energetics and the availability of excess SET solution. However, the technology provider has also noted that some SET products of energetics destruction are ionic and contribute to the conductivity of the reacting SET solution.

The headspace in the reactor-product vessel is monitored continuously for traces of agent in the gas stream. If agent is present in the off-gas, the gas is held and reworked to ensure complete destruction.

Hold-test-release provisions for the abrasive reactor, the munition tumbler, and the rotary-plow mixer ensure that all metal parts and dunnage are tested for decontamination before release for packaging and disposal.

Feed Streams

Tables 9-11 and 9-12 list the feed streams entering the Teledyne-Commodore system when processing VX-filled M55 rockets and HD-filled 155- mm projectiles, respectively.

Waste Streams

The Teledyne-Commodore system generates at least 17 process-waste streams for disposal. Those identified by Teledyne-Commodore are listed in Table 9-13. Tables 9-14 and 9-15 provide mass flows for the consolidated waste streams leaving the system when processing VX-filled M55 rockets and HD-filled 155-mm projectiles, respectively.

Start-up and Shutdown

Start-up and shutdown procedures for the full-scale system are still being developed.

TABLE 9-11 Process Inputs for the Teledyne-Commodore Technology Package for VX-filled M55 Rockets Processed at a rate of 20/hr

Process Inputs	Mass Flow (lb/hr)
Munitions	
Agent VX	200.0
Propellant	384.0
Other energetics	64.0
Metal parts	896.0
Abrasive	132.0
Sodium (for agent)	39.6
Sodium (for energetic)	277.7
Sodium (for decontamination)	80.0
Water (make-up)	230.2
Oxidants (for agent)	3,540.5
Oxidants (for energetic)	3,139.2
Cement	3,279.7
Dunnage	440.0
Total	12,700.9

Source: Adapted from Teledyne-Commodore, 1998b.

TABLE 9-12 Process Inputs for the Teledyne-Commodore Technology Package for HD-filled 155 mm Projectiles Processed at a rate of 100/hr

Process Inputs	Mass Flow (lb/hr)
Munitions	
HD	1170.0
Energetics	41.0
Metal parts	8,600.0
Abrasive	154.0
Sodium (for agent)	778.1
Sodium (for energetic)	22.6
Sodium (for decontamination fluid)	132.6
Oxidants (for agent)	7,575.8
Oxidants (for energetic)	398.3
Cement	150.0
Dunnage	525.0
Total	19,547.4

Source: Adapted from Teledyne-Commodore, 1998b.

TABLE 9-13 Process Waste Streams Released to the Environment

Waste Source	Waste Treatment	Final Disposal Method
Aqueous slurry of abrasives, metal cuttings, and fuze parts after SET treatment	Solidification with cement	RCRA Subtitle C landfill
Ferrous metal parts from projectiles	SET treatment to 3X	Shipment to Rock Island for 5X processing
Driving bands and nonferrous metals from projectiles after SET treatment	Separation from ferrous parts	RCRA Subtitle C landfill
Land-mine and non-PCB M55 shredded metal parts	None	Landfill
SET-treated M55 residue with PCBs	Passed through clarifier to remove fiberglass	RCRA Subtitle C landfill
Residual sodium salts from oxidation of SET-treated agent and energetics	Evaporation, filtration, dewatering, and packaging	RCRA Subtitle C landfill
Isopropanol liquid from oxidation of SET-treated GB	Integrated with off-gas	Burned for energy recovery
Ethanol liquid from oxidation of SET-treated VX	Integrated with off-gas	Burned for energy recovery
2-diisopropyl-(amino) ethane sulfonic acid liquid from oxidation of SET-treated VX	Separated and stabilized in cement	Landfill
N,N=-tetrisopropylpiperazine dihydroxide from oxidation of SET-treated VX	Unknown	Unknown
Residues containing lead from oxidation of SET-treated energetics	Stabilized in cement	RCRA Subtitle C landfill
Sodium nitrate and polymeric organics (no lead) from oxidation of SET-treated energetics	Packaged for shipment	RCRA Subtitle C landfill
Aqueous decontamination fluids containing agent	Treatment with caustic or hypochlorite; cement stabilization	RCRA Subtitle C landfill
Noncondesable off-gases	Reheated to 50% relative humidity and passed through a deep-bed carbon adsorber	Released to air
Hydrogen and hydrocarbon off-gases from agent-destruction SET and oxidation subsystems	None	Recycled to the process as supplemental fuel

Source: Adapted from Teledyne-Commodore 1997, 1998a.

EVALUATION OF THE TECHNOLOGY PACKAGE

Process Efficacy

Effectiveness of Disassembly of Munitions

Teledyne-Commodore proposes to use a completely new disassembly process based on ammonia-jet cutting. Although water-jet cutting offers many potential benefits for the demilitarization of ordnance (see Appendix G), Teledyne-Commodore proposes substituting ammonia for water in the fluid-jet cutting operation mainly because the primary treatment process is carried out in anhydrous liquid ammonia.

Teledyne-Commodore also mentions several advantages of ammonia over water. They point out that many of the hazardous materials that are washed out of

TABLE 9-14 Process Outputs for the Teledyne-Commodore Technology Package for VX-filled M55 Rockets Processed at a rate of 20/hr

Process Outputs	Mass Flow (lb/hr)
Metal parts	896.0
Stabilized abrasive	396.0
Stabilized propellant	8,593.1
Product slurry (from agent)	1,861.0
Product slurry (from energetic)	399.2
Gases (from agent)	32.2
Gases (from energetic)	3.4
Dunnage	520.0
Total	12,700.9

Source: Adapted from Teledyne-Commodore, 1998b.

munitions have very low solubility in water and contain surfactants that aid in the formation of stable emulsions. These emulsions have presented serious safety and maintenance problems in full-scale conventional-munition disassembly facilities. Anhydrous ammonia, in contrast, dissolves most of the explosive materials and forms nondetonable solutions.

Teledyne-Commodore successfully demonstrated ammonia-jet cutting at Redstone Arsenal on M60 (inert) and M61 (live) rockets, a 4.2-inch M2 inert mortar, and a 105-mm M60 inert projectile (Teledyne-Commodore, 1998c). Successful wash-out of Comp B and M28 propellant was demonstrated during the M61 test series. Teledyne-Commodore claims that ammonia-jet cutting was 25 percent faster than water-jet cutting, but no details were provided on how the comparison was made.

TABLE 9-15 Process Outputs for the Teledyne-Commodore Package for HD-Filled 155 mm Projectiles Processed at a rate of 100/hr

Process Outputs	Mass Flow (lb/hr)
Metal parts	8,600.0
Stabilized abrasive	454.0
Salts (from agent)	5,218.8
Salts (from energetic)	227.7
Water (recycled)	3,980.1
Gases (from agent)	436.5
Gases (from energetic)	0.3
Dunnage	630.0
Total	19,547.4

Source: Adapted from Teledyne-Commodore, 1998b.

Ammonia-jet cutting would require a major change in the baseline munition-disassembly areas. Specifically, Teledyne-Commodore plans to conduct these operations at room temperature and at a pressure of around 10.5 atm (140 psig), which would require pressurized enclosures for the cutting workstations.

Effectiveness of Detoxification of Chemical Agents

The technology provider has conducted more than 250 tests of portions of the proposed system on 15 different chemical agents, 9 energetic materials or compositions, and 21 different metal or process-waste (i.e., dunnage) combinations. The largest quantities of agent tested in a single batch were 1.4 pounds of HD, 1.1 lb. of HT, 1.1 lb. of VX, and 1.3 lb. of GB. The concentration of agent in the SET solution after completion of the laboratory tests was below detection limits in all cases (less than 200 ppb for HD and HT; and less than 20 ppb for VX and GB.) The technology provider conducted four tests of post-treatment with sodium persulfate: one with VX, one with TNT/HD/Lewisite, one with TNT/HD, and one with TNT/Lewisite. (Lewisite is a chemical agent but is not included in the ACWA program.)

Teledyne-Commodore has demonstrated that the SET process, followed by hydrolysis, can destroy chemical agents to a destruction efficiency of at least 99.9999 percent. However, considerably more testing and analysis will be required to determine the exact molecular composition, phase distribution, and quantity of reaction products. This lack of information compounds the difficulty of developing optimum conditions for the final oxidation step. The data in Tables 9-3 through 9-6 illustrate this problem.

The data in Table 9-3 were measured experimentally. The data in Table 9-4 were predicted by the technology provider, based partly on quantitative analysis and partly on compounds identified but not quantified by NMR analysis. Teledyne-Commodore adjusted the molecular compositions shown until they were able to get a mass balance. The committee noted several inconsistencies in the two tables. The total measured hydrogen release in Table 9-3 is 0.13 lb/100 lb GB. The technology provider predicted hydrogen release in Table 9-4 is 1.2 lb/100 lb GB, almost 10 times larger. The measured isopropyl alcohol output in Table 9-3 is 9 lb/100 lb GB; the predicted output in Table 9-4 is

16.7 lb/100 lb GB. The amounts of propane released are the same in both tables. In Table 9-3, the amount of ammonia vaporized during the SET reaction is much lower than expected, based on the heat release shown. The technology provider attributes the difference to heat losses to the vessel and surroundings. The committee is not convinced that this explains the discrepancy.

The committee expected stoichiometric agreement between NaOH and H_2. Presumably, NaOH is formed by the reaction of water with excess sodium in the SET product solution, which would yield 0.5 moles of H_2 per mole of NaOH. The technology provider hypothesizes a molar ratio of 1:1 (see Table 9-4). This higher ratio is possible because some hydrogen is generated in the SET reaction prior to hydrolysis. However, the H_2 to NaOH ratio actually measured was less than 0.5.

According to the data in Table 9-3, 112 lb water/100 lb GB was added to the SET reaction products in the hydrolysis step; and 122 lb aqueous solution/100 lb GB was generated. The committee calculated that dissolved solids in the aqueous solution would be 10 lb/100 lb GB, far less than the quantities hypothesized by the Technology Provider in Table 9-4 for NaOH and NaF, both of which the committee expected to partition largely to the aqueous phase.

The committee identified similar inconsistencies between the experimental data reported in Table 9-5 and the predicted product mix in Table 9-6 for VX. In Table 9-5, the measured quantity of hydrogen released was 0.12 lb/100 lb VX; the hypothetical quantity in Table 9-6 is 0.4 lb/100 lb VX. Moreover, the quantity of hydrogen hypothesized in Table 9-6 does not bear a reasonable relationship to the quantity of NaOH hypothesized. The quantity of ethane reported is the same in both tables. As with GB, the amount of ammonia vaporized was less than expected, based on the heat release shown in Table 9-5. Considering the quantity of water added in the hydrolysis step and the quantity of aqueous solution produced, the committee calculated a dissolved solids content of 86 lb/100 lb VX. Therefore, some of the organic compounds listed in Table 9-6 would have to partition to the aqueous phase. Teledyne-Commodore did not analyze the aqueous liquid and the solids separately.

One additional point is worthy of note. Mustard contains volatile low molecular weight chlorinated hydrocarbons that are difficult to hydrolyze or to remove with caustic or acid scrubbers or with activated carbon adsorbers. The SET/oxidation process does not specifically address the management of these compounds, and their treatment remains to be demonstrated.

Effectiveness of Decomposition of Energetic Materials

Teledyne-Commodore has not demonstrated that the SET process, followed by hydrolysis, has the capacity to decompose the energetic materials. In laboratory experiments, the condensed-phase products were generally pasty, difficult to handle, of unknown polymeric composition, and sensitive to electrostatic ignition. Moreover, the reaction products identified represent less than 1 wt. percent of the material treated. The committee was troubled that in several instances, apparently spontaneous exothermic reactions occurred. The root causes of the exotherms have yet to be identified. These results, which were obtained relatively early in the technology provider's sequence of tests, indicate that SET reactions with energetics in excessive amounts of sodium can produce sodium salts that are sensitive energetic materials. Expected sodium salts may include sodium amide, sodium azide, sodium amido-peroxide, sodium nitramidate, and possibly even sodium salts of hydrazine. Depending on the consumption rate of the solvated electron, some sodium picrate or diazo or azoxy derivatives of TNT may also be present.

In a more recent report, Teledyne-Commodore (1998b) indicates that some of these problems may be solved if the sodium content is reduced. In small-scale laboratory experiments, the accumulation of problematic precipitates was eliminated by destroying excess sodium in the SET solution with isopropanol prior to the evaporation of ammonia. However, considerable additional testing will be required to demonstrate that this procedure will be effective in a full-scale system.

Effectiveness of Final Treatment to Produce Wastes Suitable for Disposal

The SET/hydrolysis products from the treatment of GB and VX include Schedule 2 compounds that must be further treated. Teledyne-Commodore has tested sodium-persulfate oxidation as the method of secondary treatment, but only on a laboratory scale. A substantial amount of additional testing will be needed to validate the process and determine optimum conditions for full-scale operation. In the few laboratory tests that were completed, the reactions were highly exothermic and led to bubbling and rapid rises in temperature. To maintain the temperature in the range of 90 to 100°C (194 to 212°F), the sodium-persulfate solution had to be added in extremely small increments, which, of course, increased the reaction time.

The technology provider has not yet developed a complete analysis of either the gaseous-phase or condensed-phase reaction products. In the treatment of VX, the piperazine formed by SET/hydrolysis appears to undergo no change during the oxidation process. Teledyne-Commodore believes that the sodium bis-diisopropylamino ethyl mercaptide formed in the primary reaction is converted to the corresponding sulfite by the sodium persulfate. Other products are inorganic phosphates and sulfates, although some methanol may also form. In the treatment of GB, Teledyne-Commodore believes the principal oxidation products are sodium bisulfate, sodium sulfate, sodium phosphate, ammonium hydroxide, and acetone.

The SET/hydrolysis products from the treatment of energetics include inorganic cyanide at levels that exceed regulatory limits. Other products of SET/hydrolysis have not been identified. Alkaline chlorination, the method most commonly used for the destruction of cyanide in aqueous systems, is not applicable to the solutions formed in SET/hydrolysis because the hypochlorite reacts with ammonia to form nitrogen trichloride, a strong irritant. The alternative copper-catalyzed alkaline persulfate/hydrogen peroxide system developed by Teledyne-Commodore has not been adequately tested. In assessing their own efforts to date, Teledyne-Commodore concluded,

> Additional work is required to validate the oxidation process. For realistic process residues, the residues will have significantly higher concentrations, affecting the process conditions. Oxidative exotherms will also be greater in higher concentrations. The oxidation process appears approximately equally effective for the nitroaromatic compounds, but less effective for M28 propellant (Teledyne-Commodore, 1998c).

In the view of the committee, the process for the final oxidation of energetics residues is still in the early research stage.

Tests showed that hydrogen and low molecular weight gaseous hydrocarbons form during hydrolysis. In the current conceptual design, however, no provision is made for venting gases released in the hydrolysis steps in Area 200 to the gas-treatment train. This issue must be addressed if the design is taken further.

Sampling and Analysis

No standard generally accepted sampling and analysis methods are available for systems based on liquid ammonia. The condensed-phase products of SET can only be analyzed for specific chemical components after hydrolysis. The hydrolysis process changes the composition of the SET products, in addition to converting excess sodium to sodium hydroxide and producing hydrogen gas.

It may not be necessary to analyze the products of SET and the products of hydrolysis separately because hydrolysis occurs immediately after SET treatment, and the hydrolysis products are passed on to the final oxidation step. Nevertheless, the identity of hydrolysis products is still uncertain.

When chemical agents are treated, color changes provide a visual indication of the completion of the SET reduction process, and measurements of conductivity provide a quantitative indication. When energetics are treated, the persistence of the blue color characteristic of solutions of sodium in liquid ammonia can be masked by the intense colors of solutions of energetics in liquid ammonia. Measurements of conductivity may also be difficult to interpret because of the formation of inorganic nitrates, nitrites, and cyanides, which contribute to conductivity.

Gases are released in both the SET process and the hydrolysis process. Teledyne-Commodore found it difficult to collect gas samples for analysis. However, there are well developed methods for collecting gas, so this problem should be relatively easy to overcome.

Maturity

The committee is not aware of any full-scale applications of the type proposed by Teledyne-Commodore. The system is quite complex and has never been operated as a totally integrated package that includes oxidation. The system involves at least 16 unit operations. Five are for SET treatment of agents, energetics, shredded dunnage, metal parts, and fuzes; the other eleven are for hydrolysis of agents and energetics, fluid-jet cutting, oxidation of energetic residue, oxidation of agent residue, evaporation of oxidized residues, recovery of ammonia, stabilization of oxidized residues of cement, detonation of fuzes, decontamination of abrasives, and stabilization of fuzes/abrasives.

The technology provider tested SET destruction of malathion, an agent simulant, at the Teledyne-Commodore pilot-scale plant in Marengo, Ohio, where many of the unit processes and operations proposed for the ACWA program were demonstrated. The largest quantity of malathion tested in a single run was 100 pounds. The agent simulant was introduced at a rate of 125 pounds per hour. The SET solution was made up of elemental calcium in liquid ammonia. Malathion was reduced to below detectable limits. The committee notes that the Marengo facility did not include a module for testing the oxidation of residuals with persulfate.

Robustness

The SET process appears to be capable of destroying agents with a wide range of feedstock compositions, temperatures, and pressures. The reactions seem to be most sensitive to the sodium/feedstock ratio, which must be high enough to ensure complete reaction. However, based on the data the committee received, the capability of the SET/hydrolysis process to deactivate energetics does not appear to be satisfactory, and optimum operating conditions have not been established.

In tests conducted at Redstone Arsenal, Teledyne-Commodore encountered and was able to resolve several operating problems, including plugging of the liquid-sodium feed lines, which was resolved by adding an ammonia wash to dissolve solidified sodium in the piping. The technology provider plans to modify the design to maintain a uniform sodium temperature throughout the flow system. Another problem was the formation of sodium-oxide plugs in the mass-transfer system caused by small amounts of air in the sodium line. Controls were added to prevent this.

Monitoring and Control

Teledyne-Commodore uses conductivity as the main indicator that the reaction is complete. However, because some of the products of the SET reaction contribute to conductivity, additional control algorithms are being developed that incorporate temperature, pressure, and feed composition. The technology provider also intends to improve the mass-flow monitor and control system for all process constituents.

Teledyne-Commodore is not far enough along in the development of a full-scale system for the committee to assess start-up and shutdown procedures.

Even for the SET component, which has been investigated in greatest detail, the firm notes, "Due to the intrinsic reactivity of ammonia and to kinetic effects, the mixing protocol can significantly impact reaction mechanisms and pathways." In batch experiments, Teledyne-Commodore has tested two mixing protocols: one called "forward addition," in which the material to be treated is added to a premixed SET solution; the other called "backward addition," in which sodium is added to a premixed solution of the material to be treated in liquid ammonia. The sodium consumption and the products of reaction are different in the two cases. For the full-scale plant, Teledyne-Commodore proposes using a hybrid protocol called "controlled stoichiometry," in which reagents are added and mixed in "carefully controlled ratios" via a static mixer.

In the few bench-scale tests on SET/hydrolysis of energetics and oxidation of the residues from the treatment of both agents and energetics, exotherms were observed that are not yet well understood. Therefore, the possibility of runaway reactions cannot be ruled out at this time.

Applicability

The SET process has been applied at a reduced scale to a wide range of agents, energetics, agent/energetic combinations, and solid coupons contaminated with agent and energetics and is conceptually applicable to all assembled chemical weapons types in the U.S. stockpile.

Process Safety

A number of pieces of equipment and processes are unique to the SET system. These must be taken into account in an evaluation of process safety:

- an extraction subsystem:
  - — a fluid-jet cutting machine using high-pressure (2,720 atm; 40,000 psig) liquid anhydrous ammonia streams (1 gpm) and garnet abrasive for opening and accessing chemical weapons inside a pressurized containment vessel (the munitions access vessel)
  - — a high-pressure (28.2 atm; 400 psig) fluid-jet wash-out using liquid anhydrous ammonia to remove remaining agent and energetics from the munition cavities inside the munitions access vessel
  - — an abrasives-removal system for separating garnet or other abrasives from the ammonia cutting fluid
  - — a fuze-removal machine for extracting intact fuzes after agent and energetics have been removed
  - — an explosion-containment chamber in which fuzes, removed from the munitions in the munitions access vessel, are initiated to effect their destruction by detonation
  - — a system for removing munitions metals and packaging
  - — a shredder system for rocket and land-mine bodies and packaging for these items
- a destruction system:
  - — six SET reactors for the destruction of agent and energetics and the decontamination of dunnage, shredded rockets and land mines and their packaging, abrasives used in cutting, and fuze parts; the reactors use a solvated-electron solution consisting of sodium dissolved in liquid anhydrous ammonia
  - — ammonia-recovery systems for the agent and energetics SET reactors
  - — separate oxidation reactors for reaction products from the agent and energetic SET reactors
  - — water wash-out chambers to remove remaining sodium from all solids (e.g., dunnage, metal parts, fuze parts, etc.) in preparation for their disposal as waste or for shipment as 3X metal parts to a U.S. Army thermal-treatment facility

The SET process operates at ambient temperatures, using evaporation of ammonia to remove heat from the mildly exothermic reactions between the SET solution and the agent and energetics in their respective reactors. Pressures as high as 2,720 atm (40,000 psig) are used for the ammonia-jet cutting solution during disassembly operations. With the exception of the small section of piping and the intensifier pump for the fluid-cutting solution, most of the systems operate at near atmospheric pressure to 10.5 atm (140 psig). Where liquid ammonia is used, vessels will be operated at approximately 10.5-atm pressure, which is the vapor pressure of ammonia at typical ambient temperatures.

Worker Health and Safety

The Teledyne-Commodore technology package is basically an adaptation of existing technologies used for the destruction of conventional munitions and other hazardous chemicals. Assuming careful design and operation, the general technology of fluid cutting and fluid wash-out can achieve acceptable worker-safety levels. The aspects of the process that require special attention from a worker-safety standpoint are listed below:

- use of liquid ammonia-cutting fluid at 2,720 atm (40,000 psig)
- fluid-jet wash-out with liquid ammonia
- SET reaction with agent and energetics by combining the ammonia slurry containing these materials with liquid sodium in an in-line mixer
- operation of mechanical equipment in ammonia atmospheres at 10.5 atm (140 psig)
- processing of agent at elevated pressure (10.5 atm; 140 psig)
- addition of water to metallic sodium

A unique aspect of this technology from the standpoint of worker safety is the hazard associated with the primary chemicals used to destroy the agent and energetics. Liquid anhydrous ammonia boils at –33.4°C (–28°F) and becomes a toxic gas capable of burning in air. Liquid sodium (melting point = 98°C; 208°F) is a reactive and pyrophoric metal that burns violently upon exposure to air or other oxidizing media and requires special firefighting methods and materials. Sodium persulfate and hydrogen peroxide solutions are reactive chemical oxidizers as well as health hazards. (For example, because of the high reactivity and toxicity of hydrogen peroxide at concentrations above 52 percent, it is treated as a regulated substance by the Occupational Safety and Health Administration and the State of California Office of Emergency Services.)

Although all of these chemicals are widely used in industry and releases and accidents are infrequent, because of the hazardous and reactive nature of these chemicals, the SET systems must be carefully designed to ensure worker safety (e.g. from toxic gas exposures and fires). This includes minimizing the occasions when workers are in contact with the areas in which these hazardous chemicals may be present.

The disassembly processes differ from the baseline system in that fluid-jet cutting is used to access agent and energetics in all of the munitions. Indexing technology now used in the disassembly of munitions will be used to control the location of the cuts. The cutting machines in a pressurized ammonia environment may be less reliable than anticipated and may require more maintenance to achieve desired throughputs. The added maintenance would increase opportunities for worker exposure.

Most of the experience with cutting and fluid-jet wash-out has been with fluids other than ammonia (see Appendix G). The SET reactions have been well documented, but the destruction efficiencies over a prolonged period of operation with cutting operations will have to be carefully monitored.

The use of liquid anhydrous ammonia, with attendant ammonia vapors, is a significant hazard to workers because of the toxicity and flammability of ammonia (although the flammability range is only 4 to 15 percent). The committee's concern is that ammonia is not only flammable but can also become a fuel if a fire is initiated by other sources. The committee recognizes that anhydrous ammonia is a widely used industrial chemical, is the fluid of choice in large refrigeration systems, and is extensively used as a fertilizer. The widespread use of anhydrous ammonia by workers with varying levels of familiarity suggests that ammonia could also be used in the proposed systems without significant risks to worker safety.

The most significant issue is the necessity of wearing DPE suits for maintenance of the equipment in both the extraction and destruction subsystems. The equipment should be designed for easy access for the change-out and safe removal of contaminated parts. The design of tumblers and stirred reactors in the destruction subsystems will be especially important. In addition, the durability and reliability of DPE material in the presence of ammonia and residual SET solution is a concern. Although small-scale laboratory tests indicated that DPE degradation was manageable, additional tests will be required to ensure the safety of maintenance personnel.

Another safety concern is the ability of the fuze destruction (detonation) chamber to withstand a large number of explosions. The technology for fuze destruction is well established, but the area of concern is that fuzes will be handled and detonated in an ammonia-vapor atmosphere by remote control. These operations will have to be safe and the maintenance and repair requirements minimal; at the same time, high throughputs will have to be maintained.

The ammonia-agent and ammonia-energetic slurries will be mixed with liquid sodium to provide the solvated electrons for reducing chemical bonds. The long-term durability of the equipment for performing this mixing must be proven to ensure that maintenance requirements are low and opportunities for exposure are minimized.

Dunnage and other agent-contaminated materials will be shredded and then reacted with SET solution in a tumbler-type reactor. This technology has not been demonstrated in the presence of these materials. The committee is concerned that worker risk may be increased if maintenance requirements are high.

Public Safety

The committee believes that the likelihood of releases of agent or other regulated (hazardous) substances to the atmosphere or the groundwater system at the facility is small. Hold-test-release systems are used for all effluent streams except the containment ventilation system, which uses baseline air-cleaning technology. The primary cause for a release of material containing agent or other regulated substances would be a disruptive explosion. The likelihood of such an event is expected to be extremely small at the conclusion of the design process for the full-scale facility. (This design process is understood to include the completion of a quantitative risk assessment.) Ammonia at ambient temperature readily dissipates.

Human Health and the Environment

Effluent Characterization and Impact on Human Health and Environment

Teledyne-Commodore has not characterized the effluents from their total system in sufficient detail for the committee to assess potential impacts on human health and the environment. The chemical composition of the products of final oxidation is unknown, and the oxidation processes themselves are still under development. The technology provider has demonstrated that agent is unlikely to be present in any of the process effluents but has not demonstrated the absence of other chemical compounds of concern.

Gaseous Effluents. Teledyne-Commodore proposes scrubbing and recycling all ammonia vapors. Other gases produced in the system include hydrogen, low molecular-weight aliphatic hydrocarbons (propane, methane, ethane, ethylene, and propylene), and nitrogen. These gases will be passed through carbon-filter beds to remove impurities and burned for use as supplemental fuel to supply some of the energy for the process. The gases released to the environment should, therefore, consist primarily of carbon dioxide, water vapor, and nitrogen. Should the scrubbing process not remove all of the ammonia, this ammonia will pass through the carbon beds and be burned as part of the supplemental fuel, creating additional NO_x in the combustion gases. Carbon beds must be impregnated with phosphoric acid or a similar material to enable adsorption of ammonia. If this type of carbon is used, it must be tested for its ability to remove trace amounts of chemical agent.

Liquid Effluents. Teledyne-Commodore proposes recycling all wastewater for reuse in the process. Other anticipated liquid effluents include isopropanol, ethanol, diisopropyl-(amino) ethane sulfonic acid, and $2,2^1$-hydroxy diethylether. The first two will be separated in the ammonia-recovery towers and burned with the noncondensable gases as supplemental fuel. The rest will be mixed with cement and sent to a hazardous-waste landfill. The cement stabilization process for these wastes has not been tested.

Solid Wastes. The process produces many solid wastes, including metal parts, dunnage, sodium salts, polymeric materials, lead salt, and various organics. These wastes are the end products of final treatment methods that have not yet been optimized. The composition and properties of the products are, therefore, not certain and final disposal methods have not been tested.

The salt that Teledyne-Commodore anticipates will be produced in the largest quantity is sodium sulfate—5.2 lb/lb GB; 3.5 lb/lb HD; 2.4 lb/lb of M28; and additional quantities from the oxidation of VX, TNT, tetryl, tetrytol, and Comp B. Sodium sulfate in itself is not a listed or hazardous waste as defined by the EPA. However, as a product derived from the treatment of hazardous wastes (i.e., agents and energetics), sodium sulfate is subject to RCRA hazardous-waste regulations.

Sodium sulfate is quite soluble in water and may have to be stabilized prior to disposal in a landfill. However, sodium sulfate has been known to retard settling and cause spilling of the most common cement-based stabilization agents. A special cement may, therefore, have to be used to prevent leaching of the salt.

Completeness of Effluent Characterization

Final effluents from the oxidation process and the gas-treatment train have not been fully characterized. Chemical compositions and quantities are largely unknown and are likely to change as process conditions are refined.

Effluent Management Strategy

The technology provider has estimated the composition of the process effluents and has proposed disposal plans for them. The committee believes that these plans are reasonable. However, further work will be necessary to characterize the effluents accurately.

Resource Requirements

The resource requirements (including electrical power) for the proposed system are not unusual.

Environmental Compliance and Permitting

Commodore Remediation Technologies, a predecessor of Teledyne-Commodore, has received a nationwide permit under the TSCA to use solutions of calcium in liquid ammonia (known as Agent 313) to remove PCBs from contaminated soils and metal surfaces. However, the SET technology proposed here will be permitted under RCRA, which is considerably more complicated than permitting under TSCA. Furthermore, the proposed technology package has many more components than the permitted PCB system, including SET reactors for many different types of contaminants, both solids and liquids; oxidation reactors for treatment of SET residuals; and processes for preparing the products of oxidation for final disposal. Because the process as a whole (and the individual components) are unique, they will have to be permitted under Subtitle X of RCRA. Because federal and state regulatory agencies have limited experience in issuing Subtitle X permits, the permitting process will almost certainly be prolonged.

Another aspect of the process that may lead to permitting delays is the use of cleaned off-gas as a boiler fuel. Extensive testing may be required to characterize this stream to ensure its adequacy for that purpose.

STEPS REQUIRED FOR IMPLEMENTATION

The technology provider will have to take the following steps prior to implementation:

- verify products of SET/hydrolysis of agents through experimentation
- establish optimum conditions for SET/hydrolysis of agents through laboratory tests, followed by pilot-plant demonstration
- identify the unknown precipitates of SET-energetics reactions
- establish optimum conditions for the oxidation of residuals from SET/hydrolysis of both agents and energetics through laboratory tests, followed by pilot-plant demonstration
- test waste-disposal methods
- pilot test methods for decontaminating metal parts and dunnage
- revise the preliminary design for the hypothetical system, especially the interfaces between the unit processes and operations, and demonstrate the revised design at pilot-scale

FINDINGS

Finding TC-l. The use of ammonia-jet cutting in the munitions disassembly process could solve some of the problems encountered in baseline disassembly. However, the process must be thoroughly tested to address reliability and maintenance issues.

Finding TC-2. Conditions for SET destruction of agents have been reasonably well established but demonstrated only on a small scale.

Finding TC-3. Conditions for SET deactivation of energetics have not yet been determined. Moreover, energetics have not been completely deactivated in laboratory tests, which raises concerns about explosions or other upsets.

Finding TC-4. The reaction chemistry is not yet fully understood for either SET destruction of agents or SET deactivation of energetics.

Finding TC-5. The products of SET/hydrolysis of agent and energetics have not been adequately characterized. Thus, the technology provider cannot be certain that all of the SET/hydrolysis products can be oxidized by the persulfate step. Furthermore, the products of oxidation of the SET/hydrolysis products have not been adequately characterized.

Finding TC-6. The solid wastes produced by the overall process have not been characterized well enough to establish whether they are suitable for safe disposal by existing methods, such as landfill, or whether pretreatment methods (i.e., stabilization) to convert them to an acceptable form for disposal will be necessary.

Finding TC-7. The use of cleaned off-gas as a boiler fuel poses unique permitting challenges. Any process demonstration must characterize this stream to ensure that this off-gas can be permitted as a boiler fuel.

Finding TC-8. The full-scale system for hydrolysis of the SET products will differ significantly from the systems used in the reduced-scale tests. Because further chemical reactions occur during hydrolysis, the components of the full-scale system must be tested.

10

Public Acceptance of Alternative Technologies

INTRODUCTION

The Army's past failure to obtain broad public acceptance for its selected systems for the disposal of chemical weapons has contributed significantly to the schedule delays and cost increases in the baseline chemical weapons demilitarization program. Recognizing that public acceptance is critical to completing the destruction of chemical weapons, the Army's ACWA program has incorporated two innovations. The first is a public involvement process, the Dialogue on Assembled Chemical Weapons Assessment (hereinafter called the Dialogue). The Dialogue, facilitated by The Keystone Center, brings together a broad range of stakeholders to achieve consensus on methods for selecting and demonstrating alternative technologies for the destruction of chemical weapons. The second is the decision to identify and demonstrate at least two alternative technologies that may address the objections that have been raised most frequently by interest groups and the public.

This committee was asked to "gather data and analyze information on stakeholder interests at the assembled chemical weapons storage site locations..." to help the Army assess the prospects for public acceptance of alternative technologies. The committee gathered data from the following sources:

- attendance at public meetings in Richmond, Kentucky; Anniston, Alabama; and Pueblo, Colorado
- private discussions with residents and concerned citizens who attended the public meetings
- attendance at meetings of the Dialogue held during the preparation of this report
- private discussions with participants in the Dialogue
- interviews with Keystone facilitators
- discussions with state regulators in Colorado, Kentucky, and Utah
- briefings by DOD officials and managers

Analyzing the information proved to be a complex undertaking for a number of reasons. First, the committee needed to address the problematic term of "acceptance." Does it imply only tolerance, or does it mean affirmative acceptance? Is a broad public consensus necessary, or will a majority suffice? The task also required that the committee consider the diverse meanings of the term "the public" in the context of public policy. The term can be variously defined in terms of geography, level of interest or involvement in the policy debate, and policy preference. Some groups have conflicting opinions about the acceptability of alternative technologies. Which of these "publics" should be considered? The conventional use of the term "stakeholder" helps only a little, as all Americans have a stake in the safe and efficient disposal of chemical weapons. Second, very little systematic and reliable data are available on public reactions to any of the alternative technologies. Some of them are too new to have attracted widespread public attention, but even public perspectives on the technologies that are already being used for chemical demilitarization have rarely been systematically measured. Third, the acceptance of any proposed solution to a highly controversial policy issue of the kind considered here is likely to be affected as much (if not more) by the processes of public involvement as by the technological characteristics of the proposed solution.[1] Therefore, for the committee to assess public acceptance of the alternative technologies, the processes of public involvement in the selection of the alternative technologies had to be investigated first.

For these reasons, the discussion of public acceptance in this chapter does not follow the technology-by-technology approach used in the preceding chapters. Instead, this chapter begins with a brief overview of the processes by which public views of controversial policy options tend to be shaped and the role of these views in the policy-making process. The typical pattern of political debate is described, and the implications for public acceptance are addressed. Second, the discussion focuses on the development of public views on incineration, the only technology for the destruction of chemical weapons that has received broad, sustained political attention so far. The discussion includes the implications of public attitudes toward incineration for the progress of the chemical-weapons destruction program, as well as procedural and technological objections to the incineration of chemical weapons voiced by the public. Third, the committee evaluated the prospects for public acceptance of alternatives to incineration, focusing on the innovative process for public involvement by the ACWA program. The discussion addresses how the process and the basic characteristics of the proposed alternative technologies are likely to influence public acceptance and the role of environmental permitting in public acceptance of alternative technologies. The final section summarizes the committee's findings concerning public acceptance and provides general recommendations for improving the prospects for public acceptance of alternative chemical-weapons disposal technologies.

THE MEANING OF PUBLIC ACCEPTANCE

What does public acceptance mean in the context of the Army's efforts to destroy chemical weapons? Although there is broad public support for the destruction of these weapons, there are also deep disagreements over the means of that destruction. Since the mid-1980s, a number of organized groups of citizens have been mobilizing public opposition to the Army's use of incineration as the technology for destruction of chemical agents and munitions. Some of these groups have focused on the concerns of citizens living near chemical weapons storage depots, while others have broader national and international objectives (Smithson, 1994; Bradbury et al., 1994; NRC 1996b).[2] Partly because of their efforts, public meetings to obtain citizen input have often generated vocal opposition to the Army's baseline technology, particularly to the use of high-temperature, "smokestack" technologies like incineration (Bradbury et al., 1994; NRC, 1996b). Few organized groups of citizens have come forward to support incineration (although individual citizens have done so at several public meetings).

While the Army has pushed ahead with the baseline incineration approach, opponents to incineration have used multiple venues, with varying degrees of success, to express their opposition and block the implementation of chemical weapons incineration. Their efforts contributed to the passage in 1997 of Public Laws 104-201 and 104-208 mandating the evaluation and testing of alternatives to incineration for the destruction of chemical weapons. These are all clear indicators that, at least among some important segments of the public, acceptance of the Army's baseline chemical weapons destruction process has proven to be elusive.

Given the nature of the public debate over the destruction of chemical weapons, how should public acceptance be understood and evaluated? In general, when an issue is as complex as this one, there is no single public. The importance citizens attach to policy issues varies enormously (Krosnick, 1990). Some attach great importance to a particular issue and spend considerable time and resources becoming informed about it and expressing their views. Most do not. Moreover, relatively few people can devote substantial attention and resources to more than a few public policy issues. As a result, the American public consists of many relatively small "issue-publics."[3]

[1]An enormous literature demonstrates the degree to which process considerations, such as trust and fairness, influence acceptance. See, for example, Jenkins-Smith and Silva 1998; Flynn et al., 1992; and Leiss and Chociolko, 1994.

[2]Many of these groups have coordinated their efforts under umbrella organizations, such as the Chemical Weapons Working Group (CWWG), to seek alternatives to incineration.

[3]Partly because of the segmented nature of "the public" in the United States, it is notoriously difficult to "inform" the public on policy issues,

Whether or not a citizen becomes consistently involved in an issue (i.e., becomes a member of an issue-public) depends on his or her particular interests and concerns. Proximity to a chemical weapons storage depot, for example, generates interest in the related health and safety questions. The views of concerned citizens tend to be relatively sophisticated with respect to the policy issue, resistant to change, and stable over time (Krosnick 1990; Cobb and Kuklinski 1997). Members of issue-publics tend to be involved in the policy process, become active members of organized citizen groups, and express their views in public meetings or directly to political leaders. In most cases, they constitute a small fraction of the general public.

Citizens who are less involved in the issue tend to acquire relatively little information about it, rely on relatively simple heuristics, or methods, of forming their opinions (Sniderman et al., 1991), and hold opinions that are susceptible to change depending on the framing or context of the issue (Yankelovich, 1991). These citizens rarely participate directly in public debates on the issue. Nevertheless, if the issue receives significant attention in the press, public opinion can play a substantial role in shaping policy as decision makers respond to (or attempt to anticipate) broad public concerns (Page and Shapiro, 1983; Page et al., 1987; Hill and Hinton-Andersson, 1995).

RISK PERCEPTION AND POLICY DEBATES

The lay public rarely has firsthand information about potential environmental and health risks. Policy debates on controversial issues involving perceived risks tend to take the form of competing messages about the magnitude of the risk, the performance of those charged with managing the risks, and the appropriateness of competing policy options from organized interests and the responsible public agencies. When the stakes appear to be high, these messages are received and retransmitted via the news media, where they reach less involved and less informed members of the public. Given the general propensity of people to weigh prospective losses more heavily than gains (Kahneman and Tversky, 1984), members of the public tend to be more receptive to negative arguments (e.g., those against incineration) than positive ones (Cobb and Kuklinski, 1997).[4] Moreover, simple arguments with an emotionally charged message tend to weigh more heavily with less involved members of the public than "harder" or more complex arguments (Renn, 1998; Cobb and Kuklinski, 1997).[5] In the case of programs involving either incineration or managing chemical weapons, the issue falls within the "high dread" and "high uncertainty" domain of perceived risk, about which members of the public tend to be quite worried (Slovic, 1987; Krimsky and Golding, 1992; Renn, 1998). Thus, in public debates over issues for which the magnitude of the risk is an important ingredient, the playing field is tilted significantly in favor of those who argue that risks are high and who oppose the apparent risk-inducing program.

In policy debates over risk, the level of trust of the agencies charged with carrying out the potentially risky activity weighs heavily in shaping public responses (Flynn et al., 1992; Leiss and Chociolko, 1994). In general, when all else is equal, members of the public appear to give greater credence to arguments that risks are large than to arguments that risks are small (Jenkins-Smith and Basett, 1994). Moreover, when the responsible agency is not considered very trustworthy, members of the public are increasingly prone to discount reassurances of safety by that agency (Jenkins-Smith and Silva, 1998). Trust is therefore a critical ingredient in shaping public views of the risk and acceptability of a potentially hazardous program.

Another important factor is whether the public perceives the potential hazard to have been imposed upon them. The "voluntariness" of the risk has long been

even when the issue concerns health and safety. Furthermore, information about a specific policy issue must compete with the deluge of messages that bombard citizens on a daily basis. See Renn (1998) for a discussion of the information-filtering processes used by citizens.

[4]Some scholars dispute Kahneman and Tversky's (1984) claim that prospective losses will be weighed more heavily than prospective gains. For example, Cosmedes and Tooby (1994) argue that Kahneman and Tversky's results are based on the presentation of risks in probabilistic, rather than frequentist, terms. However, because the debate over the risks of incineration of hazardous materials is almost universally cast in probabilistic terms (e.g., the risk of 10^{-6}), their findings would appear to be valid in this case.

[5]The differences in how members and nonmembers of issue-publics respond to arguments may pose a dilemma for agencies charged with informing the public. The members of issue-publics tend to be more responsive to more complex arguments (involving the presentation of premises from which a conclusion is derived), while less involved and less informed citizens may view such arguments with suspicion. Thus, acceptance from one audience may come at the expense of the acceptance of the other. See Cobb and Kuklinski (1997), pp. 115–116.

identified as a moderating factor in perceptions of risk (Renn, 1998; Slovic, 1987). When a prospective program is convincingly framed to the public as a potential hazard that is imposed on them by outsiders (especially if the outsiders are not seen as particularly trustworthy), public acceptance of the program may prove to be very hard to attain. The perception of the hazard as imposed is heightened if it appears that the responsible agency is defending an option that has been decided upon prior to engaging the community (the familiar "decide, announce, and defend" approach).

Coupled with the matters of trust and perceived risk is the matter of "framing" the issue (Kahneman and Tversky, 1984). If policy activists and the news media portray an issue as one in which a risk is posed by the destruction of the chemical weapons, then the program for destruction will be singled out as a potential threat to the community. As discussed above, this is a very simple argument (e.g., incineration poses a health threat) with negative emotive content. Such messages are likely to be persuasive with less involved, less informed members of the public (Cobb and Kuklinski, 1997). If, on the other hand, the presence of the stored chemical weapons is portrayed as an ongoing threat, then the risk posed by the means of destruction contrasts the risk of continued storage. This is a more complex message and is most likely to be compelling to those who are better informed about the issue and more involved in the policy process. Over the course of the debate on the incineration of chemical weapons, the framing of the issue has been a central concern. The Army has generally argued that the risks of long-term storage are significant and that risks of incineration are modest. The Army's critics have emphasized the risks of incineration while implicitly downplaying the risks of continued storage.

In general then, gaining acceptance of the broader and less involved members of the public for a potentially hazardous activity faces an uphill battle. The population tends to be fragmented into multiple publics, ranging from intensely committed and engaged issue-publics to the publics and individuals who are relatively inactive, less informed, and less interested. This fragmentation significantly complicates the problem. Generally, it is much more difficult to reassure these publics that a potentially hazardous facility is safe than it is to raise the alarm. If trust in the implementing agency is lacking, the prospects for public acceptance are particularly low.

POLITICS AS USUAL: THE INTERACTION OF POLICY DISPUTES AND PUBLIC ACCEPTANCE

Public acceptance (or rejection) of chemical weapons disposal technologies does not occur in a vacuum. The public evaluation of alternative technologies is very likely to take place in the context of the ongoing policy debate on the disposal of chemical weapons. Therefore, the prospects for public acceptance of the alternative technologies can best be understood as part of the more general process surrounding public-policy controversies.

In most policy controversies, the shaping and implementation of programs involve multiple regulatory, judicial, and legislative arenas. The regular participants in these debates typically include: the agencies charged with carrying out the program; regulatory oversight agencies; organized interest groups; members of the press who regularly report on the issue; and the elected and appointed public officials for whom the issue is particularly salient (Anderson, 1979; Kingdon, 1984; Sabatier and Jenkins-Smith, 1993; Jenkins-Smith and Sabatier, 1994). In the course of the policy debate, opponents identify what they perceive to be problems with the proposed program and attempt to alter or block its adoption and implementation. Typically the opponents attempt to persuade political decision makers and the public of both the validity and importance of their concerns and of the negative consequences of pursuing those policies. When resources are available, opponents contest the policy in visible public settings, such as public hearings and press conferences. They may also have many opportunities to challenge the policy in administrative and judicial processes at the local, state, and federal levels of government.

In the case of the destruction of hazardous materials like chemical weapons, federal law allows states to complement federal requirements with more stringent regulations.[6] These additions tend to require demonstrations of regulatory compliance before multiple state

[6]For example, provisions of the Resource Conservation and Recovery Act (RCRA) allow states to add regulatory requirements that exceed those set forth in the federal law. See Section 6929 of Title 42 of the United State Code.

agencies, each of which may require a process of public involvement and administrative due process.[7] In a sustained policy controversy, each of the involved regulatory agencies provides an opportunity for opponents to challenge the proposed program. If administrative processes fail to generate the desired results, opponents may be able to turn to the state or federal courts.[8] The net result is often a sustained period of policy debate, in the course of which the news media have ample opportunity to transmit the positions of the program's opponents and proponents.

The extended policy debate both influences, and is influenced by, public opinion. Although research on the interactions among the mass public and political "elites" (such as representatives of organized interest groups or government officials) is far from conclusive, the available evidence suggests a reciprocal relationship in which political leaders both influence, and are influenced by, the opinions of the larger public (Hill and Hinton-Andersson, 1995; Page et al., 1987; but see Hill, 1998). Thus the extended policy debate exposes the public to the arguments of competing coalitions of agencies, interest groups, and political officials, while feedback (for example, through elections and public hearings) from segments of the public appears to influence the positions of political decision makers. In short, gaining acceptance for a controversial program from the larger public cannot be decoupled from the policy debate concerning that program among active segments of the public and government agencies.

PUBLIC REACTION TO THE ARMY'S BASELINE CHEMICAL WEAPONS INCINERATION PROGRAM

An understanding of the possible public reactions to alternative technologies for the destruction of chemical weapons can be gained by an evaluation of the Army's experience with the ongoing chemical weapons destruction program. From the early 1900s through the 1960s, chemical weapons were destroyed by burning in open pits, land burial, and ocean dumping. In 1969, a National Academy of Sciences report recommended a combination of neutralization and incineration as a means of destroying chemical weapons (NAS, 1969). By the early 1980s, the Army had identified incineration as the preferred technology. This decision was endorsed in a report by the National Research Council in 1984 (NRC, 1984). The following year the U.S. Congress passed Public Law 99-145 requiring the destruction of all U.S. chemical weapons stockpiles. The policy was further refined by an Army decision in 1988 to destroy all chemical weapons on site at the eight continental storage depots.[9]

This sequence of policy decisions, and the ongoing efforts to implement the chemical weapons demilitarization process at the eight continental depots, has pitted organized groups of citizens who oppose the incineration of chemical weapons against the Army agencies charged with their disposal.[10] Opponents have argued that the Army has "consistently" attempted to implement incineration before adequate testing and risk assessment were completed (Alailima et al., 1995). Furthermore, the opponents have argued that the incineration of chemical weapons "presents an imminent threat to public health" through the possibility of catastrophic failure, the release of small amounts of chemical agent, and the emission of trace amounts of products of incomplete combustion (PICs), such as dioxin (Alailima et al., 1995; Costner, 1993).

Coupled with these concerns about risk has been the contention that residents of communities near the storage depots have been essentially excluded from decision making about the choice of the disposal technology (Alailima et al, 1995; Futrell and Davies, 1996; Bradbury et al., 1994; NRC 1996b). As proof, members of some interest groups cite the Army's use of temporary permits to begin the operational verification tests (OVT) at the Johnston Atoll Chemical Agent Disposal System (JACADS) and to break ground at Anniston, Alabama (GAO, 1993), in which there were

[7]For example, in Indiana the permitting process for the destruction of chemical weapons agent through hydrolysis requires action by the following state and federal agencies: the Indiana Waste Management Administration, the Indiana Water Management Administration, the Indiana Air and Radiation Management Administration, and the regional office of the Environmental Protection Agency.

[8]Even if the court battle is lost, the delays and costs can be considerable.

[9]This decision was reaffirmed in 1991 in Public Law 101-520, which prohibited the Army from studying the transport of chemical weapons.

[10]The list of organizations that oppose incineration of chemical weapons is long, including both local citizens groups and organizations with national and international objectives. See Alailima et al. (1995) for a partial listing of the involved groups. Also see Smithson 1994.

no opportunities for public comment. Overall, the policy debate has been heated, involving a significant number of organized advocacy groups that have contested both the Army's and the NRC's findings that incineration can be a safe and effective technology for destruction of chemical weapons.[11] The following section identifies some of the attributes of incineration that are considered objectionable by the public and opponents to the technology. These attributes should be carefully considered in the selection of an alternative technology.

ATTRIBUTES OF INCINERATION

This committee sought public input directly from citizens in the Richmond, Anniston, and Pueblo areas and collected information on public concerns from earlier research (Bradbury et al, 1994; NRC 1996b; Brown 1993). None of these sources of information, however, provides a reliable or valid sample from which inferences can be drawn about broad public perspectives.[12] Attendees at public hearings tend to be those who have strong feelings about an issue, while those who are less interested (or who perceive the risk to be insignificant) are less likely to attend. Nevertheless, it is possible to identify some of the persistent public concerns about chemical weapons destruction technologies:

- the potential for catastrophic failure and a massive release of agent into nearby communities
- low levels of agent escaping into the air through smokestacks, with potential for long-term chronic health effects
- monitoring processes that take only periodic samples of stack emissions only identify "after-the-fact" releases of agent or other toxins ("once the alarm goes off, it's already too late")
- technologies that dispose of hazardous materials at high pressures and high temperatures
- the continued use of a chemical weapons destruction facility for other hazardous wastes once the local chemical weapons stockpile has been destroyed
- the trustworthiness of the Army to manage the chemical weapons destruction program safely
- concern that the Army will not fully inform the local community about problems with the chemical weapons destruction, if and when they arise

In general, these concerns appear to reflect the high levels of perceived risk associated with technologies that Slovic (1987) has identified as having the attributes of high dread (the worst-case outcome is catastrophic) and high uncertainty (the threat is invisible to the naked eye, the cause-effect connection to human and environmental health is not well understood, and the effects may be delayed). With respect to incineration, these fears appear to be linked to the high volumes of outputs (stack emissions) into the atmosphere that are not continuously monitored for potentially hazardous PICs that once released, cannot be reclaimed. Uncertainties about the characteristics of stack emissions (e.g., the full roster of PICs) and their potential long-term health effects also provide a ready basis for public concerns.[13]

At the same time, the available evidence suggests that public awareness of the chemical weapons incineration issue may be modest, even among people who live near the chemical weapons storage depots. In a recent telephone survey of residents of communities near the Anniston Army Depot, nearly half (49 percent) of the respondents said they either did not know of plans to destroy chemical weapons at the depot or had heard about the destruction but were "not very well informed" (Clark, 1996).[14] Less than a third of the survey respondents had ever heard about the Alabama Citizens Advisory Commission (CAC), and only 6 out of the 500 respondents said they had ever attended a CAC meeting. Other public surveys have found similarly low levels of awareness. In fact, a *majority* of the population near some chemical weapons storage

[11]See, for example, the charge by environmental groups that the NRC's 1993 report on alternative technologies for chemical weapons destruction is "politicized" (Defense Environment Alert, 1994).

[12]The Program Manager for Chemical Demilitarization is currently in the process of developing a telephone survey for eliciting systematic samples of public opinion in each of the chemical weapons storage communities. Therefore, reliable information about general public views on chemical weapons destruction technologies may be available in the future.

[13]The anti-incineration case is often based on these characteristics of incineration. See, for example, the case against the incineration of hazardous materials by Greenpeace in Costner and Thornton, 1990. But also see the evaluation of the Greenpeace argument in Chrostowski and Foster (1992) and J. Cudahy (1992).

[14]The survey was conducted for the Alabama Chemical Demilitarization Citizens' Advisory Commission in April and May of 1996 and included residents of Calhoun and Talladega Counties. No information was provided on the survey response rate.

depots is unaware of the existence of the chemical weapons stockpile (Innovative Emergency Management and Rowan and Blewitt, Inc., 1995a, 1995b, 1995c, 1995,d, 1995e). Thus, despite the Army's public relations efforts and significant local opposition to incineration by organized citizens groups in the local areas, only a small fraction of the area residents are informed and involved in the policy debate (NRC, 1996b).

Among the organized citizen groups involved in the chemical weapons demilitarization issue, opinions about the attributes of technologies are well developed (see, for example, Chemical Weapons Working Group, 1994). In addition to the concerns listed above, these groups expressed a clear reluctance to support a technology that does not permit the holding and testing of all effluents and emissions from the chemical weapons demilitarization process prior to release (hold-test-release). This concern reflects the public perception that periodic sampling and testing of the large volumes of gas emissions characteristic of incinerators will only detect system failures after the fact, when a release has already taken place.

Opportunities for meaningful input by local citizens into the process of selecting the technology is as important to members of these groups (Alailima et al., 1995; Brown 1993), as is monitoring the operation of the technology during the destruction of chemical weapons (NRC 1996b).

Discussions with representatives of some of the citizen groups that are most actively involved suggests that the following attributes are most important:

- the capability of the system to hold and test effluents prior to release
- the "transparency" of the technological process[15]
- the inclusion of specific plans for decommissioning the facility and remediating the site
- the capability of quickly and safely shutting down the facility[16]

[15]The term "transparency" in this context refers to the characteristic of a technology that permits an informed individual to observe and have a reasonable understanding of what is occurring. When transparency is absent, the technology is essentially a "black box" in which the relationship between input and outputs is unknown to the observer. In the context of public acceptance, transparency pertains to the public or a trusted representative of the public.

[16]Among this group, concerns about temperature and pressure were of relatively low importance.

PUBLIC INVOLVEMENT IN POLICY FOR CHEMICAL WEAPONS INCINERATION

Concerns about incineration have led to strong and persistent opposition to the Army's efforts to move forward with the incineration of chemical weapons. Among the most frequently cited complaints by opponents has been the Army's perceived absence of attention to local concerns and the failure to provide for meaningful public input in the selection and implementation of technologies for the destruction of chemical weapon (Alailima et al., 1995; Brown 1993; NRC 1996b). Historically, the most prominent mechanism for public involvement in the policy process has been public hearings on all federal programs that could have a significant environmental impact, as required by the National Environmental Policy Act (NEPA). Opportunities for public input were expanded in 1992 when the Congress passed PL 102-484, which called for the establishment of CACs at each of the eight continental disposal sites.[17] The nine CAC members for each site are appointed by the governors and are required to meet with a representative of the Secretary of the Army to provide "citizen and state concerns regarding the ongoing program" of chemical weapons disposal.

Despite these mechanisms for involvement, public outreach for the chemical weapons destruction program has been heavily criticized by outside activists (Smithson, 1994; Futrell and Davies, 1996; Sierra Club [Utah Chapter], 1997) by government agencies (GAO, 1991, 1993, 1997), and by agency review committees (ICF Kaiser, 1998). Reaction to the CACs has been mixed, and the vigor with which the CACs in different states have elicited input from the local populations has varied substantially.[18] The Army's efforts at public interactions have been criticized for focusing primarily on the passive provision of information (public relations). Only recently has the Army begun to seek public input about concerns at each stockpile site (public

[17]The call for CACs was first made federal law in the National Defense Authorization Act of 1993, which called for the Army to establish CACs in Kentucky, Alabama, and Maryland (the so-called "low volume states"). Later in 1993, the Secretary of the Army requested that the governors of these states form CACs and invited the governors of other states with chemical weapons stockpiles to do the same.

[18]The difficulty confronted by the CACs in some states may be structural. According to Brown (1993), the effectiveness of the CACs has been undercut by their relatively vague role (as defined by the federal legislation) and by the political nature of appointments to the CAC.

outreach). Very little emphasis has been placed on involving the public explicitly in the decision-making process (NRC, 1996b; ICF Kaiser, 1998).

Although the Army has attempted to improve its interactions with the public, the perception that public concerns have been largely ignored continues to stimulate opposition.[19] Among some opponents to the incineration of chemical weapons, the Army's public relations approach has come to be characterized as the decide, announce, and defend approach (DAD) (or simply the old way of doing business), in which the central decisions (e.g., the choice of incineration) are made prior to seeking meaningful public input. In this view, the Army's interaction with the public consists largely of public relations efforts to justify prior decisions (Alailima et al., 1995; Brown 1993).[20] At the same time, the opponents of incineration have appeared to some observers to be unwilling to give chemical weapons incineration a fair hearing (Smithson, 1994; Brown, 1993) and to have made distorted claims about incineration that have raised unwarranted fear in communities near the chemical weapons storage depots (NRC, 1996b). The result has been mutual distrust and the appearance of intractable conflict between the Army and opponents of chemical weapons incineration (Koplow, 1997).

Partly because of the problematic history of public involvement in the chemical weapons disposal decisions, opponents of incineration have made effective use of available policy venues for delaying—and in some cases halting—the Army's efforts to develop and implement baseline incineration facilities at chemical weapons storage sites.[21] Challenges in the courts have been frequent, both in state and federal jurisdictions.[22] In addition, opponents of incineration have successfully appealed to the legislatures in Kentucky and Indiana to raise the environmental requirements for permitting incinerators that would burn chemical weapons. In Kentucky (Kentucky 1988 Ky. Acts ch. 86, sec.1.) the requirement states:

> The applicant shall affirmatively demonstrate, and the cabinet shall find prior to issuance, conditional issuance, or denial of the permit, that:
>
> The proposed treatment or destruction technology has been fully proven in an operational facility of scale, configuration, and throughput comparable to the proposed facility, for a period of time sufficient to provide consistent assurance of destruction or neutralization at an efficiency of ninety-nine and nine thousand, nine hundred, and ninety-nine ten thousandths percent (99.9999%) for each substance proposed to be treated or destroyed, with the efficiency to be demonstrated as achievable during the design life of the facility under all operating conditions including during the occurrence of malfunctions, upsets, or unplanned shutdowns.

Indiana has similar requirements (Indiana 1996 IC 13-22-3-9 and IC 13-22-3-10) but has added the following stipulation:

> That monitoring data from a comparable hazardous waste facility demonstrates that there are no emissions from the comparable facility that alone or in combination with another substance present a risk of any of the following:
>
> A. An acute or chronic human health effect
> B. An adverse environmental effect

At best, these requirements would be difficult to meet and may be impossible to demonstrate given the time constraints imposed by the CWC. At a minimum, these requirements will pose formidable hurdles in the permitting process for prospective chemical weapons treatment or disposal technologies. Complicating the

---

[19]In part, the Army public relations programs have been hampered by confusion with the Chemical Stockpile Emergency Preparedness Program (CSEPP), which is responsible for planning and response to chemical stockpile incidents. CSEPP is funded by the Army but managed jointly by the Army and the Federal Emergency Management Agency (FEMA). CSEPP has been plagued by public perceptions of inadequacy (Sierra Club [Utah Chapter], 1997; GAO, 1994, 1995, 1997) and by the appearance of confusion regarding the local coordination of emergency response planning (NRC, 1996b). Problems with CSEPP have thus spilled over and generated the perception that the chemical weapons disposal program pays inadequate attention to local concerns and conditions at the stockpile sites.

[20]The perception by critics that the Army uses the DAD approach has been reinforced by Army responses to "whistle-blower" employees at the Tooele and Johnston Atoll facilities. Claims of safety failures and cost overruns by Steve Jones and Charles Oughton received substantial national media attention but were largely dismissed by the Army. Critics charge that the Army's failure to respond to problems identified by employees at Army facilities confirms that the Army will not listen seriously to citizens' concerns (Koplow, 1997).

[21]The failure to develop meaningful and successful public participation can be particularly damaging when there appear to be competing expert views on the issue (Brown 1993). In the case of chemical weapons incineration, opponents have marshaled a number of technical criticisms of the baseline incineration (e.g., Costner and Thornton, 1990; Costner, 1994; Sierra Club, 1994). Thus, to those skeptical of incineration, technical support for the Army's position appears to be merely one side in an "advocacy science" debate.

[22]US District Court-District of Utah #2:96 CV 0425C; Utah State Agency Action EPA ID# UT 521009002; Oregon Circuit Court, County of Multnomah #9708-06159; Alabama Circuit Court, County of Montgomery #CV 98 2082; Alabama State Agency Action, Docket #97-17.

TABLE 10-1 Schedule Slippages of Chemical Weapons Demilitarization

Location	Estimated Date for Completion of Operations		
	1988[a]	1992[b]	1998[c]
Utah	1994	2000	2003
Alabama	1993	2000	2005
Oregon	1993	2000	2005
Arkansas	1994	2000	2006
Colorado	1993	2000	2007
Kentucky	1992	2000	2007
Maryland	1993	1999	2004
Indiana	1993	1999	2004

[a]U.S. Army, 1988.
[b]OTA, 1992.
[c]Arthur Andersen, 1998.

process is that, in order to obtain permits to build and operate demilitarization facilities, the Army and its contractors will have to meet the regulatory requirements of *multiple* state and federal agencies. These include state agencies governing different media (air, water, and land), waste-management divisions of agencies governing different media, regional federal agencies, and emergency-management agencies. In addition, the concerns of Indian nations will have to be considered (e.g., at the Umatilla, Oregon, depot). The complexity of meeting requirements of multiple regulatory agencies is compounded because there may be multiple permits having jurisdiction over the chemical weapons.[23] Each step in the process provides another potential forum for challenging, litigating, or otherwise slowing the Army's efforts to bring a baseline incinerator on line.

Although other factors have been involved, the efforts of opponents of incineration have contributed to both delays and cost increases in chemical weapons demilitarization. Table 10-1 shows the changes in the schedule for chemical weapons disposal from 1985 to 1995. In 1985, it was anticipated that demilitarization would be completed by the end of 1994. A decade later, in 1995, the expectation was that the last chemical weapons would be destroyed by 2004. More recent estimates are even less optimistic (Arthur Anderson, 1998).

[23]In Utah, three permits are required: a storage permit, the facility incineration permit for the Tooele facility, and the research and development permit for the nearby chemical agent munition disposal system.

Schedule delays, changes due to technology development, litigation, and demanding permitting requirements are expensive, and the expected costs of chemical weapons demilitarization have risen substantially. As shown in Table 10-2, from 1985 to 1996, the expected costs of the program rose by 729 percent. This history of schedule delays and cost increases can be partly attributed to the Army's inability to obtain broad public acceptance of chemical weapons disposal options.

PUBLIC ACCEPTANCE OF ALTERNATIVE TECHNOLOGIES

Given the history of public responses to the Army's past and present chemical weapons disposal program, gaining public acceptance of alternative technologies will face many of the same hurdles that have confronted (and continue to confront) chemical weapons incineration. There will continue to be many publics, a disproportionate propensity of citizens to believe dire messages, a legacy of distrust, and a labyrinthine regulatory process fraught with opportunities for delay.

Nevertheless, the ACWA program includes two kinds of opportunities for increasing public acceptance for (or reducing opposition to) chemical weapons disposal. First, the ACWA program has adopted an innovative process for including important segments of the public in the process of identifying and evaluating alternative technologies. This public involvement process may well have a significant bearing on the acceptance of the results of the ACWA process. Second, the characteristics of the alternative technologies described in the prior chapters of this report may address some of the objections most frequently raised by members of interest groups and the public.

THE DIALOGUE

The ACWA initiative, as specified by Congress, created a unique opportunity for early and direct public involvement in the identification and demonstration of technologies for chemical weapons destruction. By requiring that the process be managed by an agency other than the Army's CSDP and that the process "identify and demonstrate not less than two alternatives to the baseline incineration process" the program was given

TABLE 10-2 DOD's Estimated Life-Cycle Costs for Chemical Weapons Demilitarization (in $ billions)

Year	Life-Cycle Cost Estimate	Dollar ($) Increase Over Previous Estimate	Percentage Increase Over Previous Estimate	Cumulative $ Increase Over 1985 Estimate	Cumulative Percentage Increase Over 1985 Estimate
1985	1.7	—	—	—	—
1986	2.0	0.3	18	0.3	18
1988	3.4	1.4	70	1.7	100
1991	6.5	3.1	91	4.8	282
1992	7.9	1.4	22	6.2	365
1993	8.6	0.7	9	6.9	406
1994	10.2	1.6	19	8.5	600
1995	11.9	0.9	8	10.2	700
1996	12.4	0.5	4	10.7	729

Source: GAO, 1993, 1997

the opportunity to sidestep the political conflict that has characterized the baseline program. In essence, the ACWA program was given an independent identity from the CSDP, and ACWA officials were given the opportunity to establish a new pattern of relationships with active interest groups, regulators, and the public.

The Army's ACWA program has taken advantage of this opportunity. The program has drawn on a wide range of experience to broaden the scope of public involvement in environmental decision making (Creighton and Aggens, 1996). These efforts have included attempts to include public input on potentially controversial decisions before the central policy choices were made so the concerns and preferences of members of the public could be factored into the policy choices in a credible and explicit manner. In that spirit, ACWA program officials initiated an ambitious process, dubbed the "Dialogue on Assembled Chemical Weapons Assessment" (hereinafter called the Dialogue), designed to include a broad range of participants in the process of technology selection and testing. The initial design of the process, and the necessary implementation of the process, were carried out by the Keystone Center, a nonprofit, neutral organization that specializes in resolving environmental and health policy disputes. The salient aspects of the ACWA Dialogue process and developments are described below.[24]

[24]This brief summary is based on Keystone (documents/undated), interviews with Keystone facilitators, observations of Dialogue meetings, and reviews of the Meeting Summaries for Dialogue meetings.

Objectives of the Dialogue Process

As described by Keystone, policy dialogues are "voluntary, interactive consensus-building processes that are designed to solve problems." They provide "an opportunity for a diverse set of individuals with different interests, perspectives, and expertise to meet face-to-face and work toward consensus solutions" (Keystone Center, undated). ACWA program officials have used this process in an attempt to achieve consensus on methods for selecting and demonstrating alternative technologies for chemical weapons destruction. Given the very aggressive schedule required under the provisions of the CWC, ACWA hoped that the Dialogue process would result in consensus—or at least reduce conflict—among interest groups, regulatory agencies, and the Army on the identification and demonstration of alternative chemical weapons disposal technologies and reduce political, legal, and regulatory obstacles to demonstrating and, perhaps, implementing alternative technologies for chemical weapons destruction. The Dialogue process was initiated to diminish the kinds of conflict, delays, and budget increases that have plagued the baseline incineration program.

Selection of Dialogue Participants

The Dialogue process is intended to provide an opportunity for discussion and consensus-building among representatives of groups and agencies involved in the policy process for the disposal of chemical weapons. Keystone facilitators attempted to ensure that (1) participants had diverse perspectives on chemical weapons

TABLE 10-3 List of Participants in the Dialogue on Assembled Chemical Weapons Assessment as of July 10, 1998

Pua'Ena Burgess	Pacific and Asia Council of Indigenous Peoples
David Christian	Serving Alabama's Future Environment
Daniel Clanton	Arkansas Dept. of Pollution Control and Environment
Ralph Collins	Kentucky Dept. for Environmental Protection
Elizabeth Cotsworth	U.S. Environmental Protection Agency
Dennis Downs	Utah Dept. of Environmental Quality
Pamela Ferguson	State of Indiana
Wm. Gerald Hardy	Alabama Dept. of Environmental Management
Douglas Hindman	Kentucky Citizens' Advisory Committee
Worley Johnson	Kentucky Citizens' Advisory Committee
Karyn Jones	Oregon GASP
Cindy King	Utah Chapter Sierra Club
Irene Kornelly	Colorado Office of Business Development
Thomas Linson	Indiana Dept. of Environmental Management
Terry Mabrey	Pueblo Chemical Depot
Brett McKnight	Oregon Dept. of Environmental Quality
Sara Morgan	Citizens Against Incineration at Newport
Jodie Neely	Blue Grass Army Depot
John Nunn	Maryland Citizens' Advisory Commission
Bob Palzer	Sierra Club Air Committee
Michael Parker	Assembled Chemical Weapons Assessment
William Pehlivanian	Assembled Chemical Weapons Assessment
Theodore Prociv	Department of Defense (Chemical Demilitarization)
George Smith	Alabama Citizens' Advisory Commission
Wesley Stites	Arkansas Citizens' Advisory Commission
Wayne Thomas	Oregon Dept. of Environmental Quality
J. Ross Vincent	Colorado Citizens' Advisory Commission
Paul Walker	Global Green USA Legacy Program
J.R. Wilkinson	Confederated Tribes of Umatilla
Craig Williams	Chemical Weapons Working Group
Suzanne Winters	State of Utah Office of Planning and Budget
Evelyn Yates	Pine Bluff for Safe Disposal

destruction technologies, (2) all of the recognized "key parties-at-interest" in the chemical weapons disposal debate were included, and (3) participants were willing to engage in the Dialogue process to seek consensus on the selection and demonstration of alternative technologies. The selected participants were drawn from a number of organizations, including CACs (members and former members), public interest groups (some of which are opposed to incineration), and officials from state and federal regulatory agencies that will be involved in the permitting and oversight of the facilities. Keystone tried to maintain a balance between a manageable-sized group and the inclusion of all current participants in the policy debates from each stockpile site.[25] The list of participants as of July 1998 and their group affiliations are shown in Table 10-3.

From the committee's observation of the Dialogue group, it was evident that a wide range of perspectives was represented. Participants ranged from those who had been actively seeking alternatives to the incineration of chemical weapons to those who perceived incineration as the most efficacious and safe way to destroy chemical weapons.

Incentives to Participate

A key challenge for public involvement processes like the Dialogue is to provide sufficient incentives for those involved in the policy debate to invest the time

[25]Keystone facilitators consulted with Army representatives, CAC members, and representatives of prominent public interest groups to compile an initial list of potential participants. As the list of participants was assembled, each new member was asked to indicate whether all relevant policy participants had been included. Additions were made until Dialogue members could not identify additional participants who were willing to join. Discussions with Dialogue members indicated general agreement that all relevant participants were included in the final list.

and energy to seek consensus with other participants who may hold opposing policy views. On both counts, the costs can be very high.[26] A willingness to enter the Dialogue process is based on the determination by all participants (including ACWA officials) that a "negotiated agreement" among the participants on alternatives to incineration would more effective than any one participant trying to influence the policy decision,[27] often referred to as a BATNA (best alternative to a negotiated agreement). As long as the BATNA was considered less desirable than a negotiated agreement through the Dialogue, members (and the Army's ACWA program officials) had sufficient incentives to participate.

The ACWA program has played a critical role in generating the trust necessary for the Dialogue process to proceed. For some Dialogue participants (such as representatives of groups that had been involved in protracted conflicts with the Army's CSDP over the baseline incineration program), trust in the Army was badly eroded. In an attempt to reestablish trust, ACWA officials have repeatedly expressed their commitment to the process,[28] provided resources for technical support to the Dialogue, and (as described below) given the Dialogue group a direct role in the technology selection process. Throughout the period that this committee observed the Dialogue process, participants indicated that the prospects of the outcome were, in their views, superior to BATNA.

[26]Regular Dialogue meetings lasted for several days. None of the non-DOD participants were paid by DOD to participate in the Dialogue. Participants who were state or federal employees (and, therefore, may have been paid through their regular jobs) spent a substantial amount of time away from their other responsibilities. Volunteers, primarily representatives of interest groups, were not compensated for their time. All Dialogue members committed substantial time and energy to the process.

[27]For the general approach underlying the Keystone's Dialogue process, see Susskind and Cruikshank, 1987. For a more general discussion of the conditions under which negotiated solutions can be expected to succeed, see Quirk, 1989.

[28]Most of the early Dialogue meetings, and many later ones, included sessions by senior ACWA officials focused on explaining the Army's commitment to the Dialogue process. For example, the summary for the May 21–23, 1997, meeting included the following account of opening remarks by Michael Parker, ACWA program manager: "Parker expressed his dedication to the dialogue process and stressed that he felt that the program's success lies in the Department of Defense doing business in a new way by incorporating stakeholder involvement at the onset of this program." A subsequent statement by senior DOD official Theodore Prociv "emphasized that the [Dialogue] program is being taken very seriously…"

Dialogue Group Process

As part of the Dialogue process, ground rules were developed and adopted for interactions within the group and interactions with the press and other outside parties. The ground rules created a process to ensure open discussions of difficult and controversial issues. Provisions called for prior notice and comment periods for news releases by participants about the Dialogue, set limits for the discussions to keep the focus on the ACWA program, and agreed that statements in meeting summaries would not be attributed to individual Dialogue members. Overall, these rules were intended to facilitate the free exchange of ideas among participants.[29]

One of the important problems facing the Dialogue is the highly technical nature of the discussions concerning the chemistry and engineering of the alternative technologies. Some Dialogue members have technical backgrounds or can rely on technical staff support, but others have little training or experience in these areas. Representatives of citizen interest groups were at a serious disadvantage. To address this problem, the ACWA program provided funding for the Dialogue to select and hire a technical consultant[30] to provide independent technical information and participate in the Dialogue oversight of the Army's evaluation of the alternative technologies. The technical consultant primarily assisted the members of citizen interest groups.

Another problem was the enormous commitment of time required for participants to develop an understanding of the alternative technologies, digest the available information about the performance of these technologies, and evaluate them based on the selection criteria. In addition, participants had to sign confidentiality agreements to obtain direct access to proprietary information provided by the technology providers. Again, the problem was most acute for the representatives of citizen interest groups. To address the problem, the Dialogue formed a four-member Citizens Advisory

[29]Because the ACWA Dialogue is publicly funded, the ground rules had to conform with the provisions of the Federal Advisory Committee Act. The ground rules were also subject to change over time, as agreed to by the Dialogue participants.

[30]The Dialogue selected SBR Technologies, Inc., of South Bend, Indiana, to provide technical support.

Technical Team (CATT)[31] to work directly with the Army's technical team and report to the citizen interest groups (as well as the entire Dialogue) to ensure that the Dialogue's input was being considered. Throughout the ACWA evaluation process, the CATT conducted daily teleconference calls, and CATT members participated directly in ACWA review and evaluation meetings. Discussions with the citizen interest group participants in the Dialogue indicate that the CATT has increased trust that the Dialogue's input is being utilized as intended.[32]

The Dialogue has been an integral participant in all of the major steps taken to date in the ACWA process, which are listed below:

- drafting of the criteria for alternative technologies that were included in the ACWA RFP (the Go/No Go criteria)
- the evaluation of proposed technologies to determine if they met the Go/No Go criteria and CLIN 0001 awards (data-gap resolution)
- identification of data gaps
- evaluation of the technologies based on the data-gap resolutions and CLIN 0002 awards (development of demonstration plans)
- evaluation of proposed demonstration work plans and CLIN 0003 awards (demonstration)
- observation of (and feedback regarding) the set-up phases of the technology demonstrations

As these steps indicate, the Dialogue has had ample opportunities to provide input into the selection of alternative technologies and to observe that their input has been seriously considered. The ACWA program has thus met the criteria for meaningful input from the Dialogue participants. In the view of the committee, the Dialogue has been a positive step toward gaining broader acceptance for alternative disposal technologies.

[31]The group is also humorously referred to as the watch-CATT. The development of the CATT was discussed at the June 16–17, 1997, meeting of the Dialogue in Lexington, Kentucky.

[32]The formation of the CATT also created some potential problems. The potential for CATT members to become a subgroup of "insiders" with disproportionate influence and access to information raised concerns both among the CATT and other Dialogue members. CATT members sought to avoid this problem by maintaining frequent two-way communications with the other Dialogue members.

Dialogue Group Outreach

One purpose of the Dialogue process is to ensure broad public involvement in the ACWA program's discussions. The Dialogue's success depends on Dialogue members maintaining communications with the communities they represent and the larger public. One of the Dialogue's continuing efforts has been to encourage participants to develop and maintain outreach efforts to the public, including public meetings (two of which were held jointly with this committee), mailings, and press releases. At this early stage, no reliable data are available to determine how well the Dialogue has succeeded in engaging the broader public.[33]

Challenges Facing the Dialogue

The ACWA Dialogue is a work in progress. ACWA and the Dialogue have accomplished a great deal so far, but the most difficult tasks (including selecting alternative technologies) lie ahead. As the ACWA process moves forward, the Dialogue will face a number of challenges, some of which simply come with the territory (this is a controversial issue, after all). Many potential problems have been identified by the Dialogue and ACWA officials and are the subject of ongoing discussions. The committee has highlighted some of the challenges below in hopes that they will be addressed constructively.

Scheduling Issues. The CWC deadline for chemical weapons destruction allows very little leeway for the development and testing of new technologies. Many of the proposed technologies, however, are relatively immature, and experience suggests that bringing them to maturity will take substantial time. If the schedule does not allow enough time for the development of promising technologies (or systems of technologies), the consensus within the Dialogue is likely to be seriously strained. Under the terms of the CWC, a country can appeal for a five-year extension of the deadline. Members of the Dialogue disagree as to whether an appeal will be necessary or would be appropriate. Regardless

[33]This assessment might be facilitated by future surveys by the Program Manager for Chemical Demilitarization, which are currently in the planning stages.

of whether an extension is sought, the issue of time for the development of alternative technologies is likely to make scheduling a contentious issue.

Future Role of the Dialogue Group. In the initial stages of the ACWA process, the role of the Dialogue was relatively clear: to provide input into the process of selecting alternative technologies. As the ACWA process moves forward, it is not clear how the role of the Dialogue will evolve. If ACWA succeeds in identifying and demonstrating promising technologies, the Dialogue may be an invaluable help in furthering public and regulatory acceptance. Taking steps now to ensure that the Dialogue can play a constructive part in implementing alternative technologies is an urgent task.

Funding for Demonstrations. At the time of this writing, ACWA has been able to fund demonstrations of three of the six alternative technologies that received CLIN 002 awards. Limited funding at this stage has created two kinds of strain within the Dialogue. First, among opponents of incineration, the Army's inability to fund what are perceived to be promising alternatives has created a concern that the Army is not seriously interested in alternatives to incineration. Second, among proponents of incineration, the limited funding has raised fears that the resources necessary for design and construction of baseline incineration facilities may be shifted to alternative technologies. Thus, the constrained funding situation could exacerbate policy disagreements and undermine the ability of the Dialogue to reach and maintain consensus.

Funding for Technical Advisors. In March 1999, funding for the CATT technical advisor was nearly exhausted. The loss of the technical advisor would have meant that the CATT (and the nontechnically trained Dialogue members) would cease to have independent technical advice, which could have eroded the trust of some Dialogue members that technical decisions reflect Dialogue input and increased the strain on the entire Dialogue process. Fortunately, the program manager for ACWA was able to identify funding for the technical advisor through September 1999. The constructive role of the Dialogue is contingent upon how future challenges like this are met.

Implications of the ACWA Dialogue for Public Acceptance

This discussion is limited to the implications of Dialogue participation on public acceptance of alternatives to incineration because the implications for public acceptance of the baseline incineration process may depend on the as yet unknown outcome of the ACWA process.[34] If one or more viable alternative technologies are identified through the ACWA process, and if the Dialogue concurs in that assessment, the organized interest groups that have opposed incineration in the past may well support the alternatives. One of the chief procedural objections to incineration—the absence of meaningful public input—will have been addressed by the Dialogue. Furthermore, representatives of several of the groups most opposed to incineration are active participants in the Dialogue.[35]

Support for a viable chemical weapons destruction technology by these groups would fundamentally change the pattern of politics in the chemical weapons program. First, rather than increasing the fears and opposition of the public, news media reports of former opponents to incineration who support an alternative, viable technology would be likely to assuage public concerns (Jenkins-Smith and Silva, 1998). Second, these same groups are unlikely to mount the kinds of administrative and judicial challenges to the alternative technologies that have plagued baseline incineration.[36] Thus, in the best of all possible outcomes—where the ACWA process identifies one or more viable

[34]For example, if the ACWA process does not identify a viable alternative to baseline incineration, and if the Dialogue concurs with that finding, organized opposition to incineration may well decrease. Alternatively, if the Dialogue does not concur with an ACWA finding that there are no viable alternatives to incineration, organized opposition to incineration could be increased.

[35]This fact has led one pro-incineration member of the Dialogue to object that the Dialogue has "institutionalized a citizen lobbying group against incineration" (Robbins, 1998).

[36]An illustration of this effect has been the relatively speedy process of permitting for the chemical weapons destruction facilities in Aberdeen and Newport. The permit for the Aberdeen facility was expedited through a "high priority permitting process" that coordinated the actions of the four agencies required to grant permits and was completed in February 1999. According to Ed Hammerberg, chief of the Regulations and Permitting Division, Indiana Hazardous Waste Management Administration, public support for the use of neutralization reduced opposition and delays substantially. In his view, "the public wanted low pressure and low temperature, and they got it" (Hammerberg, 1998).

alternatives and the Dialogue agrees—the ACWA Dialogue process is likely to increase public support for an alternative technology.

Nevertheless, Dialogue support for one or more alternative technologies certainly does not guarantee public acceptance of those technologies for chemical weapons destruction. The general conditions that make garnering public acceptance so difficult will remain, and new organized groups may arise to oppose alternatives to incineration.[37] Moreover, strains in the Dialogue (over scheduling, funding limits, and the role of the Dialogue) could undermine its future effectiveness. Thus, although the Dialogue has been a positive force for public acceptance, it should not be seen as a panacea for resolving the chemical weapons disposal logjam.

On balance, this committee considers the Dialogue an important innovation and a critical component in the ongoing ACWA process. The committee recommends that support for the Dialogue be continued and that ACWA give serious consideration now to the future role of the Dialogue.

TECHNOLOGY ATTRIBUTES AND PUBLIC ACCEPTANCE

The committee is not aware of any reliable data on public perceptions of the acceptability of alternative technologies for chemical weapons destruction. However, based on public statements from public hearings and the positions of organized interest groups, the committee arrived at several general findings.

First, objections to baseline incineration have centered on the uncertainties associated with large volumes of emissions (gas) that are released into the atmosphere without prior testing of the entire volume to ensure that no hazardous substances are present. Therefore, any alternative technology that reduces the volume of emissions and facilitates testing before release will be more likely to gain public acceptance. Note that the providers of all of the alternative technologies evaluated in this report have incorporated this characteristic into critical parts of their processes. However, all of the alternative technologies also have emissions and/or effluents that are not subjected to a hold-test-release procedure. If serious opposition is mounted, all of these technologies would be vulnerable to the same kinds of challenges that have been raised about baseline incineration.

Second, if the characteristics of the alternative technology are sufficiently similar to those of incineration, they might be confused with incineration, and public acceptance may be affected. This is particularly likely if an issue-public perceives the alternative as incineration under another guise. If these similarities are present, the uphill battle for public acceptance will become notably steeper. In the committee's judgment, the plasma-arc-based technology is most vulnerable to this kind of confusion.

Third, the "unknown risks" associated with chemical weapons destruction for all technologies have a common technical source. For baseline incineration, the perceived unknown risk is that the monitoring of emissions is not continuous. For example, monitoring and testing for dioxins and other trace organics (PICs) tends to be conducted only periodically. The existing literature (e.g., Costner and Thornton, 1990; Costner 1994) and discussions with citizen activists indicate that this concern is the underlying reason for calls for technologies with hold-test-release systems. Given that none of the technologies evaluated in this report can fully implement this strategy, addressing the risk will require that continuous monitoring be used to ensure that the entire system is operating as required.

This principle of "continuous performance assurance" is applicable to all chemical weapons destruction technologies. Public acceptance of any alternative technology will be enhanced if it can be demonstrated that it is more amenable to continuous performance assurance than its technological rivals, including incineration. By the same token, failure to demonstrate that continuous performance assurance will make any technology package vulnerable to the same kinds of objections that have been raised about incineration.

Fourth, the unknown risks associated with chemical weapons destruction have an important social aspect. As discussed earlier in this chapter, trust is a crucial ingredient in the broad acceptance of potentially hazardous programs. If the affected communities do not trust the Army or its contractors, even the best

[37]Note that proponents of incineration can use the same kinds of tactics in opposition to alternative technologies that have been used by opponents of incineration.

monitoring and testing technology will not allay public fears or generate public acceptance. In these circumstances, public acceptance may require continuous *public* monitoring of the chemical weapons destruction process, regardless of the technology. In other words, providing members of the public (or their representatives) with access to real-time monitoring data and establishing clear protocols for responding to emission violations can reassure the public that a technology is operating safely.[38]

In the presence of significant distrust, meaningful public monitoring of real-time data could be facilitated in several ways. First, providing for community representation on a public monitoring team may assure the community that its interests are fully represented in the monitoring process. The options for providing community representation vary depending on the affected community, but they could include (1) having political representatives of the community (e.g., mayors or county commissioners) name representatives to the monitoring team; (2) having organized citizens groups name representatives; or (3) a combination of the above (like the Dialogue itself). Second, public confidence in monitoring may require that independent expert observers be available for the nontechnical members of the monitoring team. Funding for these experts by DOD could facilitate access to independent observers responsive to the community (or its representatives). This arrangement would impose financial and administrative costs on DOD, but it would reassure the affected public in the face of deep-seated fears and a lack of trust in program managers. Although continuous public monitoring may not be necessary or appropriate in all cases, it may be necessary for public acceptance in the communities that are most distrustful of the Army's chemical weapons disposal program.

[38]Public monitoring will be more meaningful if the system of technologies is relatively transparent to a lay observer so that an observer can "make sense" of the process.

11

Summary, Findings, and Recommendations

This chapter summarizes the operating characteristics of the seven technology packages and then presents general findings and recommendations that have broad applicability across the technologies. (Findings specific to each technology package are presented in Chapters 3 through 9.)

SUMMARY OF THE OPERATING CHARACTERISTICS OF THE TECHNOLOGY PACKAGES

Each of the seven proposed technology packages represents a unique combination of technologies for destroying assembled chemical weapons. In Chapters 3 through 9, these packages were examined in detail. Because the information contained in those chapters is quite extensive, the committee decided it would be useful to highlight some of the key points in a general summary. Thus, Table 11-1 has been developed to summarize the fundamental operational characteristics of the seven technology packages.

GENERAL FINDINGS AND RECOMMENDATIONS

Because the munitions contain both chemical agents and energetic materials in various configurations, the destruction of assembled chemical weapons is an extremely complex undertaking. The committee has examined the packages proposed by the seven technology providers in detail and evaluated them according to the criteria set forth in Chapter 2. The following general findings and recommendations are applicable to all of them. Recommendations are listed at the end of the section, with references to the associated findings. These findings and recommendations should be considered together and not quoted out of context.

General Findings

General Finding 1. The chemistries of all four of the primary technologies, (hydrolysis, SILVER II, plasma arc, and SET) as proposed, can decompose the chemical agents with destruction efficiencies of 99.9999 percent. However, each technology package raises other technical issues that must be resolved. One of the crucial issues is the identity and disposition of by-products.

General Finding 2. The technology base for the hydrolysis of energetic materials is not as mature as it is for chemical agents. Chemical methods of destroying energetics have only been considered recently. Therefore, there has been relatively little experience with the alkaline decomposition of ACWA-specific energetic materials (compared to experience with chemical agents). The following significant issues should be resolved to reduce uncertainties about the effectiveness and safety of using hydrolysis operations for destroying energetic materials:

- the particle size reduction of energetics that must be achieved for proper operation
- the solubility of energetics in specific alkaline solutions
- process design of the unit operation and the identification of processing parameters (such as the degree of agitation and reactor residence time) necessary for complete hydrolysis

- the characterization of actual products and by-products of hydrolysis as a function of the extent of reaction
- the selection of chemical sensors and process control strategies to ensure that the unit operation following hydrolysis can accept the products of hydrolysis
- development of a preventative maintenance program that minimizes the possibility of incidents during the cleanup of accumulated precipitates

"Neutralization" (i.e., decomposition and detoxification) of chemical agents has been studied since World War I as part of the chemical weapons defense program for the protection of U.S. troops and protection in the event of accidental releases. Hydrolysis, the first approach selected, was used on a large scale to neutralize the sarin (GB) in cluster bombs destroyed at Rocky Mountain Arsenal in the 1970s.

However, the standard method of destroying explosives and propellants has been open-air burning or detonation. Because chemical methods of destroying energetics have only been considered recently, there has been relatively little experience with the alkaline decomposition of ACWA-specific energetic materials (compared to experience with chemical agents). Most of the work on base hydrolysis of TNT focused on its precipitation from "pink water" (see Appendix E). Almost all of the literature on the base hydrolysis of other energetic materials was conducted with dilute solutions that were well within the solubility limits of these materials. Even though, several undesirable products and precipitates resulted, the qualitative (rather than quantitative) understanding of these reactions suggests that the use of strong base is probably the most efficient way to ensure that hydrolysis is driven to completion.

As shown in Appendix E, the reaction of some energetics with bases is much slower than the reaction of chemical agents. In most cases, the rate of reaction is limited by the rate of dissolution of the energetic materials, which are only slightly soluble in water.

General Finding 3. The conditions under which aromatic nitro compounds, such as trinitrotoluene (TNT) or picric acid, will emulsify in the aqueous phase and not be completely hydrolyzed are not well understood. Therefore, this type of material could be present in the output stream from an energetic hydrolysis step.

In Appendix E, the products of pressurized alkaline hydrolysis of some typical propellants are shown to be dependent on the additives in the compositions. Some additives in propellants P1–P5 did not completely react. For example, diphenylamine (DPA) and centralite precipitated as solid residues or appeared as emulsions in the liquid phase. The most problematic component was found to be dinitrotoluene (DNT). In experiments performed with pure 2,4-DNT, only 7 percent of the nitrogen was found as nitrite in the liquid phase. No DNT was found in the solid residue. It is believed that the DNT was not completely decomposed and might still have been present as an emulsion in the aqueous phase (Bunte et al., 1997). Emulsified components, such as DNT and DPA, would have to be removed before any subsequent unit operations (e.g., aerobic biotreatment) could proceed. Compounds such as TNT and tetryl (both of which are present in assembled chemical weapons) as well as picric acid, nitrated phenols, or nitrated cresols (all of which could be formed during hydrolysis of the energetics in these weapons) are expected to behave in a similar fashion.

General Finding 4. The products of hydrolysis of some energetic materials have not been characterized well enough to support simultaneous hydrolysis of different kinds of energetic materials in the same batch reactor.

Lead stearate, an additive in M28 propellant, is insoluble in water at ambient temperature, but soluble in hot alcohol (Sax and Lewis, 1987). If lead stearate dissolves in hot alkaline solution, then the lead cations could combine with other anionic substrates in a batch reactor and precipitate out sensitive compounds. This possibility is supported by the results of testing on propellant P3 (Bunte, et al., 1997) discussed in Appendix E. For example, picric acid will be formed during hydrolysis of the TNT or tetryl contained in the M55 rocket bursters. If bursters and propellant are hydrolyzed simultaneously, lead from the propellant could either precipitate out or form lead picrate. In the hydrated form, lead picrate is not particularly sensitive. However, enough heat could be produced from this exothermic process to heat and dehydrate the lead picrate deposited on vessel walls. As indicated in the TNT hydrolysis section of Appendix E, dry lead picrate is an extremely sensitive explosive and is very

TABLE 11-1 Summary of the Key Process-Engineering Data for the Seven Technology Packages

	General Atomics	Lockheed Martin	Parsons/AlliedSignal	AEA	ARCTECH	Burns & Roe	Teledyne-Commodore
Disassembly							
Approach	Baseline system with modifications.	Baseline system with modifications.	Baseline system with modifications.	Baseline system with modifications.	Baseline system with modifications.	Baseline system with modifications.	High-pressure fluid-jet cutting with liquid ammonia at 2,722 atm (40,000 psig).
Modifications	Projectiles and Mortars: cryofracture of downloaded munitions to access agent.	Rockets: Water-jet to remove and break up propellant. Projectiles: Projectile bodies turned upside-down to drain agent; bodies flushed with NaOH solution.	Rockets: Water-jet cutting to remove fuze and separate motor and warhead; water-jet to wash out propellant and burster. Projectiles: water jet to wash out burster.	Rockets: six cuts; sheared-parts handling machine to segregate rocket parts.	Rockets: ten cuts; water-jets may be used for wash-out of energetics.	Number of munitions per tray limited to nine.	N/A
Primary Treatment of Agent/Energetics							
Technology	For mustard, hydrolysis at near-neutral pH. For GB, VX, and energetics, hydrolysis in NaOH solution.	For mustard, hydrolysis at near-neutral pH. For GB, VX, and energetics, hydrolysis in NaOH solution.	For mustard, hydrolysis at near-neutral pH. For GB, VX, and energetics, hydrolysis in NaOH solution.	Electrochemical oxidation with Ag(II).	For mustard, hydrolysis at near-neutral pH in presence of humic acid. For GB, VX, and energetics, base (KOH) hydrolysis in presence of humic acid.	Thermal decomposition in a plasma heated furnace. Plasma feed-gas is argon, nitrogen or carbon dioxide. Explosives initiated in an explosion chamber; residual material fed to plasma furnace.	Solvated-electron reduction by solution of sodium in liquid ammonia, followed by hydrolysis of the condensed-phase reaction products.
Operating Temperature	60 to 90°C	90 to 135°C	60 to 90°C	60 to 90°C	90°C	15,000°C in the actual plasma; gases exit at around 1100°C.	20°C
Operating Pressure	Ambient	Ambient for most cases; slightly pressurized when at 135°C case.	Ambient	Ambient	Ambient	Ambient	10.5 atm
Electric Power	Nominal	Nominal	Nominal	9.15 MW peak	Nominal	> 6.5 MW	Nominal
Processing Mode	Batch	Batch	Batch	Continuous (periodic slag removal)	Batch	Continuous and batch	Batch

Secondary Treatment of Agent/Energetics							
Technology	SCWO	SCWO using a transpiring platelet wall reactor; gas-phase chemical reduction (GPCR) of hydrolysis off-gases.	Bioremediation with AlliedSignal's immobilized cell bioreactor; liquids further treated with UV/peroxide; gases pass through catalytic oxidizer before scrubbing.	None	For mustard, oxidation with hydrogen peroxide; for VX and GB, oxidation with Fenton's reagent; no secondary treatment for energetics.	Neutralization of off-gas contamination (NO_X, SO_X) with caustic.	Hydrolysis of the sodium with water; sodium persulfate oxidation for agents; copper-catalyzed sodium persulfate/hydrogen peroxide oxidation for energetics.
Operating Temperature	650°C	780°C for SCWO; > 870°C for GPCR.	20°C for biotreatment; 450°C for catalytic oxidation.	N/A	90°C	90°C	95 to 100°C
Operating Pressure	231 atm	238 atm for SCWO; ambient for GPCR.	Ambient	N/A	Ambient	Ambient	Ambient
Electric Power	Nominal	Nominal	Nominal	N/A	Nominal	Nominal	Nominal
Processing Modes	Continuous	Continuous	Continuous	N/A	Batch	Continuous	Batch
Treatment of Fuzes	Fed to rotary hydrolyzer and heated to discharge conveyor.	Fed to hydrolysis vessels.	Sent to detonation chamber.	Dissolution in nitric acid; treatment with SILVER II.	Dissolution in a-HAX.	Detonation or burning in an explosion chamber.	Detonation in an explosion chamber.
Treatment of Metal Parts							
Technology	Rotary hydrolyzer followed by heated discharge conveyor.	Washed with caustic and reduced with hydrogen to eliminate organic matter.	Heated in superheated steam.	Cleaned with high-pressure water jets; heated in the metal parts decontamination chamber.	Washed with a-HAX; shipped for smelting.	Melted in the PWC; metal turned into ingots.	Metals shredded; treated in SET reactor; shipped for smelting.
Operating Temperature	> 540°C (1,000°F)	> 540°C (1,000°F)	> 540°C (1,000°F)	> 540°C (1,000°F)	90 to 100°C	> 1,000°C (1,800°F)	20°C
Final Condition	5X	5X	5X	5X	3X	5X	3X
Treatment of Dunnage							
Approach	Pallets, boxes, bulk: shredded, hydropulped, hydrolyzed, treated with SCWO.						
	Plastic and Rubber: sliced, cryocooled, shredded, hydrolyzed, SCWO.	Washed in NaOH solution and pyrolyzed in a GPCR reactor.	Shredded and heated in superheated steam.	Shredded, pulped, mixed with dilute nitric acid, treated with SILVER II.	Shredded and treated by base hydrolysis.	Shredded and passed through a PWC.	Shredded SET treatment.

continued

TABLE 11-1 *Continued*

	General Atomics	Lockheed Martin	Parsons/AlliedSignal	AEA	ARCTECH	Burns & Roe	Teledyne-Commodore
Feed Streams							
To primary treatment	Liquid nitrogen, sodium hydroxide, water.	Sodium hydroxide, water.	Sodium hydroxide, water, abrasives.	Nitric acid, sodium hydroxide, oxygen, calcium nitrate.	Potassium hydroxide, humic acid.	Argon, carbon dioxide, or nitrogen.	Ammonia, sodium, water, abrasives.
To secondary treatment	Air, kerosene.	Water, oxygen, isopropyl alcohol.	Microorganisms, nitric acid, air, water, dextrose, diammonium phosphate, ammonium chloride, peroxide.	N/A	Hydrogen peroxide, ferrous sulfate, cement.	Caustic	Sodium persulfate or hydrogen peroxide.
Solid Waste Streams							
HD	Filter cake consisting of sodium chloride and sodium sulfate.	Filter cake consisting of sodium chloride and sodium sulfate.	Biomass, sulfates, chlorides, ammonia; salts produced by evaporating solutions.	Salts, including Ca_2SO_4	Precipitate and residual reaction liquor solidified and stabilized; sent to a hazardous-waste landfill for disposal.	Slag sent to a landfill; solids from the brine sent to a landfill.	Stabilization in cement and sent to a landfill.
GB	Filter cake consisting mainly of sodium phosphate and sodium fluoride.	Filter cake consisting mainly of sodium phosphate and sodium fluoride.	Biomass, phosphates, fluorides, ammonia; salts produced by evaporating solutions.	Calcium phosphate	Precipitate and residual reaction liquor solidified and stabilized; sent to a hazardous-waste landfill for disposal.	Slag sent to a landfill; solids from the brine sent to a landfill.	Stabilization in cement and sent to a landfill.
VX	Filter cake consisting mainly of sodium phosphate and sodium sulfate.	Filter cake consisting mainly of sodium phosphate and sodium sulfate.	Biomass, sulfates, phosphates, ammonia; salts produced by evaporating solutions.	Salts containing calcium phosphate, calcium.	Precipitate and residual reaction liquor solidified and stabilized; sent to a hazardous-waste landfill for disposal.	Slag sent to a landfill; solids from the brine sent to a landfill.	Stabilization in cement and sent to a landfill.
Energetics	Included in above.	Salts, including sodium nitrates, sodium nitrites, acetone, acetic acid.	Biomass, nitrites, ammonia; salts produced by evaporating solutions.	Salts	Precipitate and residual reaction liquor solidified and stabilized; sent to a hazardous-waste landfill for disposal.	Slag sent to a landfill; solids from the brine sent to a landfill.	Stabilization in cement and sent to a landfill.
Liquid Waste Streams							
HD	None	None	None	Concentrated and dilute nitric acid.	Solidified and stabilized; sent to a hazardous-waste landfill for disposal.	Liquids from the PAS scrubber processed to recover salts.	Evaporation and water recycling.

GB	None	None	None	Concentrated and dilute nitric acid.	Solidified and stabilized; sent to a hazardous-waste landfill for disposal.	Liquids from the PAS scrubber processed to recover salts.	Evaporation and water recycling.
VX	None	None	None	Concentrated and dilute nitric acid.	Solidified and stabilized; sent to a hazardous-waste landfill for disposal.	Liquids from the PAS scrubber processed to recover salts.	Evaporation and water recycling.
Energetics	None	None	None	Concentrated and dilute nitric acid.	Solidified and stabilized; sent to a hazardous-waste landfill for disposal.	Liquids from the PAS scrubber processed to recover salts.	Evaporation and water recycling.
Gaseous Waste Streams							
HD	Carbon dioxide, nitrogen with traces of organic compounds and carbon monoxide, discharged through HEPA and activated carbon filters.	From GPCR, hydrocarbons, hydrogen, HCl; H_2S to scrubber; fuel gas to boiler.	CO_2, air sent to catalytic oxidizer; cooled; scrubbed; passed through carbon filter.	Off-gases from the electrolytic reactions scrubbed and released to the air.	None	Gases from the PWC scrubbed and burned in boiler.	Scrubbed and burned in boiler.
GB	Carbon dioxide, nitrogen with traces of organic compounds and carbon monoxide, discharged through HEPA and activated carbon filters.	From GPCR, hydrocarbons, hydrogen, HCl; H_2S to scrubber; fuel gas to boiler.	CO_2, gases go to catalytic oxidizer; cooled; scrubbed; passed through carbon filter.	Off-gases from the electrolytic reactions scrubbed and released to the air.	None	Gases from the boiler scrubbed and released in boiler.	Scrubbed and burned in boiler.
VX	Carbon dioxide, nitrogen with traces of organic compounds and carbon monoxide, discharged through HEPA and activated carbon filters.	From GPCR, hydrocarbons, hydrogen, HCl; H_2S to scrubber; fuel gas to boiler.	CO_2, air sent to catalytic oxidizer; cooled; scrubbed; passed through carbon filter.	Off-gases from the electrolytic reactions scrubbed and released to the air.	None	Gases from the boiler scrubbed and released to the air.	Scrubbed and burned in boiler.
Energetics	Carbon dioxide, nitrogen with traces of organic compounds and carbon monoxide, discharged through HEPA and activated carbon filters.	From GPCR, hydrocarbons, hydrogen, HCl; H_2S to scrubber; fuel gas to boiler.	CO_2, air sent to catalytic oxidizer; cooled; scrubbed; passed through carbon filter.	Off-gases from the electrolytic reactions scrubbed and released to the air.	None	Gases from the boiler scrubbed and released to the air.	Scrubbed and burned in boiler.

dangerous to handle. Therefore, the committee believes that, to avoid forming sensitive compounds such as lead picrate, hydrolysis of bursters and propellant should be performed in separate vessels.

General Finding 5. The primary chemical decomposition process in all of the technology packages produce environmentally unacceptable reaction products. Therefore, all of the packages are complicated processes that include subsequent treatment step(s) to modify these products.

General Finding 6. The waste streams of all of the ACWA technology packages could contain very small amounts of hazardous substances (besides any residual chemical agent). These substances were not fully characterized at the time of this report; therefore, all waste streams must be characterized to ensure that human health and the environment are protected. If more than one phase (gas, liquid, or solid) is present in a waste stream and, each phase should be characterized separately.

All of the alternative technology packages appear to be capable of meeting the current destruction efficiency limits for agent and hazardous materials of regulatory concern. However, they may create new pollutants that could have adverse environmental effects. Therefore, complete characterizations of the process effluents (solids, liquids, and gases) from the secondary-treatment waste streams will be essential. Characterization may require pilot-scale operation of the integrated processes before a final conclusion can be determined on environmental acceptability.

The waste streams of all of the proposed technology packages (gas, liquid, and/or solid) may contain small amounts of hazardous materials, even under normal operating conditions (this is a characteristic of virtually any industrial chemical process). To ensure that no toxic effluent is accidentally discharged, all waste streams must be monitored. In the committee's opinion, all of the packages are fundamentally capable of being monitored to ensure the protection of human health and the environment. (Although the detection and analysis of trace substances can be done to very low levels, no monitoring or analytical method can guarantee a true zero level of any known or unknown compound.)

General Finding 7. None of the proposed technology packages complies completely with the hold-test-release concept for all gaseous effluents (both process and ventilation effluents).

General Finding 8. Hold-test-release of gaseous effluents may not ensure against a release of agent or other hazardous material to the atmosphere. No evidence shows that hold-test-release provides a higher level of safety than current continuous monitoring methods for gaseous streams with low levels of contamination. Furthermore, none of the technologies provides for hold-test-release of effluents from ventilation systems that handle large volumes of gases from contaminated process areas.

In an earlier report on alternative technologies the NRC noted that,

> The risk of toxic air emissions can be virtually eliminated for all technologies through waste gas storage and certification or treatment by activated-carbon adsorption. Either of these options can be combined with methods to reduce the volume of gas emissions (NRC, 1993).

Some of the technology packages include hold-test-release steps of gaseous process effluents (1) when the effluent stream flow rates are relatively low or (2) when the effluent streams occur in batches that can be easily contained. For continuous gaseous effluent streams that have high flow rates (e.g., from SCWO units and exhaust gas from biotreatment units), elaborate designs would be required to incorporate a hold-test-release step.

The committee believes that the hold-test-release step is not a panacea for ensuring that gaseous effluents are free of agent or other hazardous materials. Some low-concentration hazardous volatile materials may adsorb onto internal tank surfaces or be absorbed into liquids or solids in holding tanks where they may escape detection. When the holding tank is vented to the atmosphere, these undetected materials may be desorbed and released to the environment. To the committee's knowledge, no experiments have been performed to demonstrate that these phenomena do not occur. Moreover, if a process upset contaminates the holding tanks, decontamination (and verification of decontamination) may present significant technical difficulties.

All of the ACWA technology packages include the baseline continuous monitoring and filtering systems (ACAMS and DAAMS) for air ventilation in process areas where contamination is expected. To add hold-test-release steps for this large volume of air flow would be very difficult and probably would entail very expensive (and impractical) design.

Continuous monitoring coupled with interlocks that shut down the process quickly if concentration limits are exceeded may be just as reliable as hold-test-release steps for protecting human health and the environment, especially for large-volume effluent streams. Whatever approach is adopted will require additional testing to demonstrate its viability and effectiveness. Tests may very well show that hold-test-release steps do not ensure safety any better than simpler continuous monitoring methods combined with robust process controls (i.e., continuous performance assurance).

General Finding 9. Solid salts will be hazardous waste, either because they are derived from hazardous waste (see Chapter 2) or because they leach heavy metals above the levels allowed by the Resource Conservation And Recovery Act Toxicity Characteristic Leaching Procedure. Stabilization—mixing waste with a reagent or reagents to reduce the leachability of heavy metals—will probably be required before the salts can be sent to a landfill. The potentially high chloride and nitrate content of these salts will make the waste difficult to stabilize, and treatability studies will be necessary to determine a proper stabilization formula.

General Finding 10. Testing, verification, and integration beyond the 1999 demonstration phase will be necessary because the scale-up of a process can present many unexpected challenges, and the ACWA demonstrations were limited in nature.

The reasons supporting this finding are discussed below. The ACWA demonstrations tested only the unit operations that DOD believed were most critical or least proven for that technology package. However, other unit operations may also require additional development before full-scale implementation can proceed.

Second, the accelerated ACWA schedule required that the demonstrations be relatively short. Thus, the longer-term reliability of the processes could not be evaluated. In addition, the duration of the demonstrations may have been too short to characterize fully the steady-state operational behavior, the buildup of trace materials in recycling loops, and problems with corrosion. Longer lasting tests with the full range of materials to be processed will be necessary for identifying the best materials of construction.

Third, the demonstrations did not include interfacing the unit operations into a complete system (i.e., when the output stream of one process step becomes the input stream of the next) when unexpected problems often arise. For example, scheduling is especially difficult to design when a batch or semibatch process (e.g., the hydrolysis reactors) is coupled with a continuous process (e.g., the SCWO reactor). Incomplete processing in one stage may cause contamination or a materials incompatibility in a later stage. Also, a bottleneck can be created if one step does not achieve the expected throughput. Therefore, for each piece of equipment, the implications of operating with input streams that are off specification, that are not moving at the design flow rate, or that are completely blocked must be tested.

Fourth, scale-up of a process is not always linear. Although the scale-up of some types of standard chemical process equipment can be straightforward, the scale-up of new equipment designs can raise problems if not all parts of the process scale in the same way. For example, many mass-transfer processes scale with length. Surface wash-out, heat transfer, and other surface phenomena scale with surface area. Homogeneous chemical reactions scale with reactor volume. When all these phenomena occur simultaneously (in a hydrolysis or SCWO reaction vessel), the different scaling properties must be accommodated in the design.

Fifth, impurities that are not detected in small-scale tests may be evident in larger scale tests. Because of the limited quantities used in the small-scale and demonstration-phase tests, trace impurities in the waste streams may not be detectable. The impurities and small amounts of intermediates that are produced in a full-scale (or near full-scale) plant are not necessarily the same as those observed during laboratory or bench-scale experiments, or even during demonstration-scale tests. Small excursions or variations in the conditions under which a reaction is run can also alter the nature or the amount of trace impurities that are produced. Because the scale of the demonstration testing is limited, the quantity of trace impurities in the reaction products may be too small for detection analysis. Operation at near full-scale would reveal trace impurities that may require process changes to mitigate or eliminate them.

Sixth, materials-handling equipment was generally not evaluated during the ACWA demonstrations. In the scale-up operations of waste treatment facilities, the materials-handling equipment, such as equipment for sizing and feeding waste, has been the weak link in the operational chain. The materials-handling equipment will have to be tested and evaluated prior to full-scale implementation to ensure that integrated facilities operate properly.

This additional testing, verification, and integration could be done in a pilot-scale facility. However, even the construction and operation of a pilot-scale facility will not necessarily ensure a trouble-free start-up of a full-scale facility. Industrial experience suggests that unanticipated problems will occur at full-scale in spite of pilot-plant experience.

General Finding 11. Although a comprehensive quantitative risk assessment (QRA), health risk assessment (HRA), and ecological risk assessment (similar to assessments performed for the baseline process) cannot be completed at this stage of process development, these assessments will have to be performed and refined as process development continues.

All of the proposed destruction systems are in the conceptual design stage, which means that many design details have not been developed. At this stage, only *qualitative* risk assessments could be done, and all of the technology providers prepared preliminary hazard analyses that qualitatively describe potential accidents.

General Finding 12. The "optimum" system for a particular chemical weapons storage depot might include a combination of unit operations from the technology packages considered in this report.

The technology packages proposed for the ACWA program address the destruction of assembled chemical weapons at the hypothetical depot described in the RFP. The actual depots under consideration have very different munition inventories. For example, the Pueblo Chemical Depot has only mustard-filled projectiles and mortars in its inventory. The Blue Grass Army Depot in Richmond, Kentucky, however, has a large inventory of M55 rockets, which contain GB or VX. Some of the components and processes in the proposed systems are very effective for one or another portion of the overall demilitarization process. Technology packages may also differ in their applicability to particular munitions and particular chemical agents.

This discussion notwithstanding, the committee's task is to evaluate the technology packages, as proposed, for the hypothetical depot. The committee did not, therefore, consider "mixing and matching" component technologies for specific sites or munitions.

General Finding 13. Some of the ACWA technology providers propose that some effluent streams be used commercially. New or modified regulations may have to be developed to determine if these effluent streams can be recovered or reused.

According to current Army standards, a solid material that has not been subjected to 5X treatment can only be disposed of in a hazardous-waste facility. If a process under consideration produces a waste stream that could be reused by, for example, a metal reclaimer or a fertilizer plant, this waste stream would have to be subjected to 5X treatment. To date, liquids and gases from chemical demilitarization processing have not been recycled or reused commercially; therefore, existing standards may have to be reexamined.

General Finding 14. An extraordinary commitment of resources will be necessary to complete the destruction of the assembled chemical weapons stockpile in time to meet the current deadline using any of the ACWA technology packages. This would demand a concerted national effort. It is unlikely that any of the technology packages could meet this deadline.

The chemical-hydrolysis destruction of bulk agents at Aberdeen Proving Ground, Maryland, and Newport, Indiana, are examples of how much time could be required to bring any of the alternative destruction systems from its present state of development to the pilot-plant stage and finally to the production stage. The schedules for the design, construction, and operation of the destruction facilities at these two sites (see Figures 11-1 and 11-2) indicate that the destruction of munitions will be completed by the end of 2004. The ACWA program is approximately three years behind the Aberdeen and Newport schedules, and the development of an acquisition design package for ACWA is

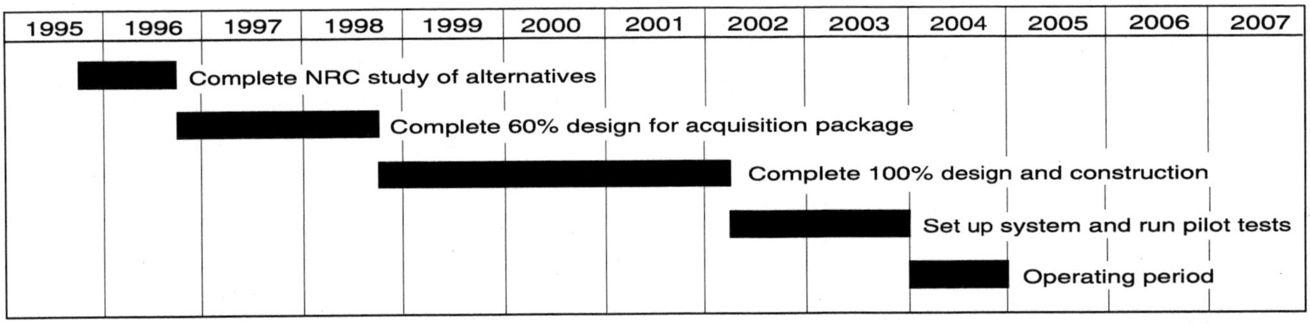

FIGURE 11-1 Schedule for the Aberdeen Chemical Agent Disposal Facility as of January 6, 1999. Source: Adapted from Pecoraro, 1999.

not likely to begin before October 1999; for Aberdeen and Newport, it began in November 1996.

The programs at Aberdeen and Newport are less complex than those required for other sites because the stockpiles at those two sites contain only bulk agents in one-ton containers. Only one agent is stored at each site, and there are no munitions to be disassembled and no energetics to be treated. Thus, the committee expects that the development cycle at other sites such as Richmond, Kentucky, and Pueblo, Colorado, could take even longer because of modifications to the disassembly process and the numerous interfaces between unit operations. In addition, the number of munitions at Pueblo suggests a much longer operating period than at Aberdeen or Newport. Therefore, meeting the April 2007 CWC treaty deadline will be very difficult. (A recent report [Arthur Andersen, 1998] concluded that the baseline incineration technology will also have difficulty meeting the April 2007 deadline. This committee did not evaluate the methodology used by Arthur Andersen to reach this conclusion.)

A "crash program" to expedite the implementation of any of the alternative technology packages is possible, of course. However, this would require significantly more financial resources than have been planned for the disposal sites. (Note that the Aberdeen and Newport designs have already been put on a fast track to conduct pilot-scale testing concurrent with the construction of full-scale facilities to reduce the time to start-up.)

General Finding 15. The Dialogue process for identifying an alternative technology is likely to reduce the level of public opposition to that technology. The committee believes that the Dialogue has been and continues to be a positive force for public acceptance of alternatives to incineration. Although the Dialogue process requires a significant commitment of time and resources, it has been a critical component of the ACWA program to date.

Reducing opposition by the general public or by organized interest groups could reduce the time and resources required to obtain state and federal permits for constructing and operating disposal facilities. For example, the speed with which the Aberdeen facility received permits can be partly attributed to the lack of public opposition (Hammerberg, 1998). The ACWA

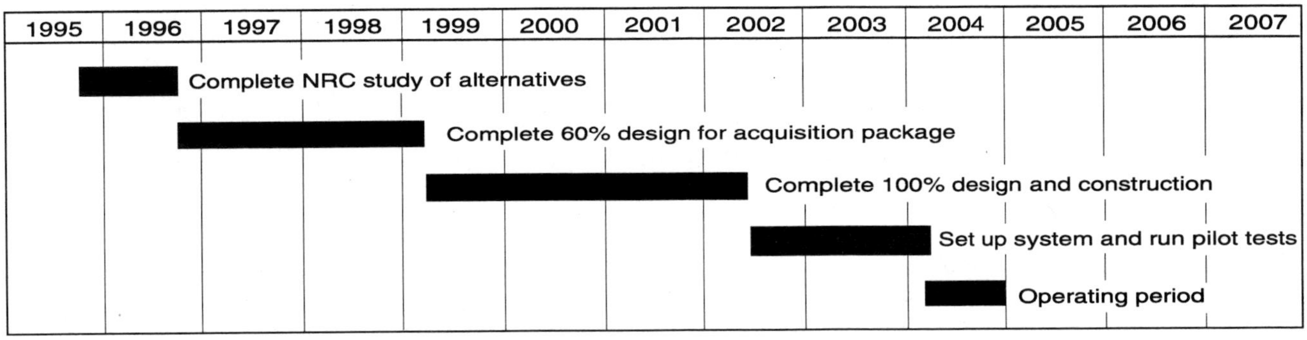

FIGURE 11-2 Schedule for the Newport Chemical Agent Disposal Facility as of January 6, 1999. Source: Adapted from Pecoraro, 1999.

process, as mandated by Congress, has provided a unique opportunity for sidestepping the kind of conflict that has impeded progress in the baseline incineration program. The Dialogue process initiated by the program manager for ACWA is a basis for building trust between DOD officials and citizen and environmental groups that have traditionally been opposed to incineration.

General Finding 16. Although the committee did not have access to scientific data on the attributes of a technology that would be most acceptable to the public, input from members of the active publics and previous research indicates that technologies with the following characteristics are likely to stimulate less public opposition:

- minimal emissions, particularly gaseous
- continuous monitoring of effluents to verify that the process is operating as designed (process assurance measurement)
- provisions for representatives of the local community to observe and participate in the process assurance measurement

General Recommendations

General Recommendation 1. If a decision is made to move forward with any of the ACWA technology packages, substantial additional testing, verification, and integration should be performed prior to full-scale implementation (see General Finding 10).

General Recommendation 2. The sampling and analysis programs at each phase of development should be carefully reviewed to ensure that the characterization of trace components is as comprehensive as possible to avoid surprises in the implementation of the selected technology (see General Finding 6).

General Recommendation 3. If a decision is made to move forward with any of these technology packages, health and safety evaluations should progress from qualitative assessments to more quantitative assessments as the process design matures. Quantitative (QRA), health (HRA), and ecological risk assessments should be conducted as soon as is practical. Early initiation of these assessments will allow findings to be implemented with minimal cost and schedule impact. (See General Finding 11.)

The QRA is a tool for managing risk in the design as it becomes increasingly well defined. In the early stages, QRAs can indicate the systems or unit operations that appear to be major contributors to risk at that stage of design development. If a pilot-facility is constructed, preliminary quantitative, health, and ecological risk assessments should be developed prior to the completion of the pilot facility design. These analyses should then be factored back into the designs and the risk assessments completed before operation of the pilot facility begins. If a full-scale facility is constructed, preliminary risk assessments for the full-scale facility should be developed prior to the completion of the facility design. The preliminary analyses should then be factored back into the full-scale design. These risk assessments should be completed before operation of the facility begins. The QRA should include assessments of public and worker risk, as well as uncertainties. The specific protocol for the HRA and ecological risk assessment will have to be determined in cooperation with state and federal permitting agencies.

General Recommendation 4. Any of these technology packages, or any component of these technology packages, should be selected on a site-specific basis. (See General Finding 12.)

General Recommendation 5. Whatever unit operation immediately follows the hydrolysis of energetic materials should be designed to accept emulsified aromatic nitro compounds, such as TNT or picric acid, as contaminants in the aqueous feed stream. (See General Finding 3.)

General Recommendation 6. Simultaneous processing of different types of energetic materials should not be performed until there is substantial evidence that the intermediates formed from the hydrolysis of aromatic nitro compounds will not combine with M28 propellant additives or ordnance fuze components to form extremely sensitive explosives, such as lead picrate. (See General Finding 4.)

General Recommendation 7. The Department of Defense should continue to support the Dialogue throughout the current ACWA program and should seriously consider the participation of the Dialogue in any follow-on programs.

References

AEA. 1997. Proposal for Assessment of Technologies for Assembled Chemical Weapon Demilitarization. Submitted in response to U.S. Army solicitation number DAAM01-97-R00031, September 1997. Bethel Park, Pa.: AEA Technology.

AEA. 1998a. Data Gaps Resolution Report. Submitted under contract number DAAM01-98-R-0006, February 1998. Bethel Park, Pa.: AEA Technology.

AEA. 1998b. Information provided by Bob Boylston, AEA, to Randy Seeker, NRC ACW Committee, August, 1998.

Alailima, K., J. Capillo, E. Crowe, and M. Tuckey. 1995. Chemical Weapons Disposal: The Citizens' Solution to the Costly Mistake of Incineration. Berea, Ky.: Kentucky Environmental Foundation.

Anderson, J. 1979. Public Policy Making (2nd ed). New York: Holt, Rinehart, and Winston.

ARCTECH and ICF Kaiser. 1998. Data Gap Resolution Technical Report: ACTODEMIL Technology for the Safe, Environmentally Friendly, and Cost Effective Demilitarization of Chemical Weapons. Submitted under Contract No. DAAM01-98-R-0001. Delivery Order #001. February 17, 1998. Chantilly, Va.: ARCTECH, Inc.

Arthur Anderson. 1998. A Schedule Risk Assessment of the Chemical Stockpile Disposal Project. Prepared for the Program Manager for Chemical Demilitarization. Aberdeen Proving Ground, Md.: Program Manager for Chemical Demilitarization.

ASME (American Society of Mechanical Engineers). 1988. Hazardous Waste Incineration: A Resource Document. New York: American Society of Mechanical Engineers.

Batz, M.L., P.M. Garland, R.C. Reiter, M.D. Sanborn, and C.D. Stevenson. 1997. Explosion and ion association chemistry of the anion radicals of 2,4,6-trinitrotoluene, 2,6-dinitrotoluene, and trinitrobenzene. Journal of Organic Chemistry 62(7): 2045–2049.

Bollog, J.M. 1992. Decontaminating soil with enzymes: an in situ method using phenolic and anilinic compounds. Environmental Science and Technology 26: 1876–1881.

Bradbury, J., K. Branch, J. Heerwagen, and E. Liebow. 1994. Community Viewpoints of the Chemical Stockpile Disposal Program. Washington D.C.: Battelle Pacific Northwest Laboratories.

Brown, M. 1993. Public Trust and Technology: Chemical Weapons Destruction in the United States. Washington D.C.: The Committee for National Security.

Bunte, G., H. Krause, and T. Hirth. 1997. Disposal of energetic materials by alkaline pressure hydrolysis and combined techniques. Journal of Propellants, Explosives, and Pyrotechnics. 22 (3): 160–164.

Burns and Roe. 1997. Assembled Chemical Weapon Assessment Program. Proposal submitted in response to U.S. Army Solicitation No. DAAM01-97-R-0031. September 15, 1997. Oradell, N.J.: Burns and Roe Enterprises, Inc.

Burns and Roe. 1998a. Data Gap Resolution Report. Submitted under Contract No. DAAM01-98-D-0007, Task 1. February 1998. Oradell, N.J.: Burns and Roe Enterprises, Inc.

Burns and Roe. 1998b. Final Demonstration Study Plan. Submitted under Contract No. DAAM01-98-D-0007, Task 2. November 1998. Oradell, N.J.: Burns and Roe Enterprises, Inc.

Burns and Roe. 1999. Plasma Waste Converter Technology. Briefing to the Committee on Review and Evaluation of Alternative Technologies for Demilitarization of Assembled Chemical Weapons, Washington, D.C., March 11, 1999.

Caton, J.E., R.T. Williams, and W.H. Griest. 1994. Characterization of insoluble fractions of TNT transformed by composting. Journal of Environmental Science and Health 29(4): 659–670.

Chemical Weapons Working Group. 1994. The Citizens' Solution for Chemical Weapons Disposal. Richmond, Ky.: Chemical Weapons Working Group.

Chrostowski, P. and S. Foster. 1992. Scientific Peer Review of Greenpeace's Position on Hazardous Waste Incineration Impacts in Its "Report on the Hazardous Waste Incineration Crisis" and "Playing with Fire." Fairfax, Va.: Clement International Corporation.

Clark, L.J. 1996. Report on Results of Community Interest Survey on the Destruction of Chemical Weapons at Anniston Army Depot. Jacksonville, Ala.: Center for Economic Development and Business Research, Jacksonville State University.

Cobb, M.D., and J.H. Kuklinski. 1997. Changing minds: political arguments and political persuasion. American Journal of Political Science 41: 88–121.

Cosmedes, L., and J. Tooby. 1994. Better than rational: evolutionary psychology and the invisible hand. American Economic Review 84: 327–332.

Costner, P. 1993. Chemical Weapons Demilitarization and Disposal: Johnston Atoll Chemical Disposal System, GB and VX Campaigns. Washington, D.C.: Greenpeace.

Costner, P. 1994. The Incineration of HD Agent at JACADS: MPF Trial Burn and LIC Demonstration Burn. Washington, D.C.: Greenpeace.

Costner, P., and J. Thornton. 1990. Playing with Fire: Hazardous Waste Incineration. Washington, D.C.: Greenpeace.

Creighton, J., and L. Aggens. 1996. Environmental Managers' Handbook on Public Involvement. Fort Belvoir, Va.: U.S. Department of Energy, Institute for Water Resources.

Cudahy, J. 1992. The Rest of the Story: Literature Research and Technical Review of the Greenpeace Report "Playing with Fire." Knoxville, Tenn.: Focus Environmental, Inc.

Defense Environmental Alert. 1994. Environmentalists Attack as "Politicized" Report Pushing Incineration. January 26, 1994. Washington, D.C.: Defense Environmental Alert.

De Frank, J.J., I.J. Fry, C.M. Foster, and J.P. Early. 1996. Biodegradation of Hydrolyzed Nerve Agent VX in Sequencing Batch Reactors. Pp. 203–209 in Proceedings of the 1996 Edgewood Research Development and Engineering Center Scientific Conference on Chemical and Biological Defense Research. Aberdeen, Md.: Edgewood Research Development and Engineering Center.

Dellinger, B., W.A. Rubey, D.L. Hall, and J.L. Graham. 1986. Incinerability of hazardous wastes. Hazardous Waste and Hazardous Materials 3(2): 139–150.

Dempsey, C.R., and E.T. Oppelt. 1993. Incineration of hazardous waste: a critical review update. Air and Waste 43: 25–73.

Department of the Army. 1997. Toxic Chemical Agent Safety Standards. DA Pam 385-61. March 31, 1997. Washington, D.C.: Department of the Army.

DOD (U.S. Department of Defense). 1998. Assembled Chemical Weapons Assessment Program Annual Report to Congress. December 1998. Aberdeen Proving Ground, Md.: Program Manager for Assembled Chemical Weapons Assessment.

Eissner, C. 1999. Assembled Chemical Weapons Assessment Program. Presentation to the Committee on Review and Evaluation of Alternative Technologies for Demilitarization of Assembled Chemical Weapons, Washington, D.C., March 11, 1999.

EPA. 1998a. Human Health Risk Assessment Protocol for Hazardous Waste Combustion Facilities. EPA530-D-98-001B. July 1998 Agency (5305W): (http://www.epa.gov/epaoswer/hazwaste/combust.htm)

EPA (Environmental Protection Agency). 1998b. Hazardous Waste Combustors; Revised Standards. Final Rule - Part 1: RCRA Comparable Fuel Exclusion. Permit Modifications for Hazardous Waste Combustion Units, Notification of Intent to Comply, Waste Minimization and Pollution Prevention Criteria For Compliance Extensions. June 19, 1998. Washington, D.C.: Environmental Protection Agency.

Erickson, L. 1997. The binding of organic contaminants to humic substances. Centerpoint 4(1): 13–14.

Flynn, J., W. Burns, C.K. Mertz, and P. Slovic. 1992. Trust as a determinant of opposition to a high-level radioactive waste repository: analysis of a structural model. Risk Analysis 12: 417–430.

Futrell, R., and C.G. Davies. 1996. Chemical Weapons Disposal Issue and Citizens' Participation: A Sociological Viewpoint. Berea, Ky.: Kentucky Environmental Foundation.

GA (General Atomics). 1997. Treatability Study: Supercritical Water Oxidation of VX Hydrolysate. Test Report 229003/A. San Diego, Calif.: General Atomics, Inc. (produced under Subcontract PS-02552 from Stone & Webster Corporation, Army Prime Contract No. DAAA-15-91-D-0005).

GA (General Atomics). 1998. Assessment of Technologies for Assembled Chemical Weapon Demilitarization Data Gap Resolution Report. February 17, 1998. San Diego, Calif.: General Atomics, Inc.

GAO (General Accounting Office). 1991. Chemical Weapons: Stockpile Destruction Cost Growth and Schedule Slippages are Likely to Continue. GAO/NSIAD-92-18. November 20, 1991. Washington, D.C.: General Accounting Office.

GAO. 1993. Chemical Weapons Destruction: Issues Affecting Program Cost, Schedule and Performance. GAO/NSIAD-93-50. January 21, 1993. Washington, D.C.: General Accounting Office.

GAO. 1994. Chemical Weapon Stockpile: Army's Emergency Preparedness Program Has Been Slow to Achieve Results. GAO/NSIAD-94-91. February 22, 1994. Washington, D.C.: General Accounting Office.

GAO. 1995. Chemical Weapons: Army's Emergency Preparedness Program Has Financial Management Weaknesses. GAO/NSIAD-95-94. March 15, 1995. Washington, D.C.: General Accounting Office.

GAO. 1997. Chemical Weapons and Material: Key Factors Affecting Disposal Costs and Schedule. GAO/NSIAD-97-118. March 11, 1997. Washington, D.C.: General Accounting Office.

GAO. 1999. Environment Protection: Research and Development of Plasma Arc Technology Used to Treat Hazardous Waste. GAO/NSIAD-99-51R. February 16, 1999. Washington, D.C.: General Accounting Office.

Getman, G. 1998. Personal communication from Gerry Getman, Teledyne-Commodore, to Joan Berkowitz, Committee on Review and Evaluation of Alternative Technologies for Demilitarization of Assembled Chemical Weapons, December 16, 1998.

Glassman, I. 1987. Combustion (2nd ed). New York: Academic Press.

Hammerberg, E. 1998. Telephone interview with Ed Hammerberg, Indiana Hazardous Waste Management Administration, by Hank Jenkins-Smith, Committee on Review and Evaluation of Alternative Technologies for Demilitarization of Assembled Chemical Weapons, September 17, 1998.

Hazelbeck, D.A., K.W. Downy, D.W. Ordway, A.J. Roberts, J.P. Elliot, and M.H. Spritzer. 1998. Design and Operational Data for Large-Scale HTO System. Pp. 763-767 in Proceedings of the 17th Annual IT3 Conference. Irvine, Calif.: University of California, Irvine, Office of Enviromental Health and Safety.

Hill, K.Q. 1998. Policy agendas of the president and the mass public: a research validation and extension. American Journal of Political Science 42: 1328–1334.

Hill, K.Q., and A. Hinton-Andersson. 1995. Pathways of representation: a causal analysis of public opinion-policy linkages. American Journal of Political Science 39: 924–35.

Hindman, D., I. Kornelly, B. Palzer, and P. Walker. 1999. Memorandum to Joe Klimek, Startech, Environmental Corp., from the Dialogue Citizens' Advisory Technical Team, January 21, 1999.

ICF Kaiser. 1998. Chemical Demilitarization Program Overarching Issues Assessment: Annual Report for FY 1998. Prepared for the Office of the Deputy Assistant Secretary of the Army for Chemical Demilitarization. Washington, D.C.: ICF Kaiser.

Innovative Emergency Management and Rowan and Blewitt, Inc. 1995a. Chemical Demilitarization Public Outreach: Anniston

Baseline Survey. Vol. 2: Detailed Survey Results. Aberdeen Proving Ground, Md.: U.S. Army Program Manager for Chemical Demilitarization, Public Affairs Office.

Innovative Emergency Management and Rowan and Blewitt, Inc. 1995b. Chemical Demilitarization Public Outreach: Tooele Baseline Survey. Vol. 2: Detailed Survey Results. Aberdeen Proving Ground, Md.: U.S. Army Program Manager for Chemical Demilitarization, Public Affairs Office.

Innovative Emergency Management and Rowan and Blewitt, Inc. 1995c. Chemical Demilitarization Public Outreach: Pine Bluff Baseline Survey. Vol. 1: Analysis and Recommendations. Aberdeen Proving Ground, Md.: U.S. Army Program Manager for Chemical Demilitarization, Public Affairs Office.

Innovative Emergency Management and Rowan and Blewitt, Inc. 1995d. Chemical Demilitarization Public Outreach: Pine Bluff Baseline Survey. Vol. 2: Detailed Survey Results. Aberdeen Proving Ground, Md.: U.S. Army Program Manager for Chemical Demilitarization, Public Affairs Office.

Innovative Emergency Management and Rowan and Blewitt, Inc. 1995e. Chemical Demilitarization Public Outreach: Anniston Baseline Survey. Vol. 1: Analysis and Recommendations. Aberdeen Proving Ground, Md.: U.S. Army Program Manager for Chemical Demilitarization, Public Affairs Office.

Jenkins-Smith, H., and G. Bassett. 1994. Perceived risk and uncertainty of nuclear waste: differences among science, business, and environmental group leaders. Risk Analysis 14(5): 851–856.

Jenkins-Smith, H., and P. Sabatier. 1994. The Advocacy Coalition framework for policy change: an assessment. Journal of Public Policy 14: 175–203.

Jenkins-Smith, H., and C. Silva. 1998. The role of risk perception and technical information in scientific debates over nuclear waste storage. Reliability Engineering and System Safety 59: 107–22.

Kahneman, D., and A. Tversky. 1984. Choices, values and frames. American Psychologist 39: 341–50.

Keystone Center. Undated. The Keystone Policy Dialogue Process. Keystone, Colo.: The Keystone Center.

Keystone Center. 1997. Ground Rules for Dialogue on Chemical Weapons Assessment. Keystone, Colo.: The Keystone Center.

Kingdon, J.W. 1984. Agendas, Alternatives, and Public Policies. New York: Harper Collins College Publishers.

Koplow, D. 1997. By Fire and Ice: Dismantling Chemical Weapons While Preserving the Environment. Amsterdam, Netherlands: Gordon and Breach Publishers.

Krimsky, S., and D. Golding, eds. 1992. Social Theories of Risk. Westport, Conn.: Preager Publishers.

Krosnick, J. 1990. Government policy and citizen passion: a study of issue publics in contemporary America. Political Behavior 12: 59–92.

Kumar, R.K., and Koroll, G.W. 1995. Ignitability of hydrogen/oxygen/diluent mixtures in the presence of hot surfaces. Nuclear Safety 36(1): 68–93.

Lehmani, A., P. Turq, and J.P. Simonin. 1996. Oxidation of kinetics of water and organic compounds by Silver(II) using a potentiometric method. Journal of the Electrochemical Society 143(6): 1860–1866.

Leiss, W., and C. Chociolko. 1994. Risk and Responsibility. Montreal, Quebec: McGill-Queens University Press.

Levenspiel, O. 1962. Chemical Reaction Engineering: An Introduction to the Design of Chemical Reactors. New York: John Wiley and Sons.

Lockheed Martin. 1997. Assembled Chemical Weapons Assessment. Submitted in response to U.S. Army solicitation number DAAM01-97-R00031, September 15, 1997. Albuquerque, N.M.: Lockheed Martin Corporation.

Lockheed Martin. 1997. Assembled Chemical Weapons Assessment. Submitted in response to U.S. Army solicitation number DAAM01-97-R00031, September 15, 1997. Albuquerque, N.M.: Lockheed Martin Corporation.

Lockheed Martin. 1998. Data Gap Resolution Technical Report. Submitted under Contract No. DAAM01-97-0031, Task 1. February 17, 1998. Albuquerque, N.M.: Lockheed Martin Corporation.

Lupton, S. Personal communication from Steve Lupton, Allied-Signal, to Walter May, Committee on Review and Evaluation of Alternatives to Incineration for Demilitarization of Assembled Chemical Weapons, March 30, 1998.

Lyon, R. and R. Ullrich. 1998. Supercritical Water Oxidation as an Economic Alternative to Incineration and Other Thermal Treatments. Pp 743–745 in Proceedings of the 17th annual IT3 Conference. Irvine, Calif.: University of California, Irvine, Office of Environmental Health and Safety.

NAS (National Academy of Sciences). 1969. Disposal Hazards of Certain Chemical Warfare Agents and Munitions. Washington, D.C.: Ad Hoc Advisory Committee of the National Academy of Sciences.

NRC (National Research Council). 1984. Disposal of Chemical Munitions and Agents. Committee on Demilitarizing Chemical Munitions and Agents. Washington, D.C.: National Academy Press.

NRC. 1991. National Research Council. Demilitarization of Chemical Weapons by Cryofracture: A Technical Assessment. Panel on the Current Status of the Cryofracture Process, Board on Army Science and Technology. Washington, D.C.: National Academy Press.

NRC. 1993. Alternative Technologies for the Destruction of Chemical Agents and Munitions. Committee on Alternative Chemical Demilitarization Technologies, Board on Army Science and Technology. Washington, D.C.: National Academy Press.

NRC. 1994. Recommendations for the Disposal of Chemical Agents and Munitions. Committee on Review and Evaluation of the Army Chemical Stockpile Disposal Program, Board on Army Science and Technology. Washington, D.C.: National Academy Press.

NRC. 1996a. Review and Evaluation of Alternative Chemical Disposal Technologies. Panel on Review and Evaluation of Alternative Chemical Disposal Technologies, Board on Army Science and Technology. Washington, D.C.: National Academy Press.

NRC. 1996b. Public Involvement and the Army Chemical Stockpile Disposal Program. Committee on Review and Evaluation of the Army Chemical Stockpile Disposal Program, Board on Army Science and Technology. Washington, D.C.: National Academy Press.

NRC. 1997. Review of Acute Human-Toxicity Estimates for Selected Chemical-Warfare Agents. Committee on Toxicology. Board on Environmental Studies and Toxicology. Washington, D.C.: National Academy Press.

NRC. 1998. Using Supercritical Water Oxidation to Treat

Hydrolysate from VX Neutralization. Committee on Review and Evaluation of the Army Chemical Stockpile Disposal Program, Board on Army Science and Technology. Washington, D.C.: National Academy Press.

OTA (Office of Technology Assessment). 1992. Disposal of Chemical Weapons: Alternative Technologies—Background Paper. OT-BP-0-95. Washington, D.C.: Government Printing Office.

Page, B.I., and R.Y. Shapiro. 1983. Effects of public opinion on policy. American Political Science Review 77: 175–90.

Page, B.I., R.Y. Shapiro, and G. Dempsey. 1987. What moves public opinion? American Political Science Review 81: 23–43.

Parsons-AlliedSignal. 1998. Assembled Chemical Weapons Assessment Program Demonstration Study Plan. Submitted under contract no. DAAM01-98-D-005. November 19, 1998. Pasadena, Calif.: Parsons Infrastructure and Technology Group, Inc.

Parsons-AlliedSignal. 1999a. Parsons/AlliedSignal ACWA Presentation to the NRC. Presentation to the Committee on Review and Evaluation of Alternative Technologies for Demilitarization of Assembled Chemical Weapons, Washington, D.C., March 11, 1999.

Parsons-AlliedSignal. 1999b. Biological Treatment System for Combined Chemical Agent and Energetic Hydrolysates. Report to the NRC. January 15, 1999. Des Plaines, Ill.: AlliedSignal Environmental Systems and Services.

Pecoraro, J. 1999. Alternative Technologies and Approaches Project Overview. Presentation to the Committee on Review and Evaluation of the Army Chemical Stockpile Disposal Program, Washington, D.C., January 6, 1999.

Pennington, J.C., C.B. Price, E.F. McCormick, and C.A. Hayes. 1992. Effects of wet and dry cycles on TNT losses from soil. Technical Report #EL-92 37. Vicksburgh, Miss.: U.S. Army Corps of Engineers, Waterways Experiment Station.

Prugh, R.W. 1998. Safe Design Concepts for a Hydrogen Reduction Process. Wilmington, Del.: Process Safety Engineering, Inc.

Quirk, P. 1989. The cooperative reduction of policy conflict. American Political Science Review 83: 905–921.

Renn, O. 1998. Implications of the hormesis hypothesis for risk perception and communication. BELLE Newsletter 7(1): 2–9.

Richmond, J. 1998. Assembled Chemical Weapons Assessment Program: Evaluation Update and Site Visits. Presentation to the Dialogue on Assembled Chemical Weapons Assessment, Salt Lake City, Utah, February 10, 1998.

Robbins, C. 1998. Army's huge supply of nerve gas poses unnerving questions. Wall Street Journal, May 18, 1998.

Sabatier, P., and H. Jenkins-Smith. 1993. Policy Change and Learning: An Advocacy Coalition Framework. Boulder, Colo.: Westview Press.

Sanjay, H.G., and A. Fataftah. 1997. Humic Acid: A Review of Characteristics, Properties, and Applications. Chantilly, Va.: ARCTECH, Inc.

Sax, N.I., and R.L. Lewis, Sr. 1987. Hawley's Condensed Chemical Dictionary (11th ed). New York: Van Nostrand Reinhold.

Sierra Club (Utah Chapter). 1997. Written Response to a guide to risk management policy and activities. Salt Lake City, Utah: Sierra Club.

Sierra Club. 1994. Sierra Club Analysis of Dr. Carl Peterson's Recommendations for the Disposal of Chemical Agents and Munitions. Pueblo, Colo.: Sierra Club.

Slovic, P. 1987. Perception of risk. Science 236: 280–285.

Smithson, A. 1994. The U.S. Chemical Weapons Destruction Program: Views, Analysis, and Recommendations. Washington, D.C.: Henry L. Stimson Center.

Sniderman, P., R. Brody, and P. Tetlock. 1991. Reasoning and Choice: Explorations in Political Psychology. Cambridge, U.K.: Cambridge University Press.

Susskind, L., and J.L. Cruikshank. 1987. Breaking the Impasse: Consensual Approaches to Resolving Public Disputes. New York: Basic Books.

Teledyne-Commodore. 1997. Teledyne-Commodore Solvated Electron System in Support of the Assembled Chemical Weapons Assessment Program. Proposal submitted in response to U.S. Army Solicitation No. DAAM01-97-R-0031. Huntsville, Ala.: Teledyne-Commodore.

Teledyne-Commodore. 1998a. Data Gap Resolution Technical Report. Huntsville, Ala.: Teledyne-Commodore LLC.

Teledyne-Commodore. 1998b. Letter from J. Woosley, Teledyne-Commodore, to R. Bailey, Committee on Review and Evaluation of Alternative Technologies for Demilitarization of Assembled Chemical Weapons. October 20, 1998.

Teledyne-Commodore. 1998c. Demonstration of Simulated Chemical Agent Rocket Demilitarization Using Solvated Electron Technology and Ammonia Fluid Jet Technology. Huntsville, Ala.: Teledyne-Commodore LLC.

U.S. Army. 1988. Chemical Stockpile Disposal Program Final Programmatic Environmental Impact Statement. Aberdeen Proving Ground, Md.: U.S. Army Program Manager for Chemical Demilitarization.

U.S. Army. 1996. U.S. Chemical Weapons Stockpile Information Declassified. News Release, January 22, 1996. Washington, D.C.: Office of Assistant Secretary of Defense (Public Affairs).

U.S. Army. 1997a. Assessment of Technologies for Assembled Chemical Weapon Demilitarization. Solicitation Number DAAM01-97-R-0031, July 28, 1997. Aberdeen Proving Ground, Md.: U.S. Army Chemical and Biological Defense Command.

U.S. Army. 1997b. Assembled Chemical Weapon Assessment Program Broad Agency Announcement. BAA-97-ACWA-1, August 25, 1997. Aberdeen Proving Ground, Md.: U.S. Army Chemical and Biological Defense Command.

U.S. Army. 1997c. Project Information Package for the Newport VX Chemical Agent Disposal Facility Newport Chemical Depot, Newport, Indiana. November 1997. Aberdeen Proving Ground, Md.: U.S. Army Product Manager, Alternative Technologies and Applications.

Yankelovich, D. 1991. Coming to Public Judgement. Syracuse, N.Y.: Syracuse University Press.

APPENDIXES

APPENDIX A

Description of Assembled Chemical Weapons

The U.S. chemical weapons stockpile is made up of a variety of munitions that serve different military functions. This study is concerned with assessing technologies to destroy munitions that contain both chemical agent and energetic materials (i.e., propellant and/or explosive charges) in an assembled configuration (hence, designated as assembled chemical weapons). There are three basic classes of assembled chemical weapons: (1) projectiles and mortars, (2) rockets, and (3) land mines. Each class is described below in greater detail. The specifications for the munitions were taken from the *Assembled Chemical Weapons Assessment Request for Proposal* (U.S. Army, 1997) and are summarized in Table A-1.

PROJECTILES AND MORTARS

Both projectiles and mortars are shells that are fired from guns or cannons. They have roughly cylindrical steel bodies with tapered noses and a hollow cylindrical tube, known as the burster well, running down the center of the shell. This tube holds the burster, a high-explosive charge that disperses the chemical agent upon detonation. The liquid agent itself is contained in the annular region between the burster well and the shell wall. The nose of the shell consists of either an explosive fuze or a lifting ring, depending on the type of munition. Mortars, which are typically muzzle loaded, are intended for shorter ranges than projectiles and are fired at lower velocities and higher trajectories. Despite their differences, because projectiles and mortars are comparable in design and construction, the approach to their destruction is also similar.

105-mm Projectiles

The 105-mm projectile is 105 mm in diameter (just over 4 inches) and has a mass of 16 to 18 kg. As shown in Table A-1, there are two types of 105-mm projectiles—the M60, which contains HD, and the M360, which contains GB (see Figure A-1). The burster for the M60 is smaller than for the M360 because HD has a tendency to burn rather than disperse if the charge is too powerful. The 105-mm projectiles are stored with their fuzes attached.

155-mm Projectiles

There are five types of 155-mm projectiles in the chemical stockpile—the M121, the M121A1, the M104, the M110, and the M122 (see Table A-1). All of these are 155-mm in diameter (just over 6 inches) and have a mass of 42 to 45 kg; the type of chemical agent (GB, VX, H, or HD) varies, as does the type and amount of burster material (Composition B4 or tetrytol). The bursters for HD (mustard) rounds are much smaller than those for the nerve agent rounds. A cutaway of the M121 is shown in Figure A-2. The 155-mm projectiles are stored with lifting rings in place of fuzes.

8-inch Projectiles

The 8-inch (20.32 cm) projectile, designated the M426, has a mass of more than 90 kg and contains either GB or VX (see Figure A-3). Like the 155-mm projectile, the M426 is stored with a lifting ring in place of a fuze.

TABLE A-1 Assembled Chemical Weapons in the U.S. Stockpile

Munition Type	Total Mass (kg)	Agent	Agent Mass (kg)	Burster Explosive	Burster Mass (kg)
105-mm projectile					
M60	17.6	HD	1.4	tetrytol	0.12
M360	16.1	GB	0.73	tetrytol/ Comp B4	0.50/0.50
155-mm projectile					
M121	44.1	GB	2.9	tetrytol	1.2
M121A1	44.9	GB/VX	2.9/2.7	Comp B4	1.1
M104	43.1	H/HD	5.3/5.3	tetrytol	0.19
M110	42.9	H/HD	5.3/5.3	tetrytol	0.19
M122	44.1	GB	2.9	tetrytol	1.2
4.2-in mortar					
M2	11.3	HD/HT	2.7/2.6	tetryl	0.064
M2A1	11.3	HD	2.7	tetryl	0.064
8-in projectile					
M426	90.3	GB/VX	6.6	Comp B4	3.2
Rocket[a]					
M55	25.9	GB/VX	4.9/4.5	Comp B/ tetrytol	1.5/1.5
Land mine					
M23	10.3	VX	4.8	Comp B4	0.37

[a]The M55 rocket also contains 8.75 kg of M28 double-base propellant.

Source: Adapted from U.S. Army, 1997.

4.2-inch Mortars

The two types of 4.2-inch (105-mm) mortars in the chemical stockpile are the M2 (filled with HD or HT) and the M2A1 (filled with HD), both of which have a mass of 11.3 kg. These rounds are similar to 105-mm projectiles, except the outer shell wall is thinner, and there are internal vanes in the agent cavity (see Figure A-4). The 4.2-inch mortars are stored with fuzes in place.

M55 ROCKETS

A rocket is an airborne weapon propelled by fuel and oxidizer, which is carried along during flight. The only rocket in the chemical stockpile is the 115-mm diameter M55 (see Figure A-5). This rocket is 1.98 m long and has a mass of nearly 26 kg. It consists of two sections: (1) an aluminum-alloy warhead section, which contains the chemical agent, two bursters, and the fuze; and (2) a steel motor section, which contains the propellant grain, the igniter assembly, and the nozzle and fins. The chemical agent is either GB or VX, and the bursters are either Composition B (Comp B) or tetrytol. The propellant is double-base M28 (nitroglycerin/nitrocellulose). The rocket is stored in a shipping and firing tube made of fiberglass-reinforced resin that can contain polychlorinated biphenyls (PCBs). An indexing ring on the outside of the tube

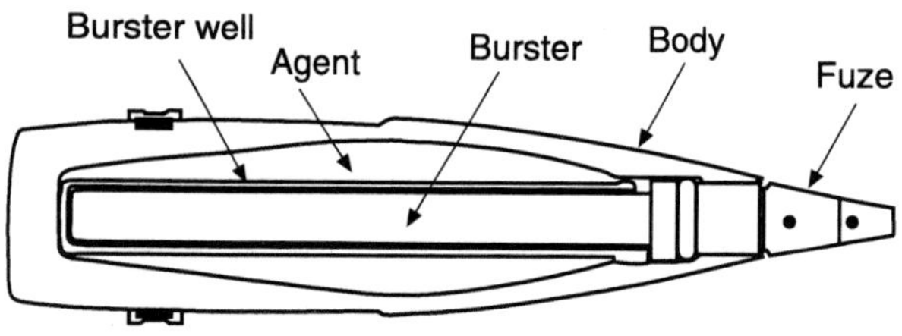

FIGURE A-1 105-mm M360 projectile. Source: U.S. Army, 1988.

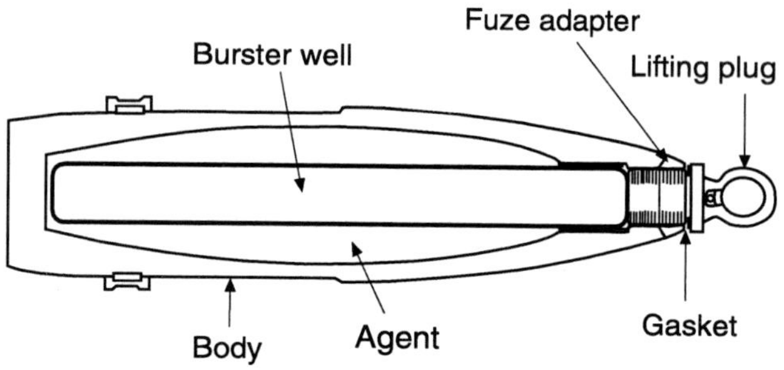

FIGURE A-2 155-mm M121 projectile. Source: U.S. Army, 1988.

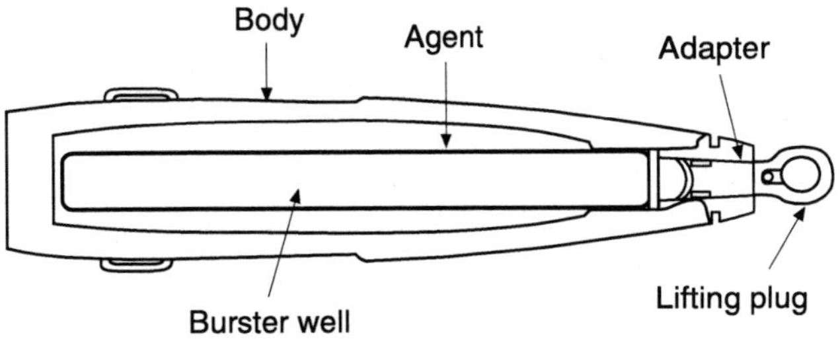

FIGURE A-3 8-inch M426 projectile. Source: U.S. Army, 1988.

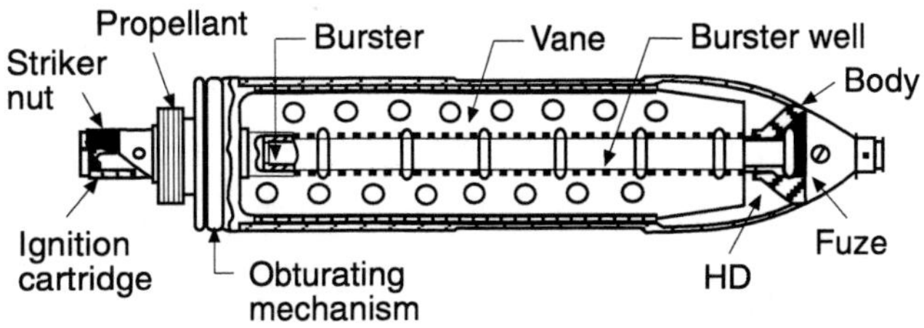

FIGURE A-4 4.2-inch M2 mortar. Source: U.S. Army, 1988.

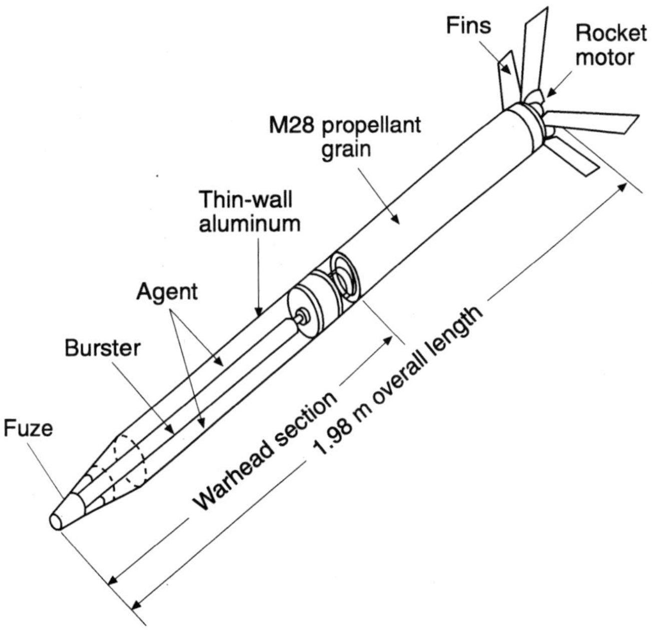

FIGURE A-5 115-mm M55 rocket. Source: SAIC, 1996.

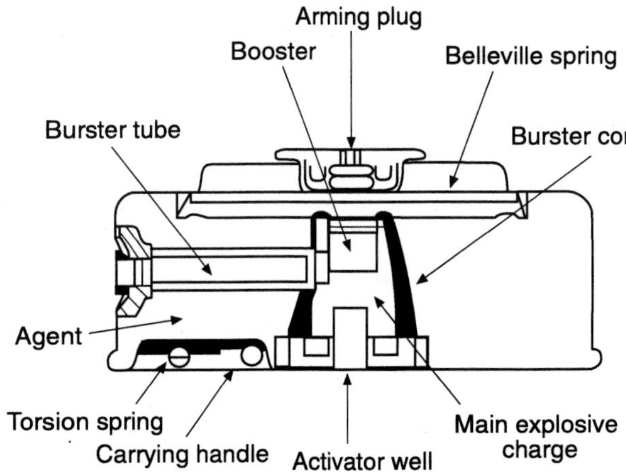

FIGURE A-6 M23 land mine. Source: SAIC, 1996.

near the front (fuze) end of the rocket identifies the front of the rocket in the shipping and firing tube. Aluminum caps seal the ends of the tube.

M23 LAND MINES

A land mine is an explosive device that is usually concealed just below the surface of the ground. When the mine is disturbed, it detonates, causing damage to nearby objects and personnel. The chemical stockpile contains only one type of land mine, designated the M23. This cylindrical mine, shown in Figure A-6 (33 cm in diameter and 13 cm high), is filled with VX and weighs 10.3 kg without the fuze. The M23 contains several explosive components, including a conical burster (Comp B4; 0.37 kg), a tubular initiator (Comp B4; 0.054 kg), a cylindrical booster (Comp A5; 0.009 kg), and a small booster pellet (tetryl; 0.003 kg). The fuze is packaged separately.

REFERENCES

SAIC (Science Applications International Corporation). 1996. Tooele Chemical Agent Disposal Facility Quantitative Risk Assessment. SAIC-96/2600, December 1996. Abingdon, Md.: Science Applications International Corporation.

U.S. Army. 1988. Chemical Stockpile Disposal Program Final Programmatic Environmental Impact Statement. Aberdeen Proving Ground, Md.: U.S. Army Program manager for Chemical Demilitarization.

U.S. Army. 1997. Assessment of Technologies for Assembled Chemical Weapon Demilitarization. Solicitation No. DAAM01-97-R-0031, July 28, 1997. Aberdeen Proving Ground, Md.: U.S. Army Chemical and Biological Defense Command.

APPENDIX B

Meetings and Site Visits

COMMITTEE MEETING 1, AUGUST 25–27, 1997

Edgewood, Maryland

NRC Participants

Committee chair: Robert Beaudet. Committee members: Richard Ayen; Joan Berkowitz, Willard Gekler, Walter May, Kirk Newman, William Rhyne, Stanley Sandler, Randy Seeker, Leo Weitzman. BAST liaison: Lawrence Delaney. NRC staff members: Robert Bailey, Bruce Braun, Douglas Bauer, Jenifer Austin, Shirel Smith.

Objectives

Welcome and introduce new members; complete administrative matters; complete discussion of composition and balance; perform historical review; tour the chemical demilitarization training facility (CDTF) to familiarize committee with the baseline program and technology; discuss preliminary evaluation criteria; review report concept, outline, and strategy; finalize objectives, location, and date of the next committee meeting.

COMMITTEE MEETING 2, OCTOBER 21–22, 1997

Phoenix, Arizona

NRC Participants

Committee chair: Robert Beaudet. Committee members: Richard Ayen; Joan Berkowitz, Hank Jenkins-Smith, John Margrave, Walter May, Kirk Newman, William Rhyne, Stanley Sandler, Randy Seeker, Leo Weitzman. BAST liaison: Lawrence Delaney. NRC staff members: Robert Bailey, Bruce Braun, Jenifer Austin, Margo Francesco.

Objectives

Receive briefing from Department of Defense (DOD) on results of the selection process; meet jointly with Dialogue members to hear their concerns and to discuss NRC/Dialogue interaction; receive briefings from technology providers that passed the DOD threshold (go/no-go) criteria; discuss and finalize technology evaluation criteria; re-examine committee composition for potential additions; organize committee into teams; finalize objectives, location, and date of the next committee meeting.

COMMITTEE MEETING 3, DECEMBER 11–12, 1997

Washington, D.C.

NRC Participants

Committee chair: Robert Beaudet. Committee members: Richard Ayen; Joan Berkowitz, Nosa Egiebor, Willard Gekler, Hank Jenkins-Smith, John Margrave, Walter May, Kirk Newman, Jimmie Oxley, William Rhyne, Stanley Sandler, Leo Weitzman. BAST liaison: Lawrence Delaney. NRC staff members: Robert Bailey, Bruce Braun, Jenifer Austin.

Objectives

Welcome and introduce two new members; complete administrative matters; complete discussion of composition and balance; receive briefing from DOD on the results of the data-gap identification process; receive detailed briefings from technology providers that passed the DOD threshold (go/no-go) criteria; receive update from Dialogue Citizens' Advisory Technical Team (CATT) members; finalize objectives, location, and date of the next committee meeting; discuss locations and dates of visits to stockpile sites and technology providers.

SITE VISIT 1, FEBRUARY 16, 1998

Huntsville, Alabama

NRC Participants

Committee chair: Robert Beaudet. Committee members: Joan Berkowitz, John Margrave, Kirk Newman. NRC staff Member: Robert Bailey.

Objectives

Receive presentations and data from technology provider (Teledyne-Commodore) regarding solvated-electron-based demilitarization system. Tour test facility.

SITE VISIT 2, FEBRUARY 25, 1998

San Diego, California

NRC Participants

Committee members: Kirk Newman, Stanley Sandler, Leo Weitzman. NRC Staff member: Robert Bailey.

Objectives

Receive presentations and data from technology provider (General Atomics) regarding hydrolysis/supercritical water oxidation (SCWO)-based demilitarization system. Tour SCWO test facility.

SITE VISIT 3, MARCH 5–6, 1998

Harwell/Porton Down, United Kingdom

NRC Participants

Committee Chair: Robert Beaudet. Committee members: Willard Gekler, Kirk Newman, Randy Seeker, Leo Weitzman. NRC staff member: Bruce Braun.

Objectives

Receive presentations and data from technology provider (AEA Technologies) regarding electrochemical-oxidation-based demilitarization system. Tour test facility.

COMMITTEE MEETING 4, MARCH 12–13, 1998

Lexington/Richmond, Kentucky

NRC Participants

Committee chair: Robert Beaudet. Committee members: Richard Ayen; Joan Berkowitz, Nosa Egiebor, Willard Gekler, Hank Jenkins-Smith, Walter May, Kirk Newman, Jimmie Oxley, William Rhyne, Stanley Sandler, Leo Weitzman. BAST liaison: Lawrence Delaney. NRC staff members: Robert Bailey, Bruce Braun, Jenifer Austin.

Objectives

Tour the Bluegrass Army Depot (BGAD) chemical weapons storage yard; discuss condition of stockpile and emergency planning with BGAD personnel; attend a Kentucky Citizens' Advisory Commission (CAC) meeting to solicit public input about proposed alternatives; review and revise first draft of report chapters; discuss the results of site visits to date; finalize objectives, location, and date of next committee meeting; finalize locations and dates of site visits.

SITE VISIT 4, MARCH 30, 1998

Des Plaines, Illinois

NRC Participants

Committee members: Richard Ayen, Nosa Egiebor, Walter May, William Rhyne. NRC staff member: Robert Bailey.

Objectives

Receive presentations and data from technology provider (Parsons-AlliedSignal) regarding hydrolysis/biotreatment-based demilitarization system. Tour AlliedSignal research facility.

SITE VISIT 5, APRIL 6, 1998

Toronto, Ontario

NRC Participants

Committee chair: Robert Beaudet. Committee members: Hank Jenkins-Smith, John Margrave, Jimmie Oxley, William Rhyne, Randy Seeker. NRC staff member: Robert Bailey.

Objectives

Receive presentations and data from technology provider (Burns & Roe) regarding plasma-arc-based demilitarization system. Tour research facility at Ontario Hydro Technologies and observe plasma waste converter (PWC) test unit.

SITE VISIT 6, APRIL 8, 1998

Pine Bluff, Arkansas

NRC Participants

Committee members: Joan Berkowitz, Nosa Egiebor, Willard Gekler, Kirk Newman, Stanley Sandler. NRC staff member: Robert Bailey.

Objectives

Receive presentations and data from technology provider (Lockheed Martin) regarding hydrolysis/SCWO/gas-phase chemical reduction (GPCR)-based demilitarization system. Tour SCWO facility for disposing of smokes and dyes at the Pine Bluff Arsenal.

SITE VISIT 7, APRIL 14, 1998

Hawthorn, Nevada

NRC Participants

Committee members: Richard Ayen, Walter May, Kirk Newman, William Rhyne. NRC staff member: Robert Bailey.

Objectives

Receive presentations and data from technology provider (ARCTECH, Inc.) regarding hydrolysis/humic-acid-based demilitarization system. Observe pilot-scale energetics-destruction system using remote camera.

COMMITTEE MEETING 5, JUNE 11–13, 1998

Pueblo, Colorado

NRC Participants

Committee chair: Robert Beaudet. Committee members: Richard Ayen; Joan Berkowitz, Willard Gekler, Hank Jenkins-Smith, John Margrave, Walter May, Jimmie Oxley, William Rhyne, Stanley Sandler, Randy Seeker, Leo Weitzman. BAST liaison: Lawrence Delaney. NRC staff members: Robert Bailey, Bruce Braun, Harrison Pannella, Jenifer Austin.

Objectives

Visit the Pueblo Chemical Depot (PCD) and meet with depot personnel to discuss site-specific problems with storage and demilitarization of munitions; receive briefing from DOD on the status of the Assembled Chemical Weapons Assessment (ACWA) program, especially status of demonstrations; receive briefing from Dialogue CATT on their site visits; receive briefing on NRC Stockpile Committee SCWO report; attend PCD summer festival to obtain public input on alternatives; discuss risk perception and ways to evaluate public acceptability; discuss the results of committee site visits since the last meeting; review and revise draft chapters, and finalize writing assignments; discuss locations, dates, and objectives of next two committee meetings.

COMMITTEE MEETING 6, AUGUST 20–21, 1998

Irvine, California

NRC Participants

Committee chair: Robert Beaudet. Committee Members: Richard Ayen; Joan Berkowitz, Nosa Egiebor, Willard Gekler, Walter May, John Margrave, Kirk Newman, William Rhyne, Stanley Sandler, Randy Seeker, Leo Weitzman. NRC staff members: Robert Bailey, Bruce Braun, Harrison Pannella, Jenifer Austin.

Objectives

Conduct annual discussion of composition and balance; receive briefing from DOD on status of ACWA program, especially progress towards demonstrations; review and revise first full-message draft of report; define steps to concurrence draft of report; set goals for next meeting.

COMMITTEE MEETING 7, OCTOBER 8–9, 1998

Woods Hole, Massachusetts

NRC Participants

Committee chair: Robert Beaudet. Committee members: Richard Ayen; Joan Berkowitz, Nosa Egiebor, Willard Gekler, Hank Jenkins-Smith, John Margrave, Walter May, Kirk Newman, Jimmie Oxley, William Rhyne, Stanley Sandler, Leo Weitzman. BAST liaison: Lawrence Delaney. NRC staff members: Robert Bailey, Bruce Braun, Harrison Pannella, Jenifer Austin.

Objectives

Receive briefing from DOD on status of ACWA program, especially progress towards demonstrations; review and revise draft of report; define steps to concurrence draft; set goals for next meeting.

SITE VISIT 8, NOVEMBER 9, 1998

Deseret Chemical Depot, Utah

NRC Participants

Committee members: Richard Ayen, Walter May. NRC staff member: Robert Bailey.

Objectives

Observe equipment (at the Chemical Agent and Munition Disposal Facility [CAMDS]) to be used in generating nerve agent hydrolysate for use in ACWA program demonstrations. Discuss system design and operation with cognizant CAMDS and contractor (Arthur D. Little) personnel.

SITE VISIT 9, NOVEMBER 12, 1998

Huntsville, Alabama

NRC Participants

Committee members: Joan Berkowitz, Nosa Egiebor, John Margrave. NRC staff member: Robert Bailey.

Objectives

Discuss questions about process chemistry with the technology provider (Teledyne-Commodore) regarding solvated-electron-based demilitarization system.

SITE VISIT 10, JANUARY 14–15, 1999

Aberdeen Proving Ground, Maryland

NRC Participants

Committee chair: Robert Beaudet. Committee

members: Joan Berkowitz, John Margrave, Jimmie Oxley, Stanley Sandler, Leo Weitzman. NRC staff members: Robert Bailey, Bruce Braun.

Objectives

Observe equipment to be used in the demonstration testing of (1) the Parsons-AlliedSignal biotreatment and catalytic oxidation unit operations for processing HD hydrolysate and (2) the Burns & Roe plasma waste-conversion system. Discuss system design and operation with cognizant DOD and contractor personnel.

SITE VISIT 11, JANUARY 25–26, 1999

Deseret Chemical Depot and Dugway Proving Ground, Utah

NRC Participants

Committee chair: Robert Beaudet. Committee members: Walter May, Leo Weitzman. NRC staff members: Robert Bailey, Bruce Braun.

Objectives

Observe equipment to be used in the demonstration testing of (1) the Parsons-AlliedSignal biotreatment and catalytic-oxidation unit operations for processing nerve agent hydrolysate; (2) the Parsons-AlliedSignal jet-cutting unit operations; and (3) the General Atomics energetics hydrolysis, metal-parts treatment, dunnage treatment, and SCWO unit operations. Discuss system design and operation with cognizant DOD and contractor personnel.

SITE VISIT 12, FEBRUARY 5, 1999

Pantex, Texas

NRC Participants

Committee members: Kirk Newman, Jimmie Oxley, William Rhyne. NRC staff member: Robert Bailey.

Objectives

Observe equipment to be used in generating energetics hydrolysate for use in the ACWA demonstrations. Discuss system design and operation with cognizant DOD, Department of Energy, and contractor personnel.

SITE VISIT 13, FEBRUARY 11–12, 1999

Pasadena, California

NRC Participants

Committee chair: Robert Beaudet. Committee members: Richard Ayen, Walter May, William Rhyne.

Objectives

Discuss questions about the latest version of the proposed process with the technology provider (Parsons-AlliedSignal).

COMMITTEE MEETING 8, MARCH 11–13, 1999

Washington, D.C.

NRC Participants

Committee chair: Robert Beaudet. Committee members: Richard Ayen, Joan Berkowitz, Nosa Egiebor, Willard Gekler, Hank Jenkins-Smith, John Margrave, Walter May, Kirk Newman, Jimmie Oxley, William Rhyne, Stanley Sandler, Randy Seeker, Leo Weitzman. BAST liaison: Lawrence Delaney. NRC staff members: Robert Bailey, Bruce Braun, Harrison Pannella, Jenifer Austin.

Objectives

Receive briefing by DOD on the results of demonstration testing to date; receive detailed briefings by technology providers that passed the DOD threshold (go/no-go) criteria detailing information generated since the selection of technology packages for demonstration; receive update from Dialogue CATT members; finalize concurrence draft of the report.

APPENDIX C

Baseline Disassembly Process

The baseline demilitarization system utilizes incineration to destroy chemical agents and energetic materials. The Army has spent considerable time and effort to develop processes for disassembling the chemical weapons prior to incineration. Many of these processes are not "incineration-specific," that is, they dismantle the munitions but do not prepare the dismantled pieces specifically for incineration. Thus, these processes could potentially be used as part of alternative disposal technologies (other than incineration).

Because several of the technology providers have incorporated the baseline disassembly process into their proposed technology packages (with some modifications), the committee included this description of the overall baseline disassembly process.[1] The storage of assembled chemical weapons is discussed first to provide background on packing configurations. Then, the procedures used to bring munitions to the chemical demilitarization facility (CDF) from the storage area and prepare them for disassembly are described. Next, the actual disassembly of each type of munition (rockets; projectiles and mortars; and land mines) is explained. This is followed by a discussion of the special treatment required for munitions that leak during storage or transport. Finally, some of the problems encountered during baseline disassembly are described. Note that the baseline process handles all types of munitions, but only one type of munition is treated at a time.

[1]Baseline disassembly is often referred to loosely as "reverse assembly" because the disassembly of projectiles, in general, reverses the steps used in their assembly. This description is not accurate for other munitions (e.g., M55 rockets), which are not reverse assembled. Therefore, the term reverse assembly is *not* used in this report to describe the overall baseline disassembly process.

STORAGE OF MUNITIONS

Assembled chemical weapons are stored on wooden pallets stacked in igloos—protective structures made of reinforced concrete. The pallet configurations vary for different types of munitions. M55 rockets, inside their fiberglass shipping and firing tubes, are stored 15 to a pallet in a 3 × 5 geometry. M23 land mines are stored in 16-gallon steel drums, three mines per drum, 12 drums per pallet. Projectiles and mortars range from eight to 24 munitions per pallet, depending on the size of the rounds.

The igloos are monitored for airborne chemical agent on a regular basis, and a small but significant number of munitions have been found to leak over time. Generally, these are vapor-phase leaks, but leaks of liquid agent have also occurred. When munitions that leak (called *leakers*) are identified, they are placed in protective containers (called overpacks), which act as a barrier to further leaks to the external environment. Leaker mines are overpacked in plastic and reloaded into drums. Leaker rockets, projectiles, and mortars are placed inside thick steel overpacks that are then repalletized for subsequent storage.

The handling and disassembly of leaker mines is essentially identical to the handling of nonleaker mines because (1) the overpack is not extensive and (2) mine processing can be easily modified to accommodate the removal of the overpack. In contrast, the handling and disassembly of leaker rockets, projectiles, and mortars is somewhat different than for nonleakers (described in the section "Processing of Overpacked Munitions"). The following description applies to nonleaker assembled chemical weapons.

TRANSPORT AND HANDLING PRIOR TO DISASSEMBLY

At the beginning of the disposal process, palletized munitions are removed from the storage igloo using a forklift and loaded into a cylindrical steel vessel known as an on-site container (ONC). Once the munitions are inside, the ONC is sealed to contain vapor leaks during transportation. A truck then takes the ONC from the storage area to the CDF. Inside the CDF, the ONC is delivered to the unpack area, where it is sampled for agent contamination prior to opening. If agent is detected, then the ONC is not opened but is instead processed as described in the section "Processing of Munitions That Leak During Transport." If no agent is detected, the ONC is opened, and the munitions are removed. The munitions are then manually separated from their packaging materials. This process is quite straightforward for projectiles, mortars, and M55 rockets, which are simply removed from their pallets. Rockets are *not* removed from their fiberglass shipping/firing (S/F) tubes.

The unpacking process for M23 land mines is more involved. First, the mine drums are removed from the pallet. Then, individual drums (containing three mines with detached fuzes and activators) are conveyed to the mine glovebox using a forklift and mechanical lift. The mine glovebox is an enclosed space that forms a boundary between the unpack area and the explosion-containment vestibule (ECV). Inside the glovebox, the drums are unpacked manually. Arming plugs are manually removed from the mines and, along with fuzes and activators, are placed into a "fuze box" identical in size and shape to a mine casing.

Once the unpack operations have been completed, the munitions (and the mine fuze boxes) are placed, one at a time, on conveyors. Rockets are oriented nose first; a special indexing ring located on the S/F tube is used by the loading machine to ensure this orientation. In contrast, projectiles are sent to the ECV base-first. The conveyors transport the munitions into the ECV and finally into the explosion-containment room (ECR), where the disassembly process begins. The ECR is separated from the ECV by blast gates and is designed to contain the effects of accidental explosions during the processing of explosively-configured munitions. Once the munitions enter the ECR, all disassembly operations take place remotely via mechanical systems.

DISASSEMBLY OF M55 ROCKETS

The disassembly process for the M55 rocket is shown schematically in Figure C-1. Once in the ECR, the rocket (in its S/F tube) is conveyed to the punch and drain station of the rocket-shear machine, which first drains the agent from the rocket by punching holes through the exterior of the S/F tube and the rocket. The agent drains by gravity and is pumped to an agent storage tank. Design requirements call for at least 95 percent of the agent to be drained from the rocket.

When draining is complete, the rocket moves via conveyor to the shear station of the rocket-shear machine, where it is cut into eight pieces by a hydraulically-driven guillotine. A combination of water and caustic spray cools the shearing blade during the cutting operation. The first cut separates the fuze from the rest of the rocket, and the other cuts break up the burster and propellant. The sheared rocket pieces drop into a hopper and are fed into the deactivation furnace system where the metal parts are decontaminated and the residual agent and energetics are destroyed. To avoid potential detonations, the feed is controlled so the fuze cannot be in the same furnace section as the burster or propellant segments.

The baseline rocket disassembly system described above produces two material exit streams: (1) liquid chemical agent; and (2) sheared rocket pieces containing metal, residual agent, burster energetics, intact fuzes, propellant, and fiberglass from the S/F tube.

DISASSEMBLY OF PROJECTILES AND MORTARS

Figure C-2 shows the disassembly process for projectiles and mortars. Inside the ECR, the projectile (or mortar) is conveyed to the projectile/mortar disassembly machine, which removes the explosive components from the munition in three steps: (1) the fuze and booster (or lifting plug) are unscrewed from the shell using a hydraulic chuck at the nose-closure removal station, (2) the fuze well cup is unscrewed using a hydraulically-driven collet at the miscellaneous parts-removal station, and (3) the burster is removed using high-pressure air at the burster-removal station. Bursters are subsequently sheared into pieces by the burster size-reduction machine, a modified version of the rocket-shear machine. The fuze, booster, and sheared

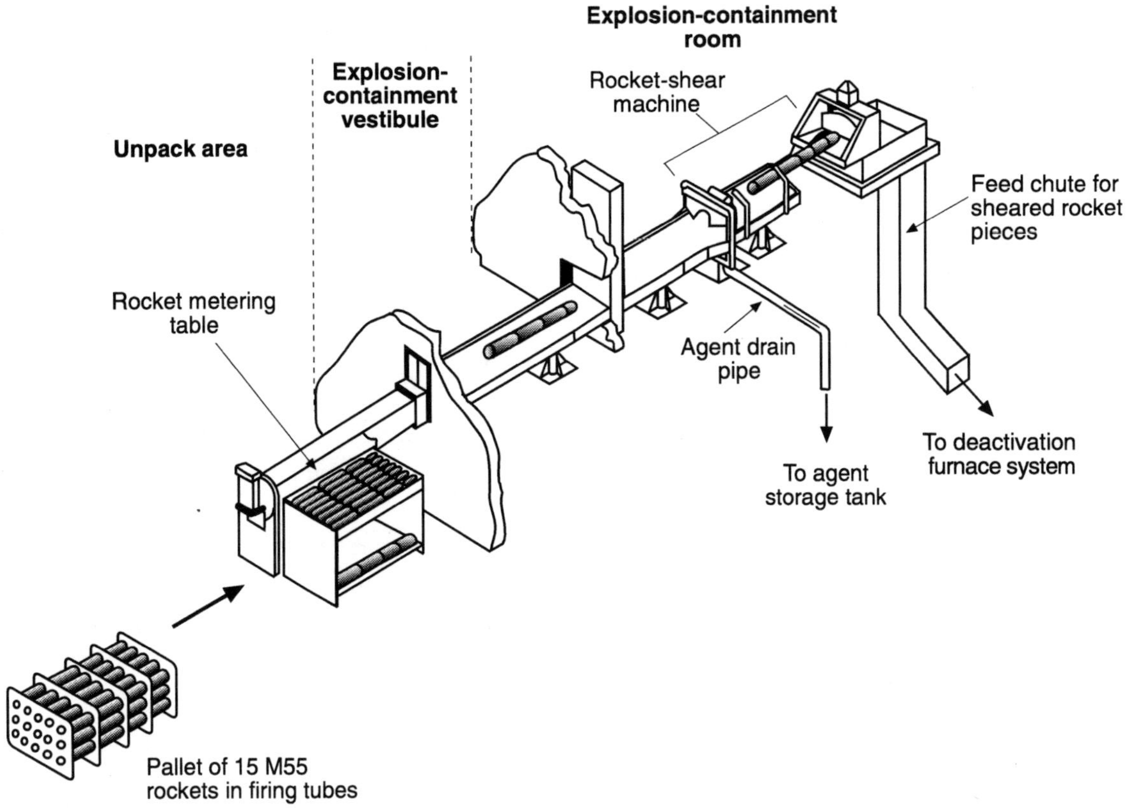

FIGURE C-1 Baseline disassembly of M55 rockets.

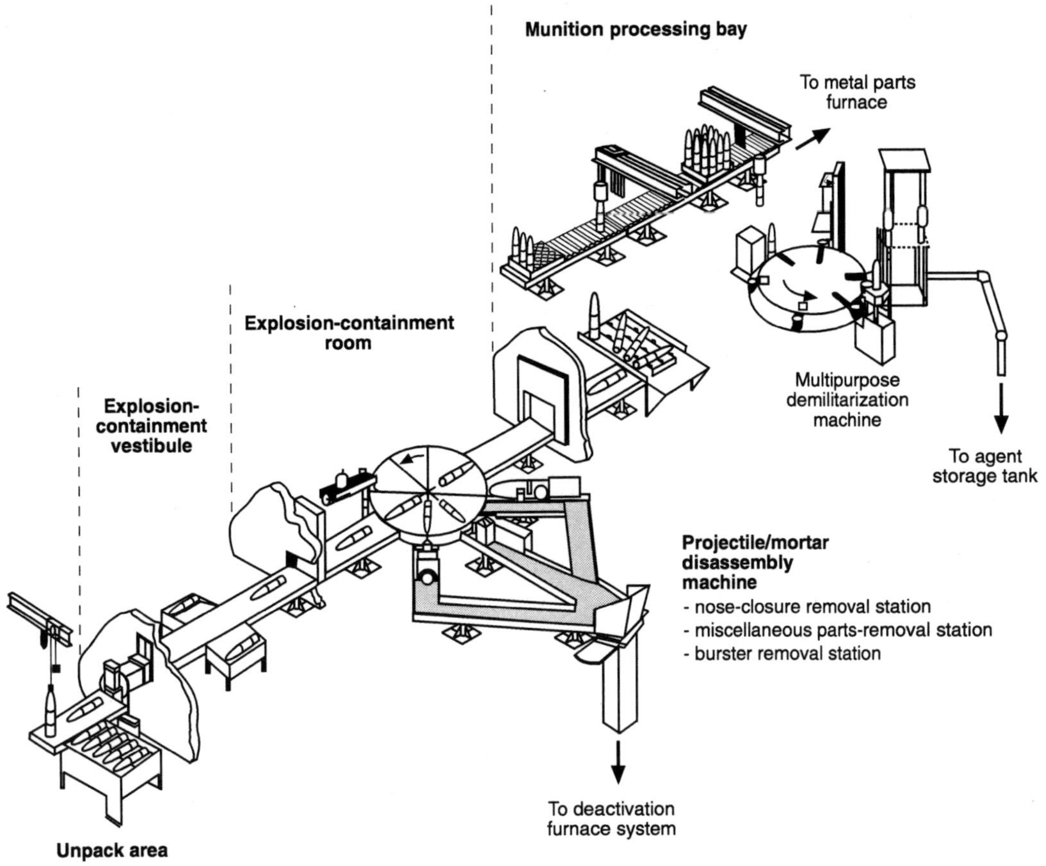

FIGURE C-2 Baseline disassembly of projectiles/mortars.

burster pieces are then fed to the deactivation furnace system for incineration.

The munition, with its explosives removed but its agent load still in place, then exits the ECR. Using conveyors and pick-and-place robots, the munition is transported to the munitions processing bay, where it is loaded into the multipurpose demilitarization machine. At the pull-and-drain station of this machine, a collet is inserted into the projectile, and the burster well is extracted. Some burster wells are welded in place, so the multipurpose demilitarization machine also has a bore station for milling out the weld or the entire well, as necessary. Once the burster well is removed (or milled out), a drain tube is inserted into the shell body, and the agent is pumped from the munition, through a strainer, and into an agent storage tank. Design requirements call for at least 95 percent of the agent to be drained from the projectile or mortar. Following draining, the burster well is crimped (to prevent reseating) and placed back into the munition body. The munition then moves on to the metal-parts furnace for thermal decontamination to a 5X condition.

The baseline projectile/mortar disassembly process produces three material exit streams: (1) liquid chemical agent; (2) energetic components, including sheared burster pieces, intact fuzes, and supplementary charges; and (3) metal munition bodies containing some residual agent but no energetics.

DISASSEMBLY OF M23 LAND MINES

The process for disassembling land mines is shown in Figure C-3. In the ECR, the mine is transported to the mine machine, which is precisely oriented to avoid the explosive components, and is punched through its side. At least 95 percent of the chemical agent is drained and pumped to a holding tank. The mine is then moved to the booster push-out station, where the central booster is removed. The mine body and booster are fed separately to the deactivation furnace system where the metal parts are decontaminated and the residual agent and energetics are destroyed.

Mine fuze boxes that enter the ECR are dumped directly into the deactivation furnace system, and the

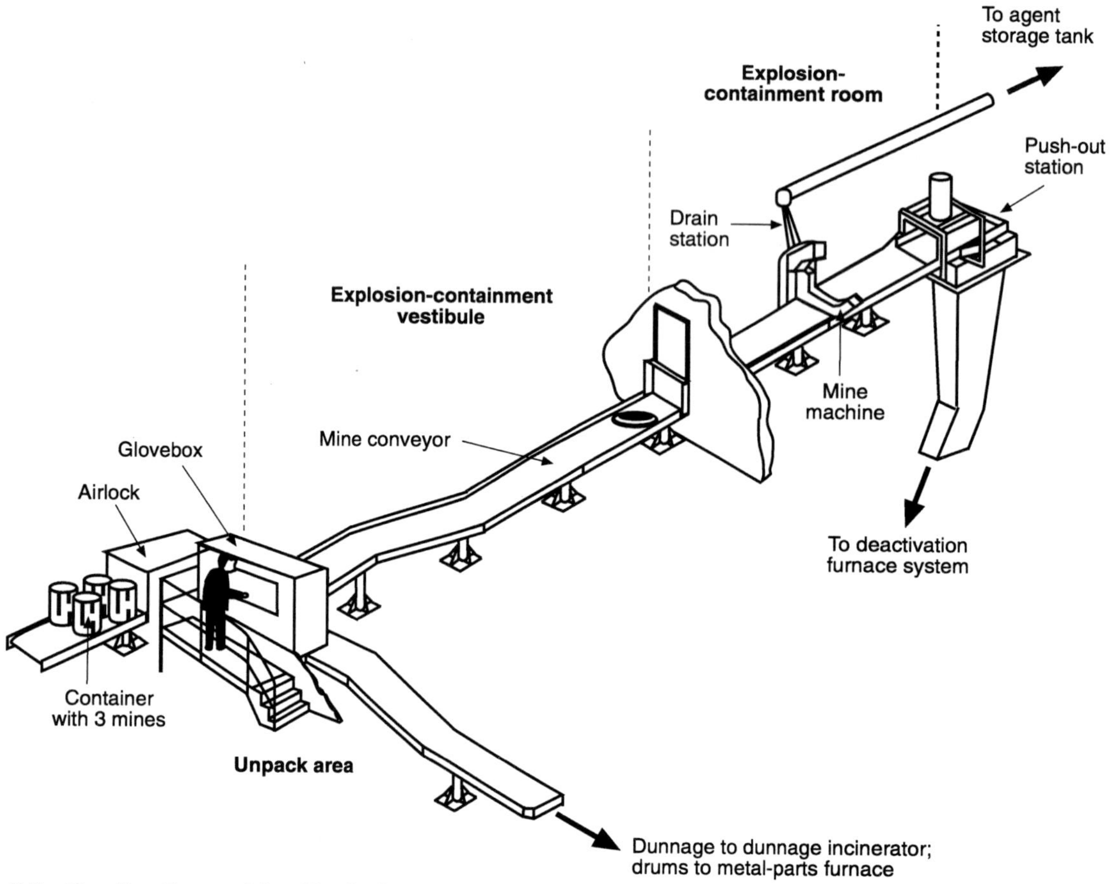

FIGURE C-3 Baseline disassembly of land mines.

fuzes, activators, and arming plugs are destroyed. A magnet is used to differentiate between the mines and the plastic fuze boxes. Empty mine drums, unloaded in the mine glovebox, are sent to the metal-parts furnace for thermal decontamination.

The mine disassembly process yields five material exit streams: (1) liquid chemical agent; (2) boosters; (3) metal mine bodies containing energetics and some residual agent; (4) plastic fuze boxes containing intact fuzes, activators, and arming plugs; and (5) steel mine drums.

DISSASSEMBLY OF OVERPACKED MUNITIONS

Leakers (rockets, projectiles, and mortars) that have been overpacked require special processing because of the increased risk of agent contamination. Like nonleakers, overpacked munitions are stored on pallets. Therefore, the procedures for removing them from an igloo and transporting them to the CDF are identical. Because of the overpacks, however, the machines normally used to load the rockets, projectiles, and mortars onto the standard transport conveyors cannot be used. Instead, the entire pallet of overpacked leakers is loaded manually onto a bypass conveyor in the unpack area and conveyed through an airlock into the ECV. There, operators in demilitarization protective ensemble (DPE) suits manually unpack the pallet, remove the rockets, projectiles, or mortars from their overpacks, and load the munitions onto the conveyors to the ECR. These munitions are unpacked in the ECV rather than in the unpack area to ensure agent containment. Once the leakers enter the ECR, the disassembly steps are identical to those already described for nonleakers.

DISASSEMBLY OF MUNITIONS THAT LEAK DURING TRANSPORT

The act of transporting munitions from the storage area to the CDF may cause some munitions to leak. These munitions are referred to as *leaking* munitions, as opposed to *leakers*, which have leaked during storage and have been overpacked. To identify leaking munitions, the internal atmosphere of the ONC is monitored prior to opening it in the unpack area. If agent is detected, the ONC is conveyed outside the facility and taken, via a separate entrance, into an area called the toxic maintenance area. This area is a "Level A" area, which means that liquid and airborne agent contamination is expected, and operators wear DPE suits. Thus, the ONC can be opened safely in this area.

After opening, the pallets are removed, and the ONC and the exterior of the munitions/pallets are decontaminated using decontamination solution. The pallet is then loaded onto a special tray and conveyed *backwards* through the facility via the processing conveyors. This means that the pallet is taken from the toxic maintenance area (on the first floor of the CDF) up to the ECV (on the second floor of the CDF), traveling opposite to the "normal" conveyor direction for nonleaking munitions. When the pallet reaches the ECV, operators in DPE suits manually unpack it and load the munitions onto the conveyor to the ECR. Once the leaking munitions enter the ECR, the disassembly steps are identical to those already described for nonleaking munitions.

PROBLEMS ENCOUNTERED DURING BASLINE DISASSEMBLY

At the Johnston Atoll Chemical Agent Disposal System (JACADS) in 1994, an M55 rocket burster initiated during shearing (SAIC, 1996). The explosion containment functioned as designed, and no agent was released from the facility. However, damage to the disassembly machines was extensive. The system operation was subsequently modified to ensure that water sprays were irrigating the cutting zone during all shearing operations.

During operational verification testing at JACADS, the multipurpose demilitarization machines (used to pull the burster wells and drain the agent from projectiles and mortars) were found to be among the least reliable of the baseline equipment. They required a large number of toxic-area entries (and even continuous manning) by personnel in DPEs (Mitre, 1993). Subsequent modifications have improved their reliability, but quantitative data on the degree of improvement were not available.

Some difficulties have also been encountered in unscrewing the nose closures on projectiles. Apparently, the pitch on the nose closure threads varies, and the projectile/mortar disassembly machines were sometimes unable to unscrew the nose closures. Most of the problems with projectile/mortar disassembly were solved by making slight modifications to the baseline disassembly equipment; however, as of 1996, between 5.1 and 8.6 percent of the projectiles processed were rejected because of some failure in the disassembly process (PMCD, 1997). (The percentage varied with the type of projectile.)

Currently, JACADS is processing mustard-filled 4.2-inch mortars. The agent in these mortars has become very thick and cannot be drained using the baseline suction approach. Therefore, after the agent cavity is opened, these rounds are being introduced to the metal-parts furnace at a reduced rate. The agent volatilizes from the mortar in the furnace and is thermally destroyed. JACADS is currently investigating a modification to its environmental permit to allow more agent per tray of mortars into the metal-parts furnace, thus, increasing the throughput rates.

REFERENCES

Mitre. 1993. Summary Evaluation of the Johnston Atoll Chemical Agent Disposal System: Operational Verification Testing. MTR93W0000036, May 1993. McLean, Va.: Mitre Corporation.

PMCD (Program Manager for Chemical Demilitarization). 1997. Presentation by PMCD representatives from JACADS to the Committee on Review and Evaluation of the Army Chemical Stockpile Disposal Program. Honolulu, Hawaii. June 23, 1997.

SAIC (Science Applications International Corporation). 1996. Tooele Chemical Agent Disposal Facility Quantitative Risk Assessment. SAIC-96/2600, December 1996. Abingdon, Md.: Science Applications International Corporation.

APPENDIX D

Agent Neutralization by Hydrolysis

Five of the seven technology packages include the destruction of chemical agents by purely chemical reactions that reduce the agents to less toxic or nontoxic products. Four of the five propose neutralization via hydrolysis with an aqueous alkali solution. *Hydrolysis* is a reaction of a target compound with water, an acid, or a base in which some chemical bond is broken in the target and OH^- or H^+ is inserted into the bond cleavage. The destruction of chemical agent via hydrolysis is often referred to as chemical *neutralization*. The military definition of *neutralize* is to render something unusable or nonfunctional. Technically, neutralization is a chemical reaction between an acid and a base to form a salt and water. Chemical agents are neither acids nor bases, however, and the use of the term neutralization for two very different processes is somewhat confusing. Nevertheless, in the literature on chemical demilitarization, the terms neutralization and hydrolysis have been used interchangeably. Therefore, unless otherwise specified, neutralization refers to the destruction of chemical agent via hydrolysis.

BACKGROUND

Agent detoxification has been an essential requirement since the introduction of mustard in World War I. Detoxification is used for decontaminating dispersed chemical agents on the battlefield, as well as in laboratories, production plants, and elsewhere. Many types of reactions can detoxify chemical agents, but only two are widely used: nucleophilic substitution (e.g., hydrolysis) and oxidation. The literature is very extensive on the neutralization of mustard but much less extensive on the neutralization of nerve agents (ERDEC, 1996, 1997). In general, work on agent decontamination has identified several problems and the need for more research in the following areas:

- Chemical agents are nonpolar compounds, but most decontaminating reagents are polar, which leads to solubility problems. HD (mustard) is very insoluble in water; VX is somewhat soluble in water at low pH but not at high pH; GB is reasonably soluble in water. Thus, the reaction takes place only at the phase interfaces, and vigorous stirring to achieve the desired reaction rates.
- Thickeners have been used in some mustard formulations, and some agents have thickened into gels by some unknown decomposition or polymerization process while in storage. (Gelling occurs most often for mustard HD and, to a lesser extent, for GB.)
- Investigations to develop decontamination methods were generally limited to room-temperature reactions because decontamination is usually performed at ambient temperatures. Higher temperatures, which are feasible for the destruction of bulk agent, were not investigated.
- Decontamination is generally carried out by flooding the surface or the liquid with a large excess of decontamination liquid (e.g., 100 to 1); for the destruction of bulk agent, this amount of liquid would create an unnecessarily large hazardous-waste disposal problem.

Major studies on chemical-neutralization (NRC, 1993; Yang, 1995) are summarized in Table D-1. Because

TABLE D-1 Examples of Large-Scale Neutralizations

Country	Agent	Quantity (tons)	Reactant	Temperature	Disposal	Period/Rate
United States	Sarin (GB)	4,000	Aq NaOH	Ambient	RCRA[a] hazardous landfill	1973–1976
Russia	Sarin, Soman, mustard	3,000	$HOCH_2CH_2NH_2$ 600 L reactor[b]	100°C	Incineration	1980–1990 @ 20t/a
	V-type[c]	30	H_3PO_4 and EG[d] 600 L reactor[b]	140°C	Incineration	1980–1990 @ 20t/a
Canada	Mustard	700	$Ca(OH)_2$ 8t batch	95°C	Incineration	1974–1976
	VX, Tabun,[e] Soman	0.3	20% KOH in MeOH	Ambient	Incineration	4 months intermittent
United Kingdom	Sarin	20	20% NaOH 250 kg per batch	Ambient	Discharged at depth into coastal sea	1967–1968
Iraq	Sarin, GF[e]	70	Aq NaOH	Ambient	Discharged to lined pit	1992–1993 @ 1 ton/day

[a]Resource Conservation and Recovery Act.
[b]A mobile reactor known as the KUASI system.
[c]A VX analog: $MeP(O)(OCH_2CHMe_2)(S(CH_2)NEt_2)$.
[d]Ethylene glycol, (EG).
[e]Tabun is $(Me_2N)P(O)(OC_2H_5)CN$; GF is MeP(O)(OR)F where R = cyclohexyl.

Source: Adapted from NRC, 1993; Yang, 1995.

the hydrolysis technologies proposed for the Assembled Chemical Weapons Assessment (ACWA) all start with either NaOH-solution (three proposals) or KOH-solution (one proposal) for nerve agents and water for mustard (four proposals), the discussion below is limited to agent detoxification with these materials.

Aqueous systems have some obvious advantages: reduced fire and other chemical hazards; a more substantial research basis than for other reagents; and a smaller volume of organic material to dispose of.

HYDROLYSIS OF GB

Basic aqueous solutions react readily with GB by the following reaction:

$$(CH_3)_2CHO{-}\overset{\overset{\displaystyle O}{\|}}{\underset{\underset{\displaystyle CH_3}{|}}{P}}{-}F + 2\,NaOH \rightarrow$$

sarin (GB)

$$(CH_3)_2CHO{-}\overset{\overset{\displaystyle O}{\|}}{\underset{\underset{\displaystyle CH_3}{|}}{P}}{-}O^- + 2\,Na^+ + H_2O + F^- \qquad (1)$$

isopropyl-methylphosphonic acid

Yang (1995) states that, "The reaction is a bimolecular displacement (S_N2 (P)) of F^- by OH^- to produce the phosphonate anion." The rate can be high; the second-order rate constant at 25°C (77°F) is 25 (molarity-sec)$^{-1}$ (Gustafson and Martell, 1962):

$$\frac{d[GB]}{dt} = -25[OH][GB]$$

where the brackets represent concentration. Assuming caustic is maintained at 0.1 molar, the concentration of GB should be reduced one million-fold in 5.5 seconds.

Sarin is soluble in water, although the exact solubility limit does not appear to have been reported. It has been reported, however, that a 1:1 volume mix of sarin and water was immediately miscible (Reeves and Macy, 1947). Alcohol (methanol or ethanol) is sometimes added to keep the GB in solution.

Excess NaOH or KOH is required to maintain the high pH. A typical impurity in GB is a diester:

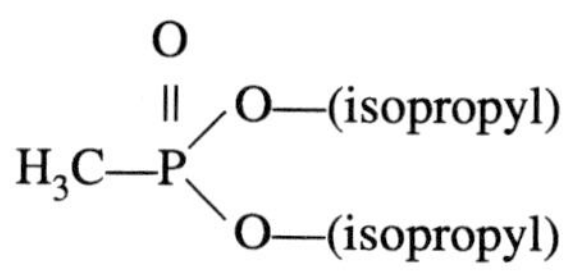

If the pH of the final solution falls below 7, this diester can react with fluoride ion to reform GB (Beaudry et al., 1993). The reaction is very slow, however.

TABLE D-2 Effect of pH on Equilibrium of Remaining GB

pH	$[OH^-]$	GB Moles/L	GB ng/L
13	10^{-1}	10^{-19}	1.4×10^{-8}
10	10^{-4}	10^{-13}	0.014
9	10^{-5}	10^{-11}	1.4
8	10^{-6}	10^{-9}	140

Excess NaOH is also required to ensure the "complete" conversion of GB. The estimated equilibrium constant, K_e, for the hydrolysis reaction is (Harris et al., 1982):

$$K_e = \frac{[IMP][F^-]}{[GB][OH^-]^2} = 10^{19}$$

where $[IMP]$ is the concentration of the isopropyl methylphosphonate (IMP) ion; $[F^-]$ is the concentration of fluoride ion; $[OH^-]$ is the concentration of hydroxyl ion; and $[GB]$ is the concentration of GB remaining. All concentrations are in gram-moles per liter. This equilibrium constant was calculated from elemental reaction rates (see Harris et al., 1982).

The effect of pH on the amount of GB remaining in solution has been calculated based on this equilibrium constant (see Table D-2). For this calculation it was assumed that the IMP and F^- concentrations were 0.1 molar each. The calculation suggests that the pH must be maintained at 9 or more for the GB to bc reduced to below the detection level. (Note: The equilibrium quoted above has been questioned [Ward, 1998a], but it is certain that GB can reform at low pH at about the levels suggested in Table D-2. An alternative explanation for the reformation of GB at low pH is the reaction of the diester impurity.) Isopropyl methylphosphonic acid, the major product of the hydrolysis of GB, is a Schedule 2 compound[1] and must be irreversibly destroyed to meet the requirements of the Chemical Weapons Convention (CWC).

Military grade GB is not very pure. Early lots (1 to 241) were manufactured to a specification of 92 percent and were distilled. Later lots (242 to 431) were manufactured to a specification of 88 percent and were not distilled (U.S. Army, 1996). Stabilizers were added to all lots to retard degradation. The first stabilizer used was tributylamine (TBA); later, diisopropylcarbodiimide (DICDI) was used. Some of the older GB has been found to contain crystalline material, believed to result from the TBA inhibitor. Some GB has been recovered, redistilled, and restabilized with DICDI. Thus, considerable variation in GB composition should be expected during the stockpile disposal program. The purity of GB at loading of weapons is reported to be between 73 percent and 93 percent.

Hydrolysis of VX (described in the next section) produces a two-phase product: a small organic layer, which is reported to result primarily from the DICDI inhibitor, remains separate from the main aqueous phase. This same behavior has recently been reported from the hydrolysis of GB (Ward, 1998b). The organic layer amounts to approximately 2 to 5 percent of the original agent volume. The flash point of this organic layer has not been reported.

During the large-scale disposal of sarin at the Rocky Mountain Arsenal in 1973–1976 (Item 1 in Table D-1), GB apparently persisted in the brine at a very low level. The brine had to be certified to have less than 2 ng/ml (2 ppb w/v) of GB before going to a dryer, where the liquid was evaporated and the solid residue was packaged to be buried. Several explanations for the persistence of GB were offered (Harris et al 1982). The one most generally accepted at present is a problem in the analytical procedure, with GB reforming during analysis. The analytical method now proposed by ERDEC is: extraction in chloroform at high pH (e.g., pH of 11), followed by gas chromatography (GC) and mass spectrometry (MS). The extraction at high pH eliminates the problem of GB reforming.

The problem was serious at the Rocky Mountain Arsenal because hydrolysis was the only process being used. A follow-up process is now required to eliminate the major Schedule 2 product (isopropl-methylphosphonic acid; see Eqn. 1). This process could be used to ensure the complete destruction of low-level residual GB.

Hydrolysis of GB has also led to a problem with formation of a solid precipitate believed to be sodium fluoride (Ward, 1998c), which has only limited solubility in water (i.e., about 4 percent). A ferrous oxide or hydroxide has also been suggested. GB hydrolysate,

[1]Schedule 2 compounds are defined as those that can be readily converted to chemical agent. Any high valence phosphorus compound with alkyl side chain in the range C_1 to C_4, is considered a Schedule 2 compound.

provided by the Chemical Agent and Munitions Disposal Systems (CAMDS) for use by the ACWA technology providers in their process demonstrations, was prepared in fairly dilute solution to avoid formation of the precipitates. The formulation used was 7 gallons of GB with 93 gallons of 5 percent NaOH solution in water. (This represents roughly 10 percent excess of NaOH solution in water and resulted in a pH between 11 and 12.)

The hydrolysate will probably be determined to be toxic when animal toxicity studies are completed (by analogy with VX-hydrolysate, below). Thus, it should be handled with care as a toxic material.

HYDROLYSIS OF VX

The hydrolysis of VX proceeds much more slowly than that of GB. In dilute solution, at 22°C (72°F):

$$\frac{d[VX]}{dt} = -5.2 \times 10^{-3}[OH][VX]$$

The rate constant shown is 4,800 times smaller than that of GB. A reduction in concentration by a factor of one million would require more than 7 hours. The activation energy for the reaction is given as 15.0 kcal/mol (at pH = 12). Thus, increasing reaction temperature from 22°C to 90°C, for example, would increase the reaction rate by a factor of 117; the reaction time would be correspondingly reduced. The temperature chosen for hydrolysis of VX at Newport is 90°C (194°F).

The rates quoted above are for VX in solution. However, VX has very limited solubility at high pH. Therefore, the reaction rate will be limited by the mass-transfer rate between the two phases. As reaction proceeds, the products (see Figure D-1), which are soluble in water, increase the solubility of the remaining VX. In development work performed for Newport, vigorous stirring was used to disperse the two phases.

The major reactions in the neutralization process are shown in Figure D-1. The overall reaction requires more than 2 moles of NaOH per mole of VX to neutralize the ethyl methylphosphonic acid (EMPA) and methyl phosphonic acid (MPA), as well as the thiolamine, and to maintain a high pH. VX exhibits behavior similar to GB: if the pH of the product solution is allowed to drop, VX will reform. Approximately 90 percent of the VX reacts directly (by the top reaction), breaking the P-S bond to form the sodium salt of EMPA and the bis (isopropyl) amino ethyl thiol ("thiolamine"). However, about 10 percent of the VX reacts by a second channel in which the P-O bond is cleaved to form the intermediate compound shown as EA2192, which is itself highly toxic but will react slowly to form MPA and more thiolamine. EA2192 is hydrolyzed at least ten times more slowly than VX. About six hours at 90°C (194°F) are required for adequate destruction.

Most of the products are soluble in aqueous NaOH, but a small residue, arising mainly from a stabilizer added to the original VX, remains as an insoluble organic layer less dense than water. The organic layer is flammable with a flash point of 53°C (127°F) (ERDEC, 1998). This should be recognized as a hazard in carrying out the reaction at 90°C (194°F).

Military grade VX is 90 to 95 percent pure; a typical analysis is shown in Tables D-3 and D-4. The agent analysis was obtained by GC with a thermal conductivity detector. The other organic components were

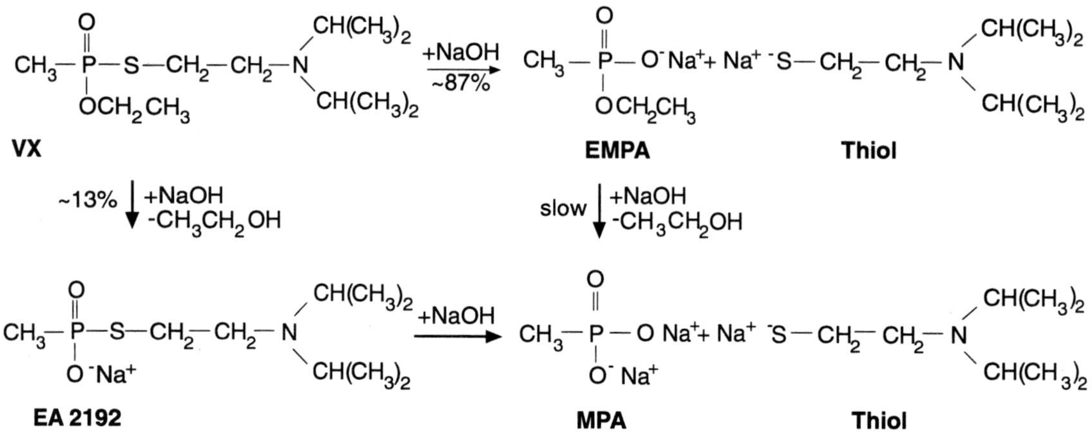

FIGURE D-1 Primary reactions involved in VX hydrolysis. Source: Yang et al., 1992; NRC, 1996.

TABLE D-3 Results of the VX Ton Container Survey Program (Organics)[a]

Organic Compounds	Mole %
VX	95.700
S, S′ bis(2-diisopropylaminoethyl) methylphosphonodithiolate	0.135
Dimethyl ketone (acetone)	0.004
Diisopropylamine	0.103
N, N-diisopropylmethylamine	0.003
Diisopropylcarbodiimide (stabilizer)	1.190
N, N-diisopropylethylamine	0.004
O-ethyl methylethylphosphinate	0.021
1, 3-diisopropylurea	0.004
Diethyl methylphosphonate	0.041
2-(diisopropylamino)ethane thiol	0.065
O, O-diethyl methylphosphonothiolate	0.098
O, S-diethyl methylphosphonothiolate	0.046
2-(Diisopropylamino)ethyl ethyl sulfide	0.081
Diethyl dimethylpyrophosphonate (pyro)	0.072
O, O-diethyl dimethylpyrophosphonothioate	0.124
O-(2-diisopropylaminoethyl) O-ethylmethylphosphonate (QB)	0.170
1, 2-bis (ethyl methylphosphonothiolo)ethane	0.450
Unknowns	0.500
Total	100.000

[a]Analyzed by gas chromatography/mass spectroscopy (GC/MS).

Source: U.S. Army, 1997a.

analyzed by GC with an electron-impact mass selective detector. Metal analysis was performed with inductively-coupled plasma/atomic emission spectroscopy (ICP/AES). Mercury was analyzed by cold-vapor atomic absorption.

The largest "impurity" listed is DICDI, which was added to the VX as a stabilizer (antioxidant). Its concentration can be as high as 5 percent. The other impurities shown are present at large parts-per-million (ppm) concentrations; many impurities in the parts-per-billion (ppb) range would probably be found with more sensitive analytic techniques.

TABLE D-4 Results of the VX Ton Container Survey Program (Metals)[a]

Metal	Analytical Result (ppm)
Arsenic	6.700
Chromium	1.200
Lead	0.370
Selenium	3.600
Zinc	4.400
Mercury	0.130
Copper	0.500
Iron	17.000
Magnesium	4.200
Calcium	28.000
Barium	0.031

[a]Analyzed by inductively coupled plasma (ICP) spectrometry.

Source: U.S. Army, 1997a.

The aqueous phase contains most of the product, about 95 percent; the organic layer has approximately 5 percent of the total. The compositions of the two product phases will, of course, be quite different. A typical composition for the total mixture (obtained by "homogenizing" the two phases) is shown in Table D-5. The analysis was based on ^{13}C and ^{31}P nuclear magnetic resonance spectrometry (NMR).

The major products are those expected from the reaction sequence given in Figure D-1, together with nonreacted impurities expected from the initial composition. Materials present at very low concentrations (e.g., ppm range or less) are not, in general, differentiated in the analysis. The exceptions are VX, shown in Table D-6 to be remaining to the ppb range, and EA 2192, in the ppm range.

The hydrolysis product has an unpleasant odor, which can be mitigated by the addition of bleach after the hydrolysis reaction is complete. This may or may not be desirable, depending on the subsequent treatment planned for the material.

TABLE D-5 Analysis of Homogenized VX Hydrolysate after 240-Minute Reaction Time[a]

Compound	Formula	Mole %
EMPA (Ethyl methlyphosphonic acid)	$OP(CH_3)(O\ CH_2\ CH_3)\ (OH)$	41.5
MPA (Methylphosphonic acid)	$OP\ (CH_3)(OH)_2$	4.6
EMPSH (O-Ethyl methylphosphonothioic acid)	$SP(CH_3)(O\ CH_2\ CH_3)(OH)$	0.3
Other ^{31}P	Not applicable	0.2
RSH (Diisopropylaminoethanethiol)	R′ - SH	43.4
RSSR (Bis [diisopropylaminoethyl] disulfide)	R′ - SS′R′	0.4
RSR (Bis [diisopropylaminoethyl] sulfide)	R′ - S - R′	0.7
RSCCSH (2 – [diisopropylaminoethylthio] ethanethiol	$R1 - S1 - CH_2\ CH_2SH$	1.3
DIPA (Diisopropylamine)	$HN[CH(CH_3)_2]_2$	0.6
CDI (Dicyclohexylcarbodiimide)	$C_6H_{11} - N = C = N - C_6H_{11}$	2.7
Other RS-compounds	Not applicable	1.9

[a]Analyzed by ^{13}C and ^{31}P nuclear magnetic resonance (NMR) spectroscopy.

Source: U.S. Army, 1997b.

VX was hydrolyzed at CAMDS to supply hydrolysate for the ACWA technology provider demonstrations. This hydrolysate had a higher concentration than the GB hydrolysate. Thirty-three gallons of VX were mixed with 67 gallons of 20 percent NaOH solution in water to produce the hydrolysate.

The primary purpose of the VX hydrolysis is to reduce the material toxicity. Animal tests have been carried out to confirm the reduced toxicity. The intravenous LD_{50} value of VX for mice was reported to be 0.0141 mg/kg. The hydrolysis product was 4 to 6 orders of magnitude less toxic. The characteristic signs of nerve agent attack were not observed with the hydrolyzed material. The remaining toxicity is due to the salt and organic content of the samples.

The major phosphorus products in the hydrolysate are considered to be Schedule 2 compounds; they must be irreversibly destroyed to meet the requirement of the CWC.

In summary, the hydrolysis reaction can reduce the toxicity of the original VX by a large factor, 5,000-fold to 25,000-fold. The product retains a low level of toxicity and will require further treatment before being released to the environment.

TABLE D-6 Residual VX and EA 2192 Concentrations from 12-Liter Reactor Tests

Reaction Time (minutes)	Homogenized Samples	
	VX[a]	EA 2192[b]
30	< 20 ppb	65 ppm
120	< 20 ppb	< 24 ppm

[a]Analyzed by GC/ion trap mass spectrometry (ITMS).
[b]Analyzed by ^{31}P NMR spectroscopy.

Source: U.S. Army, 1997b.

HYDROLYSIS OF HD

The mustard stored at Aberdeen, Maryland, will be detoxified by hydrolysis with hot water (90°C; 194°F) followed by the addition of caustic.

The overall desired reaction is:

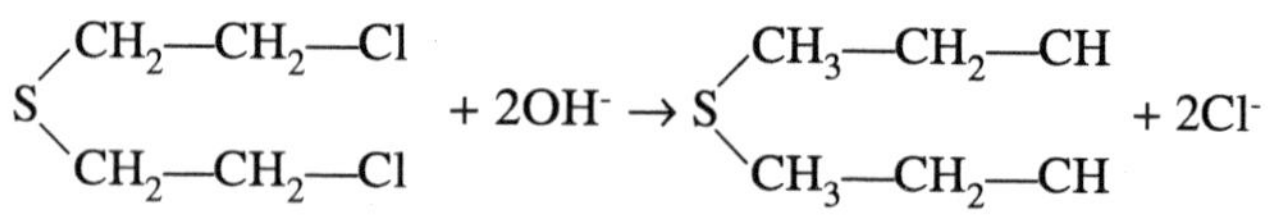

This overall reaction follows a complex series of steps, however, leading to several products. In addition, military grade mustard is usually impure (as low as 80 percent purity), and the impurities contribute to the complexity of the neutralized mix.

There are several reasons for the two-step reaction sequence (hot water hydrolysis followed by addition of caustic). Hydrolysis with water is preferred because caustic produces vinyl compounds, which are considered toxic and resist further treatment. The products of the aqueous hydrolysis are strongly acidic, and the pH must be adjusted for the subsequent treatment proposed for Aberdeen (biodegradation). Finally, the added caustic will react with a small amount of sulfonium salts, which are themselves toxic.

A typical analysis of liquid mustard is shown in Table D-7 (based on GC/MS analysis). There are probably many more components present at lower

TABLE D-7 Typical Composition of HD Agent[a]

Compound	Mole %[c]
HD	91.38
Q[b]	6.08
2-chloroethyl 4-chlorobutyl sulfide	0.86
1, 4-dithiane	0.81
1, 2-dichloroethane	0.35
Bis 3-chloropropyl sulfide	0.18
2-chloropropyl 3′-chloropropyl sulfide	0.18
2-chloroethyl 3-chloropropyl sulfide	0.14
1-chloropropyl 2-chloroethyl sulfide	0.02
1, 4-thioxane	< 0.01

[a]Analyzed by GC/MS.
[b]Q is the following cyclic sulfonium compound:

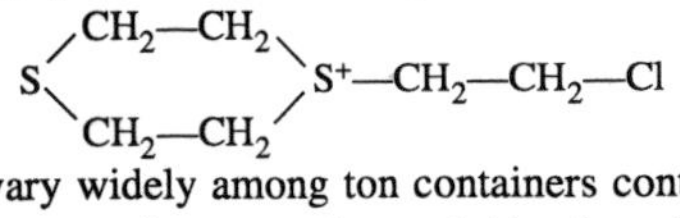

[c]Impurities may vary widely among ton containers containing HD. For example, in a survey of ton containers at Aberdeen (U.S. Army, 1996), the mole percent of 2-chloroethyl 4-chlorobutyl sulfide ranged from 0 to 2.32, for trichlorethylene it varied between 0 and 0.02.

Source: U.S. Army, 1996.

concentrations (e.g., at the ppm level). The mustard also contains small concentrations of metals (see Table D-8) the exact forms of which are not known.

There is generally a residual "heel" in mustard containers—a gel that will not flow but that can be washed out. The heel can amount to more than 10 percent of the stored agent. Analyses of three heel samples are given in Table D-9. The heel also contains metals, including a large amount of iron (e.g., > 10,000 ppm as iron sulfide).

Clean out of the ton containers containing HD leads to a small vapor stream that requires treatment prior to release. An analysis of a vapor sample is shown in Table D-10. Similar vapors should be expected when cleaning out mustard-filled weapons.

TABLE D-8 Concentration of Metals in HD Agent[a]

Metal	Content (ppm)
Aluminum	< 13.0
Antimony	< 9.0
Arsenic	18.9
Barium	< 0.3
Beryllium	< 0.2
Bismuth	< 12.0
Cadmium	< 3.0
Calcium	9.6
Chromium	< 2.0
Cobalt	< 2.0
Copper	91.8
Iron	5,035.0
Lead	< 7.0
Magnesium	< 0.4
Manganese	0.6
Nickel	< 4.0
Phosphorus	32.6
Selenium	< 14.0
Silicon	110.9
Silver	< 4.0
Sodium	< 16.0
Sulfur	264,420.0
Thallium	< 13.0
Thorium	< 11.0
Tin	< 6.0
Vanadium	< 3.0
Zinc	< 0.7

[a]Analyzed by ICP spectrometry.

Source: U.S. Army, 1996.

Mustard is relatively insoluble in water, so the hydrolysis reaction must occur at the interface. High-shear mixing has been used to speed up the reaction. The rate-determining step of the reaction with water was found to be the formation of an intermediate cyclic ethylene sulfonium ion:

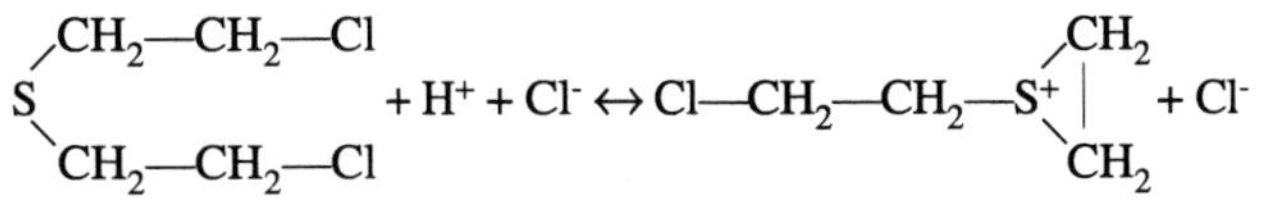

TABLE D-9 Complex Organic Compounds in HD Heel by NMR (mole percent)[a]

Sample #	Location	HD	Cyclic Sulfonium Ion[b]	1, 4-Dithiane	Other
C-03-04-HH-1936	Top of heel	53	42	—	5
C-03-05-HH-1936	Bottom of heel	33	60	1	6
C-03-06-HH-1936	Middle portion	14	86	—	—

[a]Analyzed by ^{1}H and ^{13}C NMR.
[b]The cyclic sulfonium ion is

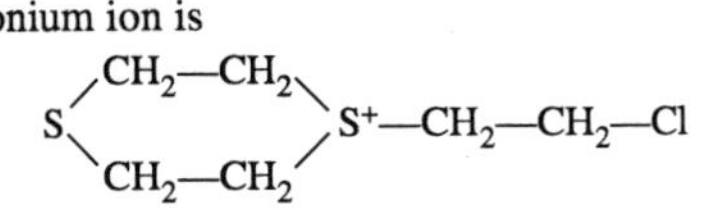

Source: U.S. Army, 1996.

TABLE D-10 Content of HD Agent and Volatile Organic Compounds in Initial Ton Container Vapor[a]

Compound	Concentration (mg/L)
1, 2-dichloroethane	6.000
Bis-(2-chloroethyl) sulfide (HD)	0.830
1, 4-dithiane	0.600
2-Chlorobutane	0.210
Tetrachloroethene	0.083

[a]Analyzed by GC/MS.

Source: U.S. Army, 1996.

This intermediate then reacts further, and a series of materials other than the expected thiodiglycol are formed. A suggested reaction pathway is shown in Figure D-2.

In early work on mustard hydrolysis, HD could not be detoxified by hydrolysis. The rate of reaction was slow and was controlled by the rate of mass transfer. As shown in the reaction scheme of Figure D-2, sulfonium ion aggregates H-TG, CH-TG, and H-2TG were formed and could be stable; also, H-2TG was believed to be quite toxic. All of these observations were based on low-temperature hydrolysis with moderate stirring.

With very vigorous high-shear mixing and high temperature (90°C; 194°F), very complete hydrolysis, primarily to the desired thiodiglycol, has been achieved. Work at ERDEC using hot water for hydrolysis with NaOH solution added at the end of the reaction has reduced mustard to less than 20 ppb (the detection level); 99 percent of the mustard was converted to the thiodiglycol.

The organic compounds in the hydrolysate at different times during the reaction—after agent addition to the hot water, and after subsequent addition of NaOH—are shown in Table D-11. (Only the major components were analyzed; there are undoubtedly many others at ppm-levels). The analysis of Table D-11 does not show the chlorinated hydrocarbons present in the original mustard, which go through the hydrolysis reaction essentially unchanged.

The hydrolyzed product remains toxic to some animal species. For example, fathead minnows (a freshwater species) showed an EC_{50} of 12 percent vol./vol.

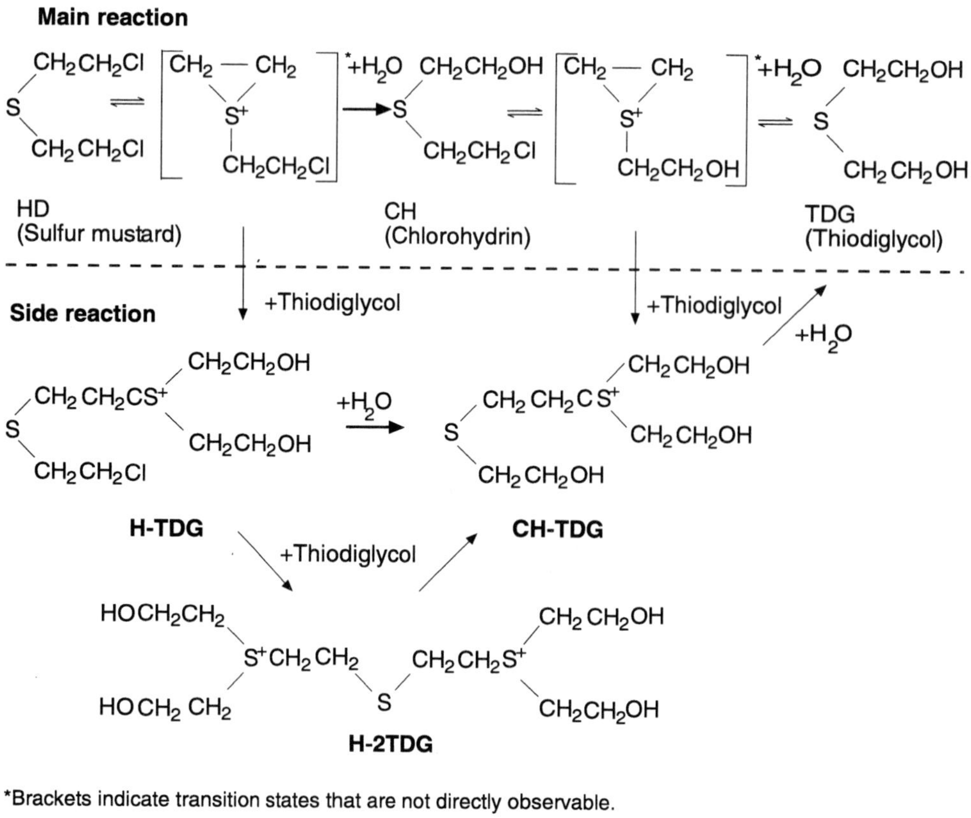

FIGURE D-2 Reversible formations of the sulfonium ion aggregates in the hydrolysis of mustard. Source: Yang et al., 1992; NRC, 1996.

TABLE D-11 Organic Compounds in HD Hydrolysate (Mole Percent)

	Analysis of Residuals	
Compound	79 Minutes after End of HD Feed	63 Minutes after End of NaOH Addition
TDG	90.70	91.20
CHTG[b]	1.70	0.05
Q-OH[c]	2.50	3.40
1, 4-dithiane	0.50	0.80
Glycol	0.50	0.50
$C_6H_{12}SO_2$ compounds	2.10	2.20
Acetone	0.02	Not detected
Other	2.00	1.80

[a]Analyzed by ^1H NMR.
[b]CHTG = $HOCH_2\ CH_2\text{-S-}\ CH_2\ CH_2\text{-S}^+\text{-(}\ CH_2CH_2OH)_2$
[c]Q-OH = $HO\ CH_2\ CH_2\text{-S-}\ CH_2\ CH_2\text{-S-}\ CH_2\ CH_2OH$

Source: U.S. Army, 1996.

(i.e., a concentration of 12 vol. percent hydrolysate in water led to the death of half the test population.) Sheepshead minnows, a saltwater species, had a higher tolerance: EC_{50} of 80 percent vol./vol. The difference in the tolerance level of the two species is probably due to the high salt content of the hydrolysate.

The hydrolysis product is not acceptable for direct discharge and must be further treated. The major product (the thiodiglycol) is considered a Schedule 2 compound and must be irreversibly reacted to meet the requirements of the CWC.

Recent work on HD has shown that hydrolysis at high temperature (90°C; 194°F) with vigorous mixing can successfully destroy the mustard. The product is not acceptable for discharge, however, without further treatment.

REFERENCES

Beaudry, W.T., J.H. Buchanan, D.K. Rohrbaugh, J.B. Samuel, L.L. Szafraniec, and J.R. Ward. 1993. Analysis of Decontamination Solution of G-agents to Detect Reformation of Agents. ERDEC-TR-005. January 1993. Aberdeen Proving Ground, Md.: Edgewood Research, Development and Engineering Center.

ERDEC (Edgewood Research, Development and Engineering Center). 1996. HD Hydrolysis/Biodegradation Toxicology and Kinetics. ERDEC-TR-382. December 1996. Aberdeen Proving Ground, Md.: Edgewood Research, Development and Engineering Center.

ERDEC. 1997. Neutralization and Biodegradation of Sulfur Mustard. ERDEC-TR-388. February 1997. Aberdeen Proving Ground, Md.: Edgewood Research, Development and Engineering Center.

ERDEC. 1998. Material Safety Data Sheet, Revised August 1998. Aberdeen Proving Ground, Md.: Edgewood Research, Development and Engineering Center.

Gustafson R.L., and A.E. Martell. 1962. Reaction of Sarin with NaOH. Journal of the American Chemical Society 84: 2309–2316.

Harris, J.M., K. DeBruin, J.E. Gebhart, B.C. Garrett, and T.M. Prociv. 1982. GB-NaOH Neutralization Mechanism and Analytical Procedures Evaluation. DRXTH-SE-CR-82161. May 1982. Arberdeen Proving Ground, Md.: U.S. Army Toxic and Hazardous Material Agency.

NRC (National Research Council). 1993. Alternative Technologies for the Destruction of Chemical Agents and Munitions. Board on Army Science and Technology. Washington, D.C.: National Academy Press.

NRC. 1996. Review and Evaluation of Alternative Chemical Disposal Technologies. Board on Army Science and Technology. Washington, D.C.: National Academy Press.

U.S. Army. 1996. Chemical Reaction Analysis Report: Literature Search and Analysis of M28 Propellant Decomposition and Potential Reactions of GB and VX Nerve Agents with Propellant. Aberdeen Proving Ground, Md.: Program Manager for Chemical Demilitarization.

Reeves, A.M., and R. Macy. 1947. Hydrolysis of Agents GB and GE. TGIR No. 369. Project A1.13. Washington, D.C.: U.S. Army.

Ward, R. 1998a. Personal communication from R. Ward, Edgewood Research, Development, and Engineering Center, to Walter May, National Research Council Committee on Review and Evaluation of Alternative Technologies for Demilitarization of Assembled Chemical Weapons, July 15, 1998.

Ward, R. 1998b. Personal communication from R. Ward, Edgewood Research, Development, and Engineering Center, to Walter May, National Research Council Committee on Review and Evaluation of Alternative Technologies for Demilitarization of Assembled Chemical Weapons, July 20, 1998.

Ward, R. 1998c. Personal communication from R. Ward, Edgewood Research, Development, and Engineering Center, to Walter May, National Research Council Committee on Review and Evaluation of Alternative Technologies for Demilitarization of Assembled Chemical Weapons, November 16, 1998.

U.S. Army. 1996. Ton Container Decontamination and Disposal Program Demonstration Report: A Second Mustard Ton Container (TC Serial #D94102). Aberdeen Proving Ground, Md.: Program Manager for Chemical Demilitarization.

U.S. Army. 1997a. Project Information Package for the Newport VX Chemical Agent Disposal Facility. Aberdeen Proving Ground, Md.: Program Manager for Chemical Demilitarization.

U.S. Army. 1997b. Neutralization of VX Nerve Agent with Sodium Hydroxide: 12-liter Mettler Test Report. Aberdeen Proving Ground, Md.: Program Manager for Chemical Demilitarization.

Yang Y.-C. 1995. Chemical Reactions for Neutralizing Chemical Warfare Agents. Chemistry and Industry 9: 334–337.

Yang Y.-C., J.A. Baker, and J.R. Ward. 1992. Decontamination of chemical warfare agents. Chemistry Review 92: 1729–1743.

APPENDIX E

Neutralization of Energetic Materials by Hydrolysis

BACKGROUND

Because most energetic materials are synthesized in acidic media (or salts of acids), they are vulnerable to hydrolysis. The hydrolysis of energetic materials, such as 2,4,6-trinitrotoluene (TNT), has been of some interest since the mid-1800s (Wilbrand, 1863; Hepp, 1882; Meyer, 1889). However, the work has been sporadic and the results often inconclusive. Limited reviews of hydrolysis reactions for energetics can be found in the following general references: Urbanski, 1964, 1965, 1967, 1984; Feuer, 1969, 1970. Recently, the Navy has published a review of alkaline hydrolysis of energetic materials pertinent to assembled chemical weapons (ACWs) (Newman, 1999).

Because open burning or open detonation (OB/OD) of energetic materials has become increasingly unpopular, more environmentally acceptable alternative technologies have been proposed. Base hydrolysis has recently been proposed for the demilitarization and/or destruction of some types of energetic materials. For example, several investigators have shown that base hydrolysis combined with supercritical water oxidation may be effective for neutralizing pressed explosive composites containing 1,3,5,7-tetraaza-1,3,5,7-tetranitrocyclooctane (HMX) (Spontarelli et al., 1994; Flesner et al., 1994; Skidmore et al., 1995; Bishop et al., 1996). Cannizzo et al. (1995) have demonstrated base hydrolysis of a composite propellant containing HMX, ammonium perchlorate, nitrate esters, and aluminum powder. Borcherding and his coworkers (Borcherding and Wolbach, 1995; Borcherding, 1997, 1998) have performed base hydrolysis on at least three types of composite propellant and have claimed, "the products of the process will be treatable in conventional waste treatment facilities." In September 1997, a 200 lb/day hydrolysis-treatment facility became operational and replaced routine OB/OD disposal of waste composite propellant at a site near San Jose, California.

Base hydrolysis decomposes energetic materials to organic and inorganic salts, soluble organic compounds, and various gases. The base usually attacks all of the functional groups of the energetic material. In the past, these reactions were usually conducted at ambient conditions with dilute concentrations of energetic material and base. Now, however, they are conducted at elevated temperature (between 60°C [140°F] and 155°C [311°F]) and often at elevated pressure (up to 14 atm) in a strong base solution (pH $\geq$ 12). In some cases, base hydrolysis has been conducted at supercritical conditions (Larson et al., 1998). Typically, higher temperature and pressure are used to increase the solubility of reactants and to prevent clogging of the reactor. Sodium hydroxide, potassium hydroxide, ammonium hydroxide, and sodium carbonate have been used. These reactions are exothermic and must be carefully controlled by a method that can immediately recognize the onset of a thermal excursion and prevent a runaway exothermic reaction. The reactor must also be designed to contain the maximum credible explosive event.

Depending on the specific conditions and materials, several reaction routes are possible that may produce undesirable precipitates. Energetic materials, such as nitrate esters, nitroaromatics, and nitramines, usually decompose to form nitrates, nitrites, ammonia, nitrogen, hydrogen, organic acids, and formaldehyde. Intermediates formed during the hydrolysis of some energetic materials can provide reaction pathways to other nitro derivatives and addition products that are still energetic. Current work indicates this occurs preferentially with aromatic nitro compounds and usually when the energetic material is in excess.

The solid particles in hydrolysis must be small enough to expose adequate surface area for expedient solvation and subsequent reaction. In most of the work on hydrolysis documented in the literature, the energetic material has been in the form of molding powder, small pellets, or small chunks from demilitarization operations. The energetic material has then been dissolved, and hydrolysis has been performed well within solubility limits. For the Assembled Chemical Weapons Assessment (ACWA), the energetic material is still in ordnance, so the first challenge is to remove it and reduce it to an appropriate size. The second challenge is to determine safe operating parameters for hydrolysis of a heterogeneous feedstock that may include energetic materials, metals, and contaminants. The third challenge is for technology providers to design process equipment and develop process-control protocols that minimize the accumulation of precipitates and minimize damage in the event of an accident. Finally, the technology providers must scale up a robust process that can meet the Army's requirements for various weapon types and processing rates. The scalability of operating parameters for base hydrolysis of energetic materials is not universal and will depend on the specific chemical munitions being demilitarized.

Base hydrolysis has been proposed by several technology providers for the destruction of energetic materials in the ACW inventory. These materials (which make up fuze components, leads, boosters, bursters, and rocket motors) include: of 1,3,5-triaza-1,3,5-trinitrocyclohexane (RDX); Composition B (Comp B) (a 60/40 gravimetric blend of RDX and TNT plus 1 wt. percent wax added as a desensitizer); 70/30 tetrytol (a 70/30 gravimetric blend of tetryl and TNT); and M28 double-base propellant (which contains nitrocellulose [NC] and nitroglycerin [NG]). These materials fall into four classes: aromatic nitro compounds (such as TNT), aromatic nitramines (such as tetryl), heterocyclic nitramines (such as RDX), and nitrate esters (such as nc and ng).

AROMATIC NITRO COMPOUNDS

Several references can provide a review of the reactivity of aromatic nitro compounds (Feuer, 1969; Urbanski, 1964, 1984). It is well documented that heterolytic substitutions occur through electrophilic and nucleophilic attacks from cations and anions, respectively. Homolytic substitutions involve the reaction of uncharged free radicals. Base hydrolysis is usually a heterolytic substitution as a result of a nucleophilic attack of the alkaline anion. However, there is some evidence that homolytic substitutions can occur simultaneously.

The reactions of aromatic nitro compounds with bases are often accompanied by intense color changes. Violet, red, or brown colors can be observed depending on the alkali system. Much of what is now known about these reactions has been attributed to Jackson (Jackson and Gazzolo, 1900), Meisenheimer (Meisenheimer, 1902), and Janovsky (Janovsky and Erb, 1886), whose fundamental work with nitrobenzene around the turn of the century identified some of the reversible reactions that form reactive intermediates. Several possible resonance structures can be quite energetic, including Jackson-Meisenheimer complexes, Janovsky complexes, and other sigma (σ) complexes. Russell and others have demonstrated how nucleophilic attack of a strong base (B^-) on aromatic nitro compounds can simultaneously produce Jackson-Meisenheimer complexes, a radial anion, and a free radical (Urbanski, 1964, 1984; Russell et al., 1964). Obviously, these species will undergo further reaction with available substrates. Depending on the reagents used to generate the anion and the reaction conditions, several anionic sigma complexes are possible (Strauss, 1970).

One of the most common aromatic nitro compounds is TNT (molecular weight 227.13 g/mole, melting point 81°C [178°F]). TNT is resistant to reaction with acids, but like all nitro compounds, it reacts easily with bases. Because the solubility of TNT in water is only about 6.0×10^{-4} M at 25°C (77°F) (DA, 1984), hydrolysis of TNT has often been performed in an alcohol or a cosolvent system.

In 1975, Hammersley (1975) published a survey of alkali reactions with TNT that affirms the conclusions of Bernasconi (1971) and of Gold and Rochester (1964). Hammersly was probably unaware of the work of Cuta and Beranek (1974). Many others have also performed relevant work in this area (Blake et al., 1966; Caldin and Long, 1955; Shipp and Kaplan, 1966; Shipp et al., 1972; Schaal and Lambert, 1962). Most of the investigations attempt to identify species that cause color changes or verify the formation of specific products. However, all of these investigations were performed to study the "pink-water" problem[1] at production facilities. Their main focus was to find ways to precipitate TNT from dilute solution and prevent the discharge of the pink contaminants to the environment. Because destruction of TNT was not the goal, the quantitative chemical reaction kinetics for the alkaline decomposition of TNT were not determined.

The base hydrolysis of TNT is probably an extremely complex series of reactions (Newman, 1999). A plausible overall mechanism for the alkaline decomposition of TNT can be constructed from the work cited above. The reaction pathways and products differ if either the base or the TNT is in excess.

With excess base, the general mechanism is:

$$TNT + RO^- \leftrightarrow TNT^-$$
$$TNT^- \rightarrow A + NO_2^-$$
$$TNT^- \rightarrow B + C + NO_2^-$$
$$B \rightarrow NH_3 \uparrow + HCN \uparrow + HNO_2$$
$$C \rightarrow D$$
$$A \rightarrow E \downarrow$$

With excess TNT, the general mechanism is:

$$TNT + RO^- \leftrightarrow TNT^-$$
$$TNT + TNT^- \leftrightarrow JC^-$$
$$TNT + JC^- \rightarrow \text{Addition Compounds} \downarrow$$

where,

TNT^- = family of MC^- resonance structures known as an anion of TNT
MC^- = Jackson-Meisenheimer complex
JC^- = Janovsky complex
A = alkali salts of dinitrocresol (these may precipitate)
B = picric acid
C = isomers of dinitrophenol (these may precipitate)
D = alkali salts of phenols
E = alkali nitronate salts (precipitate)

[1]Water on military bases was sometimes contaminated with trace amounts of TNT, which caused the water to take on a pink color.

Quantitative characterizations and prioritizations of the reaction pathways (or determination of the branching ratios) have not been completed. However, the current level of qualitative understanding is perhaps sufficient to realize that engineering practices can probably restrict the domain of possible products. The use of excess strong base is probably the most efficient means of ensuring that hydrolysis is driven to completion. However, elevated temperature and pressure must be carefully monitored to prevent spontaneous and violent exothermic reactions. A careful review of processing procedures is also warranted to safeguard personnel who may have to handle precipitates and other products of incomplete hydrolysis. This is especially true during emergency shutdowns and restart procedures. The ACWA technology providers should identify and demonstrate these engineering practices before full-scale hydrolysis of TNT is deemed safe and effective.

AROMATIC NITRAMINES

One of the most common aromatic nitramines is tetryl (2,4,6-trinitrophenylmethylnitramine) (Urbanski, 1967; DA, 1984; DA, 1991; Kaye and Herman, 1980). Tetryl is not very soluble in water. The molecular weight of tetryl is 287.1 g/mole, and its melting point is about 130°C (266°F). Tetryl forms a eutectic mixture at 58.3°C (137°F) with 57.9 wt. percent TNT (Urbanski, 1964).

Tetryl is resistant to attack by weak acids but will react with concentrated acid or any alkaline solution. As early as the 1880s, van Romburgh, Mertens, and Franchimont established a mechanism for the hydrolysis of tetryl (van Romburgh, 1889; Mertens, 1886; Franchimont and Backer, 1913). The nitramino group is hydrolyzed when tetryl is boiled in a dilute solution (2 wt. percent) of sodium carbonate, sodium hydroxide, or potassium hydroxide. The products are 2,4,6-trinitrophenol (picric acid), methylamine, and nitrous acid.

The kinetics for the hydrolysis of tetryl have not been documented. However, the rate of hydrolysis of tetryl is slower than that of TNT (Newman, 1999). The actual mechanism involved with complete alkaline decomposition is not known either but is probably very similar to those that occur in the hydrolysis of TNT. An anion resonance complex should be expected. This anion complex should be the result of a fast intermediate step, and its formation may complicate the family of possible hydrolysis products. If the tetryl is in excess, hydrolysis should yield insoluble Janovsky complexes and addition products that are potentially energetic. An excess of strong base should, however, drive hydrolysis to completion.

A plausible overall mechanism for the alkaline decomposition of tetryl can be hypothesized from the work previously cited for TNT. With excess base, the mechanism should be:

$$\text{tetryl} + RO^- \leftrightarrow \text{tetryl}^-$$
$$\text{tetryl}^- \rightarrow A + NO_2^-$$
$$\text{tetryl}^- \rightarrow B + C\uparrow + NO_2^-$$
$$B \rightarrow NH_3\uparrow + HCN\uparrow + HNO_2$$
$$A \rightarrow D\downarrow$$

With excess tetryl, the expected mechanism is:

$$\text{tetryl} + RO^- \leftrightarrow \text{tetryl}^-$$
$$\text{tetryl} + \text{tetryl}^- \leftrightarrow JC^-$$
$$\text{tetryl} + JC^- \rightarrow \text{Addition Compounds}\downarrow$$

where,

- tetryl^- = hypothetical resonance structures that behave as an anion of tetryl
- JC^- = Janovsky complexes of tetryl
- A = intermediate alkali salts (these may precipitate)
- B = picric acid
- C = methylamine
- D = alkali nitronate salts (precipitate)

More work is necessary to determine the reaction pathways and the branching ratios. However, as with TNT hydrolysis, the current level of understanding is perhaps sufficient to realize that engineering practices can probably restrict the domain of possible products.

A feasibility study has been performed by the Naval Air Warfare Center, Weapons Division, at China Lake, California, to measure the hydrolysis rates with excess NaOH of TNT and tetryl residue in simulated burster sections (GA, 1998). The NaOH concentration was varied from 12 wt. percent to 20 wt. percent (or from 3.4 M to 6.1 M). The results were reported as linear hydrolysis rates in millimeters per hour (mm/hr) as a function of temperature. The linear hydrolysis rate is a strong function of temperature.

One set of experiments was conducted in laboratory glassware, and the results are given in Table E-1. These data show that the hydrolysis rate for alkaline decomposition of tetryl increases with temperature and concentration of NaOH solution, as expected. However, the rate is only 16 grams per hour (g/hr) for tetryl in 6.1 M NaOH at 105°C (221°F). In similar tests at 95°C (203°F) in 3.4 M NaOH, the hydrolysis rate of TNT is at least 2.7 times faster.

The second set of experiments were performed in a simulated rotary hydrolyzer (see Chapter 6) in which the energetic materials were immersed in caustic 66 percent of the total test time. The results for these bench-scale rotary hydrolyzer tests are shown in Table E-2. In general, the hydrolysis rates are lower than those reported in Table E-1 for the laboratory-scale experiments. It was concluded that the mixing action for the laboratory-scale experiments was better than for

TABLE E-1 Hydrolysis Rates Obtained at Laboratory Scale

Test #	NaOH Concentration (M)	Explosive	T (°C)	Linear Hydrolysis Rate (mm/hr)	Gravimetric Hydrolysis Rate (g/hr)	Comments
2	3.4	tetryl	89	1.5	0.48	vigorous agitation
9[a]	3.4	tetryl	95	5 to 6	1.6 to 1.9	gentle agitation
1	3.4	tetryl	100	27	8.7	moderate agitation
11	6.1	tetryl	100	34	11	gentle agitation
8	6.1	tetryl	105	50	16	gentle agitation
6[a]	3.4	TNT	95		5.3	gentle agitation
5	3.4	TNT	95		40	vigorous agitation

[a]Tests 9 and 6 can be compared to assess relative hydrolysis rates for tetryl and TNT in 3.4 M NaOH.

Source: GA, 1998.

TABLE E-2 Hydrolysis Using a Bench-Scale Rotary Hydrolyzer

Test #	NaOH Concentration (M)	Explosive	T (°C)	Linear Hydrolysis Rate (mm/hr)	Gravimetric Hydrolysis Rate (g/hr)	Length of Tube (in)
3	3.4	tetryl	87	2.1	0.68	2
3	3.4	tetryl	87	1.85	0.60	4
3	3.4	tetryl	87	1.82	0.59	4
4	3.4	tetryl	95	7.4	2.4	2
4	3.4	tetryl	95	6	1.9	4
4	3.4	tetryl	95	7–8	2.3–2.6	10
4	3.4	tetryl	95	7–8	2.3–2.6	14
7	6.1	tetryl	101	18	5.8	10
13	6.1	tetryl	101	20	6.5	10
7	6.1	tetryl	101	19	6.1	14
13[a]	6.1	tetryl	101	20	6.5	14
2[a]	6.1	TNT	101		54	14

[a]Tests 13 and 12 can be compared to assess relative hydrolysis rates for tetryl and TNT in 6.1 M NaOH.

Source: GA, 1998.

the bench-scale rotary hydrolyzer sections (GA, 1998). This result may have ramifications for process design and scale up. No attempts were made during this feasibility study to identify the hydrolysis products for either tetryl or TNT.

In summary, the hydrolysis of tetryl pellets is not rapid, even in excess base and at reasonably high temperatures. The hydrolysis rate of tetryl is also much slower than that of TNT, which may prompt technology providers to raise the operating temperatures. However, tetryl will start to decompose to picric acid above 120°C (248°F) (Urbanski, 1967; DA, 1984), so attempts to increase the alkaline decomposition rate by performing base hydrolysis at temperatures above 120°C should be avoided, because these reactions may be difficult to control. Because hydrolysis is mass-transfer limited, reducing the chunk size of the tetryl feedstock and inducing vigorous agitation in excess base are probably better ways to increase the rate of hydrolysis once a threshold temperature and base concentration have been achieved. In any case, the technology providers must identify and demonstrate processing conditions for the base hydrolysis of tetryl before it is deemed safe and effective at full scale.

HETEROCYCLIC NITRAMINES

The most common heterocyclic nitramines are RDX and HMX (DA, 1984; Kaye and Herman, 1980; Meyers, 1981; DA, 1996a; DA 1996b). The molecular weight of RDX is 222.1 g/mole, and its melting point is about 190°C (374°F); the molecular weight of HMX is 296.2, and it melts at about 277°C (531°F). Because none of the munitions in the ACW inventory contains HMX, this discussion focuses on the hydrolysis of RDX.

RDX is manufactured in the United States by the Bachmann process, which produces high quality RDX that may contain from 4 to 17 wt. percent HMX as a by-product (Urbanski, 1967; DA, 1996a). RDX is practically insoluble in water.

In 1940, Somlo performed a feasibility study to de termine if RDX could be decomposed by alkali hydrolysis (Urbanski, 1967; Somlo, 1940). He reported that the heterogeneous decomposition of RDX in 1 M solution of NaOH could be accomplished in about 10 hours at 60°C (140°F). He also reported that increasing the concentration of the sodium hydroxide solution to 10 M and using intense stirring could reduce the time to about four hours. The decomposition products were nitrates, nitrites, organic acids, ammonia, nitrogen, formaldehyde, and hexamethylenetetramine.

A mechanism for the homogeneous alkaline decomposition of RDX has been reported by Epstein and Winkler (1951), Jones (1954), Hoffsommer et al. (1977), and Croce and Okamoto (1979). The detailed mechanism, a series of four reactions, was derived from base hydrolysis of RDX (3×10^{-5} to 6×10^{-4} M) in sodium methoxylate (≤ 0.5 M) between 19°C and 45°C (66°F and 113°F):

$$RDX + RO^- \rightarrow A + ROH + NO_2^-$$
$$A + RO^- \rightarrow B- + ROH$$
$$B^- + RO- \rightarrow C^{2-} + ROH$$
$$C^{2-} \rightarrow D^- + NO_2^-$$

where R = CH_3 W. H. Jones (1954) suggested that A was the product of HNO_2 elimination and might be 1,3,5-triaza-3,5-dinitrocyclohexene-1. He also suggested structures for the other hypothetical intermediates B, C, and D, which are all various resonance structures.

Jones also proposed the following equation for the rate of alkaline decomposition (*dx/dt*), which depends on the rate constants for the first and fourth reactions:

$$dx/dt = k_1 (a - 3x_1)(b - x_1) + k_2 (x_1 - x_2)$$

where,

- a = initial molarity of RO^-
- b = initial molarity of RDX
- k_1 = the rate constant for the first reaction
- k_2 = the rate constant for the fourth reaction
- x_1 = molarity of nitrite from the first reaction
- x_2 = molarity of nitrite from the fourth reaction

Jones also proposed the following expressions to determine the disappearance of RDX and RO^-:

$$- dR/dt = k_1 R (a - 3b + 3R) \quad \text{where } R = [RDX]$$
$$- dZ/dt = k_1 Z (3b - a + Z) \quad \text{where } Z = [RO^-]$$

TABLE E-3 Observed First-Order Rate Constants for Hydrolysis in Excess NaOH Solution (8.048×10^{-2} M) at 25 °C

% Completion	Time (min)	RDX Concentration $\times 10^5$ (M)	$k_1 \times 10^4$ (s^{-1})[a]
0.0	0.00	7.6	3.5
15.8	7.98	6.4	2.6
17.1	11.88	6.3	3.5
28.9	16.05	5.4	3.8
36.8	19.96	4.8	3.1
40.7	27.91	4.5	2.8
42.1	31.97	4.4	3.1
48.7	35.99	3.9	3.1
52.6	40.05	3.6	3.0
76.3	80.06	1.8	2.9
82.9	99.96	1.3	3.1
[a]100[b]	232.80[b]	0.1[b]	3.1

[a] $k_1 = 1/t \ln ([RDX]_o/[RDX]_t)$
[b] Calculated using the average first order rate constant from this experiment, k_1=3.1.

Source: Hoffsommer et al., 1977.

Epstein and Winkler (1951), Hoffsommer et al. (1977), and Croce and Okamoto (1979) have offered and verified a simplified expression for the reaction kinetics for RDX hydrolysis. The initial reaction is a proton abstraction from the acidic methylene hydrogens located between two adjacent nitramine groups, whereby nitrous acid is liberated. This initial second-order reaction is the rate-limiting step. Epstein and Winkler reported second-order rate constants for base hydrolysis of RDX in acetone with aqueous NaOH (≤ 0.6 N). The relative concentrations of the solute and solvent comprising these solutions were chosen to prevent the precipitation of solids and ensure the miscibility of the alkaline solution with acetone.

Hoffsommer et al. also reported second-order rate constants for hydrolysis of RDX ($\leq 2 \times 10^{-4}$ M) in NaOH solutions (from 0.02 to 0.5 M) at temperatures of 25°C, 35°C, and 45°C (77°F, 95°F, and 113°F). The reaction kinetics were pseudo first order with excess hydroxyl ion:

$$- d[RDX]/dt = k_2 [NaOH][RDX] \approx k_1 [RDX]$$

The observed first-order rate constants for the aqueous, homogeneous, alkaline hydrolysis of RDX in excess base are provided in Table E-3. Second-order rate constants were calculated from the first-order rate constants using the expression $k_2 = k_1 / [OH^-]$. Second-order rate constants are given in Table E-4. The formation rate of nitrite ion during the alkaline hydrolysis of RDX at 45°C (113°F) is presented in Table E-5. The variation in hydrolysis products as a function of hydroxide-ion concentration is given in Table E-6.

Hoffsommer et al. (1977) stated that the formation of both the formate ion ($HCOO^-$) and formaldehyde (CH_2O) indicated a ring opening of RDX-h-5 by hydroxide-ion attack on carbon followed by a series of

TABLE E-4 Second-Order Rate Constants for Hydrolysis of RDX in NaOH Solutions from 0.02 to 0.25 M

Temperature (°C)	Number of Kinetic Experiments	$k_2 \times 10^3$ ($M^{-1} s^{-1}$)
25.0	6	3.9 ± 0.2
35.0	6	14 ± 1
45.0	5	48 ± 3

Source: Hoffsommer et al., 1977.

TABLE E-5 Formation of Nitrite Ion during Hydrolysis of RDX (7.07×10^{-5} M) at 45°C (113°F) with Excess NaOH Solution (6.82×10^{-2} M)

Time (min)	% RDX Hydrolysis	+ [NO2-] $\times 10^5$ (M)	Mole Ratio +[NO2-] / -[RDX]
1.38	23.4	1.61	0.97
2.93	43.3	3.32	1.08
4.32	56.7	4.59	1.14
6.17	69.8	5.70	1.16
8.17	79.4	6.54	1.17
41.00	100	8.35	1.18

Source: Hoffsommer et al., 1977.

TABLE E-6 Analysis of RDX Hydrolysis Products with Different Hydroxide Concentrations (Mole ratio of product formed per RDX hydrolyzed)

$[OH^-]$	$NO2^-$	N_2	NH_3	N_2O	$HCOO^-$	CH_2O	H_2
Weak, 0.1 M	1.1	0.12	0.9	1.2	0.7	1.1	0
Strong, 19 M	2.1	0.7	1.6	0.4	1.6		0.2

Source: Hoffsommer et al., 1977.

bond cleavages to form NH_3, N_2, and N_2O. The formation of some H_2 is probably due to OH^- attack on CH_2O under Cannizzo's conditions (Cannizzo et al., 1995; Walker, 1964). No evidence of the nitrate formation observed by Somlo (1940) was found.

Dell'Orco investigated the hydrolysis of Comp B (a blend of RDX and TNT) at the Los Alamos National Laboratory (LANL) under the Joint Department of Defense/Department of Energy (DOD/DOE) Munitions Technology Program (Dell'Orco et al., 1995). Simple experiments in alkaline decomposition were performed with 1.5 M NaOH solutions at both laboratory and pilot scale. These results are summarized in Table E-7. Two reaction times were reported. One represents the amount of time at or above 81°C (178°F); the other represents the total time from start to finish of the experiment. Because these experiments were performed beyond the solubility limits of Comp B, the hydrolysis rate was slower than expected.

Color changes from orange-red to dark brown and black were observed during these experiments. A significant exotherm occurred at 81°C (178°F) as the TNT in the Comp B melted and the rate of reaction increased. This event was accompanied by vigorous bubbling of off-gas. The reaction temperature was reported as self-limited because the boiling point of the solution was about 91°C (196°F). The reported products of reaction included precipitates, gases, and liquid-phase species. Precipitates included unreacted RDX and HMX. The gas contained nitrous oxide (N_2O) and ammonia (NH_3). The organic products found in the aqueous phase included "formate" (presumably the anion and salts or esters of formic acid) and "hexamine" (apparently hexamethylenetetramine as reported in Somlo [1940]). Ion chromatography results for anionic products from the hydrolysis of Comp B are given in Table E-8. The table

TABLE E-7 Summary of Alkaline Decomposition Experiments for Comp B Performed by LANL

Scale	[NaOH] (M)	T (°C)	Time of Reaction (min)	% Completion	Gravimetric Hydrolysis Rate (g/hr)
1 g Comp B in 10 ml of NaOH solution[a]	1.5 1.5	85 33–81	15 27	98.5 98.5	3.9 2.2
100 g Comp B in 1 liter of NaOH solution[a]	1.5 1.5	>81 33–81	90 210	>99.9 >99.9	66.7 28.6
Two 711 g billets in 14 liters of NaOH solution[a]	1.5 1.5	>81 28–83	120 300	95.7 95.7	676 270

[a]The concentration of Comp B in these experiments is calculated to be 0.44 M.

Source: Dell'Orco et al., 1995.

TABLE E-8 Anionic Products from the Hydrolysis of Comp B

Anion	Shown as a Percentage of C or N in Comp B
formate	30% of C
acetate	1.2% of C
nitrite	25% of N
nitrate	0.2% of N

Source: Dell'Orco et al., 1995.

lists only the aqueous-phase products and is, therefore, an incomplete material balance. No effort was made to quantify gas-phase products or to specify final products.

HMX is present as a contaminant in RDX manufactured in the United States. The solubility of HMX in water is only about 5 ppm at 20°C (68°F). Some investigations have been performed to determine the chemical-reaction kinetics and the products of HMX alkaline decomposition (Epstein and Winkler, 1951; Croce and Okamoto, 1979; Spontarelli et al., 1994). The rate-limiting step is a bimolecular elimination reaction producing nitrous acid. The reaction kinetics are second order, and according to Epstein and Winkler's data, the hydrolysis of HMX is more than 300 times slower than that of RDX at 15.5°C (60°F).

Bishop et al. (1996) have developed a reaction-rate model for the hydrolysis of HMX. Their second-order rate constant for the base hydrolysis of HMX in sodium carbonate (1.01.5 M) was fit to an Arrhenius equation. The experimentally determined values for activation energy and pre-exponential factors are given below in the expression for the reaction-rate constant.

$$\ln k = -(92.4 \pm 0.7 \text{ kJ/mol}) / RT + 20.8 \pm 0.8 \, (\ln s^{-1})$$

In this equation, the effect of temperature is substantial; however, the rate constant (k) is low, and the alkaline decomposition is slow. Bishop et al. claim that this model can be used in the design of reactors and in scale up to production.

Because the hydrolysis of HMX is mass-transfer limited, investigators have resorted to using higher temperatures and pressures to increase the reaction rate. Bishop et al. have recently developed a solid-liquid mass-transfer model to describe how faster rates can be achieved for the hydrolysis of HMX by using higher temperatures and vigorous agitation (Bishop et al., 1998). The hydrolysis products are given in Table E-9.

TABLE E-9 Product Analysis for Sodium Carbonate Hydrolysis of HMX Powder[a]

	% Total Carbon and Nitrogen in HMX (average of ten experiments)
Carbon in solution	
Inorganic carbon	0.5 to 1.5
Organic carbon	about 99
Carbon-bearing species	
Formate (aqueous)	44.00
Acetate (aqueous)	21.00
Carbon dioxide (gaseous)	0.86
Carbon monoxide (gaseous)	0.74
Carbon	34.00
Nitrogen-bearing species	
Nitrite (aqueous)	20.00
Nitrate (aqueous)	0.05
Nitrogen (gaseous)	4.10
Nitrous oxide (gaseous)	59.00
Ammonia (gaseous + aqueous)	25.00

[a]The uncertainty in the values presented is ± 15 percent.

Source: Bishop et al., 1998.

In summary, the alkaline decomposition of RDX and HMX is better understood than the alkaline decomposition of either TNT or tetryl. However, there are still uncertainties regarding the removal and particle-size reduction of energetic materials in the ACW inventory that will constitute the feed stream for the hydrolysis unit. Insufficient size reduction may adversely impact the effectiveness and scalability of processing parameters and may increase safety concerns, such as clogging of the reactor and exothermic runaway at full scale. Therefore, the ACWA technology providers must demonstrate processing conditions for realistic feeds before the alkaline decomposition of RDX, Composition B, or HMX can be deemed safe and effective at full scale.

NITROCELLULOSE AND OTHER NITRATE ESTERS

Information on nitrate esters and their use in various explosive and propellant applications is available from several sources (e.g., Miles, 1955; Urbanski, 1965, 1984; Lindner, 1980; DA, 1984; Fedoroff and Sheffield, 1962, 1972, 1974; Meyers, 1981), many of which include substantial information about characteristics and reactivity. The physical properties of nitrate esters and their affinity to undergo base hydrolysis or alkaline decomposition are described briefly below.

Nitrate esters are classified as primary, secondary, or tertiary, depending on whether they were derived from primary, secondary, or tertiary alcohols. The complicated base hydrolysis of nitrate esters is now reasonably well understood. It involves the cleavage of both C–O and O–N bonds. The hydrolysis reactions can be summarized by the following set of expressions (Urbanski, 1984; Baker and Neale, 1954, 1955; Baker and Heggs, 1955; Anbar et al., 1954):

1. Nucleophilic attack on carbon (S_N1 and S_N2):
   $HO^- + RCH_2 - O - NO_2 \rightarrow$
   $RCH_2 - OH + NO_3^-$
2. Nucleophilic attack on nitrogen (S_N2):
   $HO^- + RCH_2 - O - NO_2 \rightarrow$
   $[RCH_2 - O + HONO_2] \rightarrow RCH_2 - O - H + NO_3^-$
3. Nucleophilic attack on α-hydrogen (E2):
   $HO^- + RCH_2 - O - NO_2 \rightarrow$
   $RCH = O + NO_2^- + H_2O$
4. Nucleophilic attack on β-hydrogen (E1 and E2):
   $HO^- + R\,CH_2CH_2 - O - NO_2 \rightarrow$
   $RCH = CH_2 + NO_3^- + H_2O$

The hydrolysis of both primary and secondary mononitrate esters is slow. In general, the hydrolysis of primary nitrate esters proceeds mostly by substitution reactions 1 and 2 above. However, the hydrolysis of secondary nitrate esters is dominated by the elimination reactions 3 and 4.

NC and NG are the two nitrate esters of greatest concern in the ACWA program. NC is notoriously unstable at elevated temperatures. At 125°C (257°F), NC decomposes to CO, CO_2, H_2O, N_2, and NO. At 50°C (122°F), the rate of decomposition of NC is approximately 4.5×10^6 wt. percent per hour, increasing by a factor of about 3.5 with each 10°C (18°F) increase in temperature. Dry NC burns very rapidly and may detonate when acted upon by a sufficient stimulus if the NC is present in large quantities or if it is confined. NC is a dangerous material to handle dry because of its sensitivity to friction, static electricity, impact, and heat. The higher the nitrogen content, the more sensitive NC tends to be. Therefore, NC is always shipped wetted with water or alcohol. However, even NC with 40 percent (vol.) water can detonate if it is confined and acted upon by a sufficient stimulus (Lindner, 1980).

NC is not a single homogeneous product. Rather, it is a family of very similar compounds containing mono-, di-, and tri-nitrocellulose. The molecular weight of NC depends on the nitrogen content and can be calculated using the following expression.

$$\text{molecular weight of NC} = 324.2 + [\%\ N / 14.14]\ 270$$

Military grades of NC contain more than 12 percent N. Four grades of NC are specified for military use; however, it is often blended to meet the performance requirements of specific weapons. The cast double-base propellant grain, M28, specified for the M55 rocket contains both NC and NG. However, the actual nitrogen content for the NC is not specified (DA, 1966).

NC is soluble in acetone but insoluble in water. The higher the nitrogen content, the less soluble NC is in ether-alcohol mixtures (Sax and Lewis, 1987), and alkaline solutions. Therefore, the residence time required for alkaline decomposition of NC (14 percent N) is expected to be longer than for NC (12 percent N). Because NC is more soluble in ammonia than in aqueous alkaline solution, some researchers recommend ammonia hydrolysis, or "ammonolysis," for destroying NC (Cannizzo et al., 1995; Morgan et al., 1994; Melvin, 1997; GA, 1997). The alkaline decomposition of NC in aqueous NaOH solution is mass-transfer limited.

Kenyon and Gray (1936) performed a quantitative study of the alkaline decomposition of NC in which they examined the time necessary for dry NC fibers to disappear in NaOH solutions. Although their time measurement technique is not extremely reliable, it does indicate a relative order of magnitude required to perform the operation on a laboratory scale. They evaluated the effect of base concentration and temperature on hydrolysis rate. Their data set for the alkaline decomposition of NC at 30°C (86°F) is given in Table E-10 and at 60°C (140°F) in Table E-11. They concluded that alkaline decomposition of NC proceeds faster at higher temperature and in stronger base. They also observed that small amounts of carbon dioxide were products of decomposition and that large amounts of the nitrates were reduced to nitrites. An average of about 66.6 percent of the nitrogen in NC was reduced to

TABLE E-10 Alkaline Decomposition of NC (12.2% N) at 30°C (86°F)

Mass of NC (g)	Volume of NaOH Solution (cc)	NaOH Concentration (%)	Time Required for Decomposition (hrs)
10	100	20 (or 6.09 M)	2.58
10	200	10 (or 2.77 M)	11.13
10	400	5 (or 1.32 M)	23.33
10	800	2.5 (or 0.64 M)	170.90
10	2000	1.0 (or 0.25 M)	245.00

Source: Kenyon and Gray, 1936.

either nitrates or nitrites during alkaline decomposition at 30°C (86°F) and about 71 percent at 60°C (140°F).

NG is a pale yellow viscous liquid, also known as glyceryl trinitrate or trinitroglycerin, with a molecular weight of 227.1 g/mole. It is soluble in almost all organic solvents and is sufficiently soluble in water (0.18 g/100 g H_2O at 20°C [68°F]) to cause a contamination problem in a process waste stream. When acid free, NG is stable, but it is very sensitive to impact. NG decomposes at temperatures above 60°C (140°F) to form nitric oxides that catalyze further decomposition. Moisture increases the rate of decomposition. Unconfined NG will burn without transitioning to a detonation; however, it will detonate if confined.

Even though it is well known that NC and NG can be destroyed by alkaline decomposition in the laboratory, the application of this method to actual propellant formulations will have to be tested further to determine its feasibility as a demilitarization technique. NG readily plasticizes NC to form a gel that will solidify and provide useful mechanical properties, making it attractive as a rocket propellant. Unfortunately, this plasticizing effect makes alkaline decomposition of double-base propellants much more difficult than merely performing base hydrolysis of its components. Preliminary experiments indicate that base hydrolysis of gun propellants under normal conditions (at standard temperature and pressure) requires several hours for decomposition. During hydrolysis of a double-base propellant, NG can be expected to be hydrolyzed first via extraction from the NC. However, the conditions that favor this separation are severe. The double-base propellant must be in small pieces, and the temperature and/or pressure must be relatively high.

Bunte et al. (1977) have demonstrated a more rapid chemical destruction of double-base propellants using high-pressure alkaline hydrolysis. To destroy a large quantity of obsolete ordnance in Germany, researchers were asked to identify an "environmentally friendly" technology. Bunte, Krause, and Hirth have advocated alkaline hydrolysis at about 150°C (302°F) and 20 atm, which they claim results in the same hydrolysate products as under normal conditions but 8 to 10 times faster. The five actual gun and rocket propellant formulations used in their experiments are presented in Table E-12 as P1 through P5.

A 180-cm^3 autoclave (OD = 70 mm, ID = 48 mm) batch reactor was used to assess the products of alkaline decomposition of these propellants. The experimental conditions included a propellant-to-water weight ratio of 1 to 10, and a propellant-to-NaOH weight ratio of 2.3 to 1. About 120 ml of this reaction

TABLE E-11 Alkaline Decomposition of NC (12.2% N) at 60°C (140°F)

Mass of NC (g)	Volume of NaOH Solution (cc)	NaOH Concentration (%)	Time Required for Decomposition (hrs)
10	100	20 (or 6.09 M)	0.05
10	200	10 (or 2.77 M)	0.23
10	400	5 (or 1.32 M)	0.83
10	800	2.5 (or 0.64 M)	8.00

Source: Kenyon and Gray, 1936.

TABLE E-12 German Propellant Formulations Used in Pressured Alkaline Decomposition Experiments

Name	P1 (% by wt)	P2 (% by wt)	P3 (% by wt)	P4 (% by wt)	P5 (% by wt)
NC	97	80	57.5	57	56
NG		14	26.7	28	26
DNT		2	8.5	11	9
Centralite		2	2.9	3	2.9
DPA	2				
Graphite	1	1			
Potassium salt		1			
Lead			1.4		
Dibutylphthalate					4.5
Vaseline			1	1	1

NC = nitrocellulose (% N not given)
NG = nitroglycerin
DNT = 2,4-dinitrotoluene
DPA = diphenylamine

Source: Bunte et al., 1997.

mixture was put into the autoclave, and the autoclave was evacuated and then filled with helium to a pressure of about 20 atm. The reactor was heated at 10°C/min to 150°C (302°F) (15 minutes), and this temperature was held for one hour. The hydrolysis is exothermic, and small temperature increases (on the order of about 10°C [18°F]) were reported.

Because NC is the major component of the P1 through P5 propellants, the products of NC decomposition were identified. The main products are nitrite and nitrate in the aqueous phase, along with small-chain organic acids (Urbanski, 1984; Kenyon and Gray, 1936; Bunte et al., 1997). Using ion-exclusion chromatography, the small-chain organic acids were identified as malonic acid, succinic acid, glutaric acid, formic acid, acetic acid, and propionic acid. These products are all biodegradable, and Bunte et al. recommend treating them with a two-step bioremediation operation, anoxic denitrification and normal aerobic treatment.

The propellant samples were originally in the form of sticks and cylinders that were ground down to a particle size of about 1–5 mm. The results for the decomposition of the propellant P2 are summarized in Figures E-1 through E-4. The results for all five propellants are generalized in Table E-13.

As Figure E-1 shows, more than 50 wt. percent of the nitrogen in P2 is decomposed to gaseous products,

TABLE E-13 Results for the Pressured Alkaline Decomposition of Propellants P1–P5

	% of the Total Carbon and Nitrogen in the Propellant				
Residue	P1	P2	P3	P4	P5
Solid residue containing DPA, graphite, centralite, lead oxide, dibutylphthalate	2–3	3–12	10	5–16	10
Liquid residue containing NO_2, NO_3, DNT emulsion DPA emulsion, COD, HCO_3			40–70		
Gaseous residue containing N_2, N_2O, NO, CO_2, CO, CH_n			20–50		

Source: Bunte et al., 1997.

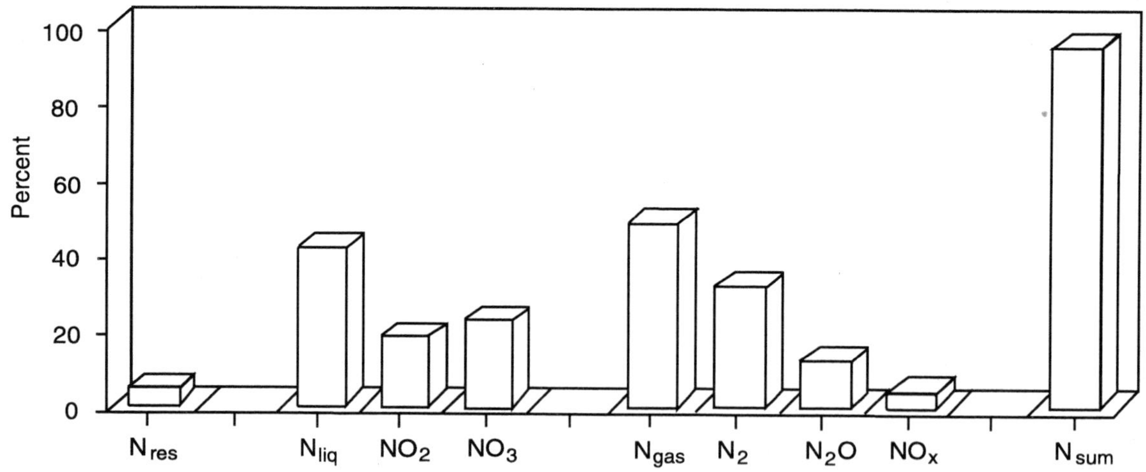

FIGURE E-1 Conversion of nitrogen (N) during pressurized alkaline decomposition of propellant P2.

which include mostly elemental nitrogen and nitrous oxide. The amount of NO_x is less than 10 wt. percent. More than 40 wt. percent of the nitrogen in P2 is decomposed to liquid products, which include nitrates and nitrites. Less than 5 wt. percent of the nitrogen in P2 is decomposed to a solid residue. The material balance for the carbon in P2 is uncertain, as shown in Figure E-2. The carbon content of the P2 propellant is transformed into about 8 wt. percent CO_2, less than 1 wt. percent CO, and less than 0.1 wt. percent, of methane. Depending on the specific additives in the propellant, between 6 and 10 wt. percent of the carbon can be found in the solid residue. The main products in the aqueous solution are carboxylic acids, carbonates, and bicarbonates. The chemical oxygen demand (COD) in the liquid phase is between 50 and 60 percent carbon. This is the measure of the amount of oxidizable compounds present in water. Because this value does not differentiate stable from unstable organic matter, it does not necessarily correlate with the biochemical oxygen-demand value.

The formation of gaseous products is strongly influenced by reaction conditions. The quantity of gaseous products can be reduced if stronger alkaline solutions are used (e.g., propellant-to-NaOH weight ratio of 1:1). Figures E-3 and E-4 show the conversion results for nitrogen and carbon, respectively, for the alkaline decomposition of propellant P5 using propellant-to-NaOH weight ratios of 2.3:1 (weaker alkaline solution) and 1:1 (stronger alkaline solution). In Figure E-3, the gaseous products are reduced from about 20.3 wt. percent to 2.6 wt. percent by going to the stronger alkaline

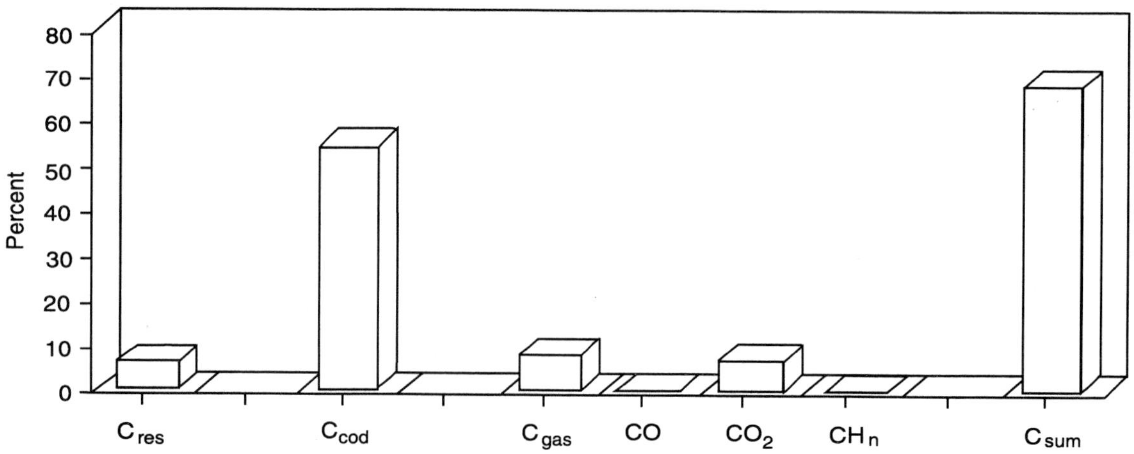

FIGURE E-2 Conversion of carbon (C) during pressurized alkaline decomposition of propellant P2.

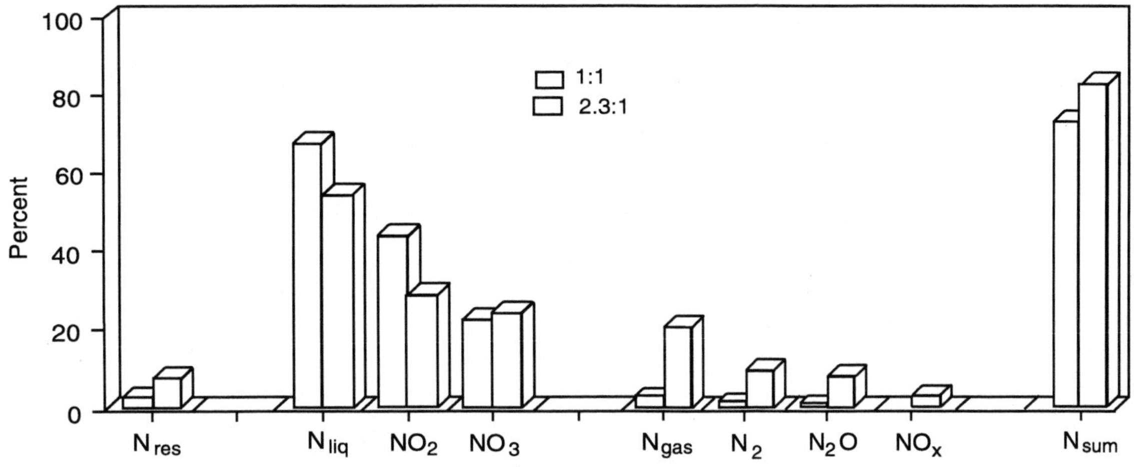

FIGURE E-3 Conversion of nitrogen (N) during pressurized alkaline decomposition of propellant P5 under different alkaline conditions.

solution. NO_x formation is also virtually eliminated. In Figure E-4, the CO_2 has been reduced from 4.5 wt. percent to less than 0.1 wt. percent. The carbon content in the solid residue is also reduced from 21.9 percent to 9.4 wt. percent. The COD value increased from 71 percent to 87 wt. percent, indicating more oxidizable components in the liquid phase.

The products of pressurized alkaline hydrolysis for propellants P1–P5 depend on the additives in the compositions. Components that do not completely react, such as diphenylamine (DPA) and centralite, will precipitate as solid residue. However, some DPA appears as an emulsion in the liquid phase. The most problematic component in this evaluation is dinitrotoluene (DNT). In additional experiments performed with pure 2,4-DNT, only 7 percent of the nitrogen was found as nitrite in the liquid phase (Bunte et al., 1997), and none was found in the solid residue. Therefore, it is believed that DNT is not completely converted and may still be present as an emulsion in the aqueous phase. Emulsified components, such as DNT and DPA, must be removed before aerobic bioremediation can proceed.

These results have been described in substantial detail because they impact the proposed approaches for demilitarizing assembled chemical weapons. Some technology providers propose using base hydrolysis of all of the energetic materials from ACW ordnance in a single batch reactor during a demilitarization campaign

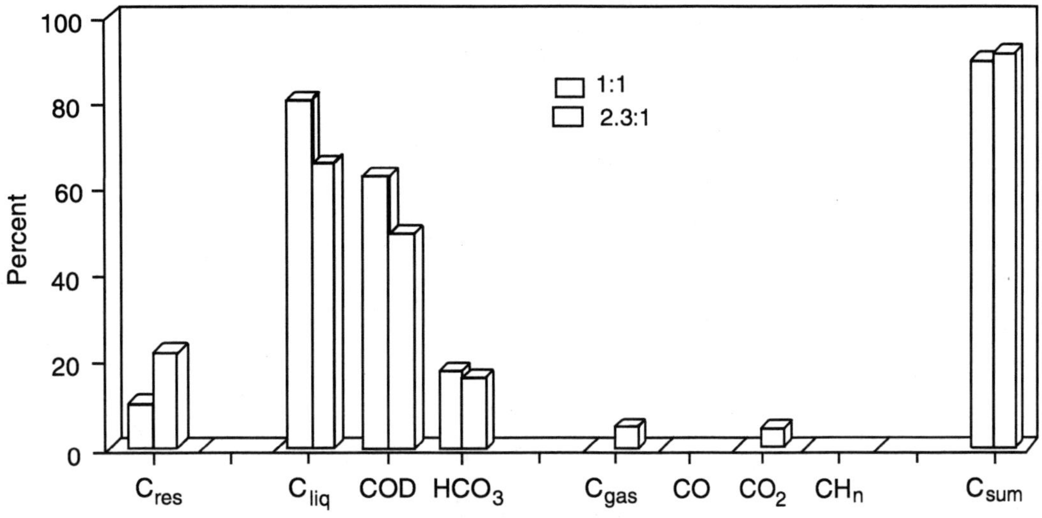

FIGURE E-4 Conversion of carbon (C) during pressurized alkaline decomposition of propellant P5 under different alkaline conditions.

(Lockheed Martin, 1997; Parsons, 1997). Because the M55 rockets contain both aromatic nitro compounds and nitrate esters, simultaneous hydrolysis on bursters and rocket motors could be problematic. First, the dissociated components of the M28 propellant could react with TNT intermediates during hydrolysis. Second, the decomposition of emulsified components or intermediates might require very severe operating conditions or extremely long residence times.

Every M55 rocket contains about 19.3 pounds of M28 propellant. Assuming that access to the grain is adequate and that particle size reduction is sufficient to obtain hydrolysis results similar to those presented for propellant P5, the NC and NG will be hydrolyzed. However, the remaining components will either be precipitates or will exist as an emulsion in the aqueous or liquid phase.

Lead stearate, an additive in M28 propellant, is insoluble in water at ambient temperature but is soluble in hot alcohol (Sax and Lewis, 1987). Hence, the committee is concerned that it may dissolve in hot alkaline solution, in which case the lead cations could combine with other anionic substrates in a batch reactor and precipitate out sensitive compounds. This possibility is supported by the results for propellant P3. For example, picric acid will be formed during hydrolysis of the TNT or tetryl contained in the M55 rocket bursters. If bursters and propellant are hydrolyzed simultaneously, lead from the propellant could either precipitate out or form lead picrate. In the hydrated form, lead picrate is not particularly sensitive. However, enough heat could be produced from this exothermic process to heat and dehydrate the lead picrate deposited on vessel walls. As indicated in the section on TNT hydrolysis, dry lead picrate is an extremely sensitive explosive and is very dangerous to handle. Therefore, the committee believes that, to avoid forming sensitive compounds such as lead picrate, hydrolysis of bursters and propellant should be performed in separate vessels unless further testing demonstrates the process is safe.

It is uncertain whether slightly soluble components are eventually decomposed or emulsified if they are allowed to undergo prolonged base hydrolysis. Bunte et al.(1997) report that alkaline hydrolysis of centralite, DPA, DNT, and TNT may require supercritical hydrolysis and oxidation (374°C [705°F] and 218 atm [3204 psi]) to be completely decomposed. In any event, the technology providers will have to demonstrate that the hydrolysate product from base hydrolysis of energetic materials is an appropriate feedstock for the next unit operation. Some of these emulsions, such as TNT, tetryl, and 2-nitrodiphenylamine, may not be good candidates for biotreatment.

SUMMARY

This review of the hydrolysis of energetic materials important to the ACWA program, based on available literature, shows that the base hydrolysis of energetic materials extracted from obsolete ordnance is not a mature technology. There is relatively little experience with the alkaline decomposition of ACWA-specific energetic materials (compared to experience with chemical agents). Most of the base hydrolysis described in the literature on other energetic materials was conducted in dilute solutions and within solubility limits. Under these conditions, several undesirable products and precipitates can result. Data in the literature relevant to the hydrolysis of energetic materials from assembled chemical weapons indicates that hydrolysis is relatively slow unless the reaction conditions are severe (elevated temperature, pressure, and pH) because these materials are not very soluble in aqueous systems and the hydrolysis of solids is mass-transfer limited. To overcome this problem, investigators have resorted to using strong alkaline solutions, high temperatures, and even high pressures. In general, the relative hydrolysis rates of the energetic materials of interest to the ACWA program are expected to be (from faster to slower):

$$\text{NG} > \text{TNT} > \text{tetryl} \geq \text{RDX} > \text{HMX} > \text{NC}$$

One of the most important unresolved issues is that the family of products is not understood well enough to support simultaneous hydrolysis of different kinds of energetic materials in the same batch reactor. This technique should be disallowed until there is substantial evidence that intermediates from the hydrolysis of aromatic nitro compounds will not combine with M28 propellant additives or fuze components to form extremely sensitive explosives or other dangerous precipitates. An associated concern is that the conditions under which aromatic nitro compounds, such as TNT or picric acid, will emulsify in the aqueous phase and not be completely hydrolyzed are not understood. Any unit operation that immediately follows hydrolysis of energetic material should be designed to accept emulsified aromatic nitro compounds, such as TNT or picric acid, as contaminants in the aqueous feed stream.

The principal problem for ACWA technology providers is to develop and demonstrate a practical production process based on previously unrelated chemical-engineering unit operations and emerging technologies to satisfy destruction efficiency requirements. These chemical-engineering unit operations are: (1) sectioning of munitions to gain access to the energetic materials, (2) reducing the size of energetic materials extracted from ordnance sections to expedite solvation and reaction, and (3) decomposing a heterogeneous feed stream consisting of energetic materials, metal, and contaminants via alkaline decomposition. The hydrolysate produced must be an appropriate feed stream for the next unit operation, which may be another emerging technology (supercritical water oxidation and biotreatment have been proposed).

An associated problem for the ACWA technology providers is to develop sound engineering and managerial practices in the event of incomplete alkaline decomposition of the heterogeneous feed stream. Products and precipitates of incomplete hydrolysis of energetic materials can still be energetic and toxic, and some emergency shutdowns during processing are likely considering the overall complexity of the process.

The current level of understanding is, perhaps, sufficient to indicate that engineering practices can probably restrict the domain of possible reaction products. The use of strong base is probably the most efficient way to ensure that hydrolysis is driven to completion. Ammonia hydrolysis for the alkaline decomposition of some energetic materials could be considered. This would decrease the solubility problem and minimize the need for severe reaction conditions to increase throughput rates.

REFERENCES

Anbar, M., I. Dostrovsky, D. Samuel, and A.D. Yoffe. 1954. Esters of inorganic oxyacids. Part I. The mechanism of hydrolysis of alkyl esters, ^{18}O being used as a tracer, and its relation to other reactions in alkaline media. Journal of the Chemical Society, 1954, Vol. III: 3603–3611.

Baker, J.W., and T.G. Heggs. 1955. Hydrolytic decomposition of esters of nitric acid. Part V. the effects of structural changes in aralphyl nitrates on the S_N and E_{CO} reactions. Journal of the Chemical Society, 1955, Part I: 616–630.

Baker, J.W., and A.J. Neale. 1954. Hydrolytic decomposition of esters of nitric acid. Part III. kinetic evidence for ion exchange between anionic reagents and the solvent in the carbonyl elimination reaction ($E_{CO}2$). Journal of the Chemical Society, 1954, Part III: 3225-3233.

Baker, J.W., and A.J. Neale. 1955. Hydrolytic decomposition of esters of nitric acid. Part IV. acid hydrolysis, and the effects of change in the nucleophilic reagent on the S_N and E_{CO} reactions. Journal of the Chemical Society, 1955, Part I: 608–615.

Bernasconi, C.F. 1971. Kinetic and spectral study of some reactions of 2,4,6–trinitrotoluene in basic solution. I. deprotonation and janovsky complex formation. Journal of Organic Chemistry 36 (12): 1671–1679.

Bishop, R.L., C. Skidmore, R.L. Flesner, P.C. Dell'Orco, T. Spontarelli, K.J. Uher, J.F. Kramer, and D.A. Bell. 1996. Base Hydrolysis Kinetics of HMX-Based Explosives Using Sodium Carbonate. Report LA-UR-96-1818. Los Alamos, N.M.: Los Alamos National Laboratory.

Bishop, R.L., R.L. Flesner, P.C. Dell'Orco, T. Spontarelli, and S.A. Larson. 1998. Modeling of the base hydrolysis of HMX powder and HMX based plastic bonded explosives using sodium hydroxide. Pp. 320–331 in Proceedings of the Fourth Conference on the Life Cycles of Energetic Materials. Los Alamos, N.M.: Los Alamos National Laboratory.

Blake, J.A., M.J.B. Evans, and K.E. Russell. 1966. Kinetic studies of proton transfer from phenols to trinitrobenzyl anion. Canadian Journal of Chemistry 44: 119–124.

Borcherding, R. 1997. An alternative to open burning treatment of solid propellant manufacturing wastes. Waste Management 17 (2/3): 135–141.

Borcherding, R. 1998. An alternative to open burning treatment of solid propellant manufacturing waste. Pp. 87–92 in Proceedings of the Fourth Conference on the Life Cycles of Energetic Materials. Los Alamos, N.M., Los Alamos National Laboratory.

Borcherding, R., and D. Wolbach. 1995. Treatment of solid propellant manufacturing wastes, base hydrolysis as an alternative to open burning. Pp. 185–193 in Chemical Propulsion Information Agency Publication 626. Columbia, Md.: Chemical Propulsion Information Agency.

Bunte, G., H. Krause, and T. Hirth. 1997. Disposal of energetic materials by alkaline pressure hydrolysis and combined techniques. Journal of Propellants, Explosives, and Pyrotechnics 22(3): 160–164.

Caldin, E.F., and G. Long. 1955. The Rates and Equilibria of Some Reactions of Trinitrotoluene and Related Sustances from –80 to 20°C. Proceedings of the Royal Society of London 226A: 263–285.

Cannizzo, L.F., G.L. Mower, L.R. Huntsman, W.R. Achatz, and W.W. Edwards. 1995. Hydrolysis of the energetic materials present in a composite modified double base solid rocket propellant. Journal of Energetic Materials 13(3–4): 331–356.

Croce, M., and Y. Okamoto. 1979. Cationic micellar catalysis of the aqueous alkaline hydrolyses of 1,3,5-triaza-1,3,5-trinitrocyclohexane and 1,3,5,7-tetraaza-1,3,5,7-tetranitrocyclooctane. Journal of Organic Chemistry 44 (13): 2100–2103.

Cuta, F., and E. Beranek. 1974. Spektrophotometrische und

Chromatographische Untersuchung der Reaktion von 2,4,6-Trinitrotoluol mit Natriumhydroxid. Collection of Czechoslovakian Chemical Communications 39: 736–741.

DA (Department of the Army). 1996a. Detail Specification MIL-DTL-398D, RDX (Cyclotrimethylenetrinitramine). Dec. 12, 1996.

DA. 1996b. Detail Specification MIL-DTL-45444C, HMX (Cyclotetramethylenetetranitramine). Nov. 26, 1996.

DA. 1991. Military Specification.Tetryl (Trinitrophenylmethylnitramine), MIL-T-339C, Notice 2. November 4, 1991. Washington, D.C.: Department of the Army.

DA. 1984. Military Explosives. TM 9-1300-214. September 1984. Washington, D.C.: Department of the Army.

DA. 1966. Military Specification. MIL-P-60071A(MU), Propellant Grain, M28. May 10, 1966, as amended by Amendment 1 of June. 15, 1966.

Dell'Orco, P., R. Flesner, T. Spontarelli, J.A. Sanchez, and J.F. Kramer. 1995. Base hydrolysis and hydrothermal processing of composition B-3 explosive. Pp. 55–74 in Los Alamos National Laboratory Report LA-12960-PR, Vol. 1. Los Alamos, N.M.: Los Alamos National Laboratory.

Epstein, S., and C.A. Winkler. 1951. Studies on RDX and related compounds. VI. the homogeneous hydrolysis of cyclotrimethylenetrinitramine (RDX) and cyclotetramethylenetetranitramine (HMX) in aqueous acetone, and its application to analysis of HMX in RDX. Canadian Journal of Chemistry 29: 731–733.

Fedoroff, B.T., and O.E. Sheffield. 1962. Picatinny Arsenal Technical Report (PATR) 2700. Encyclopedia of Explosives and Related Items. Vol. 2. AD 422747. Dover, N.J.: U.S. Army Ordnance Corps.

Fedoroff, B.T., and O.E. Sheffield,. 1972. Picatinny Arsenal Technical Report (PATR) 2700. Encyclopedia of Explosives and Related Items. Vol. 5. AD 768063. Dover, N.J.: U.S. Army Ordnance Corps.

Fedoroff, B.T., and O.E. Sheffield,. 1974. Picatinny Arsenal Technical Report (PATR) 2700. Encyclopedia of Explosives and Related Items. Vol. 6. ADA 011845. Dover, N.J: U.S. Army Ordnance Corps.

Feuer, H. (ed). 1969. The Chemistry of the Nitro and Nitroso Groups. Vol. 1. New York: Interscience Publishers.

Feuer, H. (ed). 1970. The Chemistry of the Nitro and Nitroso Groups. Vol. 2. New York: Interscience Publishers.

Flesner, R. L., T. Spontarelli, P.C. Dell'Orco, and J.A. Sanchez. 1994. Base Hydrolysis and Hydrothermal Processing of PBX-9404 Pp. 121-143 in Emerging Technologies in Hazardous Waste Management VI. D.W. Tedder and F.G. Pohland, eds. Atlanta, Ga: American Academy of Environmental Engineers.

Franchimont, A.P.N., and H.J. Backer. 1913. Etude spectrographique sur la coloration des picryl-alkylnitramines avec les alcalis. Recueil des travaux chimiques des Pays-Bas 32: 325–333.

GA. 1997. Assessment of Technologies for Assembled Chemical Weapon Demilitarization. Submitted in Response to U.S. Army Solicitation number DAAM01-97-R00031. San Diego, Calif.: General Atomics.

GA. 1998. Assessment of Technologies for Assembled Chemical Weapon Demilitarization Data Gap Resolution Report. February 17, 1998. San Diego, Calif.: General Atomics.

Gold, V., and C.H. Rochester. 1964. reactions of aromatic nitrocompounds in alkaline media. Part V. solutions of methyl picrate and 1,3,5-trinitrobenzene in aqueous sodium hydroxide in the absence of light. Journal of the Chemical Society, 1964, Part II: 1710–1716.

Hammersley, V. L. 1975. Historical and Experimental Studies of Alkali and Trinitrotoluene Reaction. WQEC/C 75-192. Crane, Indiana: Naval Weapons Support Center.

Hepp, P., 1882. Ueber Trinitroderivate des Benzols und Toluols. Justus Liebigs Annalen der Chemie 215: 344–354.

Hoffsommer, J.C., D.A. Kubose, and D.J. Glover. 1977. Kinetic isotope effects and intermediate formation for the aqueous alkaline homogeneous hydrolysis of 1,3,5-triaza-1,3,5-trinitrocyclohexane (RDX). Journal of Physical Chemistry 81(5): 380–385.

Jackson, C.L., and F.H. Gazzolo. 1900. Certain coloured substances derived from nitro-compounds. Journal of The Chemical Society, 78(1): 433–434.

Janovsky, J.V., and L. Erb. 1886. Zur Kenntniss der directen Brom- und Nitrosubstitutionsproducte der Azokörper. Article #447. Berichte der Deutschen Chemischen Gesellschaft 19B: 2155–2158.

Jones, W.H. 1954. Mechanism of the homogeneous alkaline decomposition of cyclomethylenetrinitramine: kinetics of consecutive second- and first-order reactions. a polarographic analysis for cyclotrimethylenetrinitramine. Journal of the American Chemical Society 76: 829–835.

Kaye, S.M. and Herman, H.L. 1980. Encyclopedia of Explosives and Related Items, Vol. 9. Dover, N.J.: U.S. Army Armament Research and Development Command.

Kenyon, W.O., and H.L. Gray. 1936. Alkaline decomposition of cellulose nitrate. I. quantitative studies. Journal of the American Chemical Society 58: 1422–1427.

Larson, S.A., G.R. Brewer, D.M. Harradine, C.E. Polston, L.A. Le, R.L. Bishop, P.C. Dell'Orco, and R.L. Flesner. 1998. Pilot-scale pressurized base hydrolysis of HMX plastic-bonded explosives. Pp. 312–319 in Proceedings of the Fourth Conference on the Life Cycles of Energetic Materials. Los Alamos, N.M.: Los Alamos National Laboratory.

Lindner, V. 1980. Explosives and Propellants. Pp. 561–671 in Kirk-Othmer Encyclopedia of Chemical Technology. New York: John Wiley & Sons.

Lockheed Martin. 1997. Assembled Chemical Weapons Assessment. Submitted in Response to U.S. Army Solicitation number DAAM01-97-R00031, September 1997. Albuquerque, N.M.: Lockheed Martin Corporation.

Meisenheimer, J. 1902. Ueber Reactionen aromatischer Nitrokörper. Justus Liebigs Annalen der Chemie 323: 205–213.

Melvin, W.S. 1997. Recovery and reuse of rocket propellants. Pp.193–201 in Proceedings of the 2nd Aerospace Environmental Technology Conference. NASA Conference Publications, No. 3349. Washington, D.C.: National Aeronautics and Space Administration

Mertens, K.H. 1886. Die Nitrirung von Di- und Monomethylanilin mit verdünnter Salpetersäure. Article 438. Berichte der Deutschen Chemischen Gesellschaft 19B: 2123–2127.

Meyer, V. 1889. Ueber Ringschliessung unter Abspaltung einer Nitrogruppe aus dem Benzolkern. Article 65. Berichte der Deutschen Chemischen Gesellschaft 22A: 319–323.

Meyers, R. 1981. Explosives (2nd ed). Weinheim Germany: Verlag Chemie.

Miles, F.D. 1955. Cellulose Nitrate. New York: Interscience Publishers.

Morgan, M.E., D.H. Mitchell, P.M. Stevens, W.S. Melvin, P.G. Schirk, O.J. Manar, and N.W. Rizzardi. 1994. Critical Fluid Demilitarization and Ingredient Reclamation Technology. Paper 33. Pp. 1–11 in Proceedings of the Propulsion and Energetics Panel (PEP) 84th Symposium on Environmental Aspects of Rocket and Gun Propulsion. Aalesund, Norway: Advisory Group for Aerospace Research and Development (AGARD), North Atlantic Treaty Organization (NATO).

Newman, K.E. 1999. A Review of Alkaline Hydrolysis of Energetic Materials: Is it Applicable to Demilitarization of Ordnance? Report IHTR 2167. Indian Head, Md.: Naval Surface Warfare Center Indian Head Division.

Parsons. 1997. Proposal for the Assessment of Technologies for Assembled Chemical Weapon Demilitarization. Submitted in response to U.S. Army solicitation number DAAM01-97-0031. September 1997. Pasadena, Calif.: Parsons Infrastructure and Technology Group, Inc.

Russell, G.A., E.G. Janzen, and E.T. Strom. 1964. Electron-transfer processes. I. The scope of the reaction between carbanions or nitranions and unsaturated electron acceptors. Journal of the American Chemical Society 86: 1807–1814.

Sax, N.I., and R.L. Lewis, Sr. 1987. Hawley's Condensed Chemical Dictionary (11th ed). New York: Van Nostrand Reinhold.

Schaal, R., and G. Lambert. 1962. Propriétés acido-basiques du méthanol, I etude des solutions diluée de méthoxydes alcalins. Journal de chimie Physique et de physico-chimie biologique 59: 1151–1163.

Shipp, K.G., and L.A. Kaplan. 1966. Reactions of a-substituted polynitrotoluenes. II. The generation and reactions of 2,4,6-trinitrobenzyl anion. Journal of Organic Chemistry 31: 857–861.

Shipp, K.G., L.A. Kaplan, and M.E. Sitzmann. 1972. Reactions of a-Substituted Polynitrotoluenes. III. 2,4,6-Trinitrobenzyl anion as a nucleophile at aromatic carbon. Journal of Organic Chemistry 37(12): 1966–1970.

Skidmore, C., P. Dell'Orco, R. Flesner, J. Kramer, and T. Spontarelli. 1995. Chemical Destruction of HMX-based Explosives with Ammonium Hydroxide. Report LA-13003, UC-741. Los Alamos, N.M.: Los Alamos National Laboratory.

Somlo, F. 1940. Einige Bemerkungen über die Beständigkeit von Hexogen und Penthrit gegen Alkali. Zeitschrift für das gesamte Schieß und Sprengstoffwesen 35: 175–177.

Spontarelli, T., G.A. Buntain, R.L. Flesner, J.A. Sanchez, and P.J. Unkefer. 1994. An Engineered System using Base Hydrolysis for Complete Disposal of Energetic Materials. Pp. 55–62 in Proceedings of the ADPA International Symposium on Energetic Materials Technology. Orlando, Fla.: American Defense Preparedness Association.

Strauss, M.J. 1970. Anionic sigma complexes. Chemical Reviews 70(6): 667–712.

Urbanski, T. 1964. Chemistry and Technology of Explosives. Vol. 1. New York: Pergamon Press.

Urbanski, T. 1965. Chemistry and Technology of Explosives. Vol. 2. New York: Pergamon Press.

Urbanski, T. 1967. Chemistry and Technology of Explosives. Vol. 3. New York, NY: Pergamon Press.

Urbanski, T. 1984. Chemistry and Technology of Explosives. Vol. 4. New York: Pergamon Press.

van Romburgh, P. 1889. Synthèse de la trinitrophènylmèthylnitramine. Recueil des travaux chimiques des Pays-Bas 8: 215–216.

Walker, J.F. 1964. Formaldehyde 3rd ed. ACS Monograph Series, edited by F.M. Beringer. New York: Van Nestrand Reinhold.

Wilbrand, J. 1863. Notiz über Trinitrotoluol. Justus Liebigs Annalen der Chemie 128: 178–179.

APPENDIX F

Supercritical Water Oxidation

Two of the technology providers, Lockheed Martin and General Atomics, propose using supercritical water oxidation (SCWO) for the final destruction of the hydrolysate from both agents and energetics. This appendix provides a general description of SCWO and lists the findings of the 1998 National Research Council report on SCWO of VX hydrolysate (NRC, 1998), which are considered to be applicable to this study.

BASICS

The foundation of SCWO technology is that the fluid properties of water change dramatically above its critical temperature (374°C; 705°F) and pressure (218 atm; 3,204 psi). Supercritical water functions like a dense gas with special properties, such as high organic solubility, complete miscibility with permanent gases and organics, high diffusivity, low viscosity, and a low dielectric constant similar to that of nonpolar liquids. Inorganic salts become almost completely insoluble under supercritical conditions. These properties make supercritical water an excellent oxidation medium for the destruction and mineralization of most organic compounds to simple compounds, such as carbon dioxide, nitrogen, and water.

In the SCWO process, dissolved organics, oxygen (or other oxidants), and water are reacted above the critical temperature and pressure of water. Oxidation of the organics is spontaneous, and complete mineralization to carbon dioxide and water is achieved in a few seconds of residence time. Although the SCWO of organics is exothermic, a supplementary source of fuel is normally required for the oxidation of organics in dilute aqueous solutions. A simplified flow sheet for a typical SCWO waste-treatment system is presented in Figure F-1.

HYDROLYSATES OF AGENTS AND ENERGETICS

Testing by the Army and Army contractors has shown that SCWO can achieve high destruction efficiencies for the organic constituents in VX hydrolysate (NRC, 1998). Testing has also demonstrated that SCWO is capable of destroying chemical agent (GA, 1997). For example, the SCWO oxidation of the nerve agent GB proceeds according to the following chemical equation:

$$\begin{array}{l} \qquad\quad F \\ \qquad\quad | \\ H_3C - P = O + 6.5\ O_2 \rightarrow HF + 4\ CO_2 + 0.5\ P_2O_5 + 4.5\ H_2O \\ \qquad\quad | \\ \qquad\quad OCH(CH_3)_2 \end{array}$$

In the presence of caustic, which is a component of agent hydrolysate, the HF formed during oxidation of GB will react with sodium hydroxide to form NaF salt and water as follows:

$$HF + NaOH \rightarrow NaF + H_2O$$

Because SCWO of organics is nonspecific (i.e., organic material is oxidized indiscriminately, regardless of the source), the hydrolysate from agent neutralization and energetics deactivation can either be combined or processed separately in the SCWO reactor.

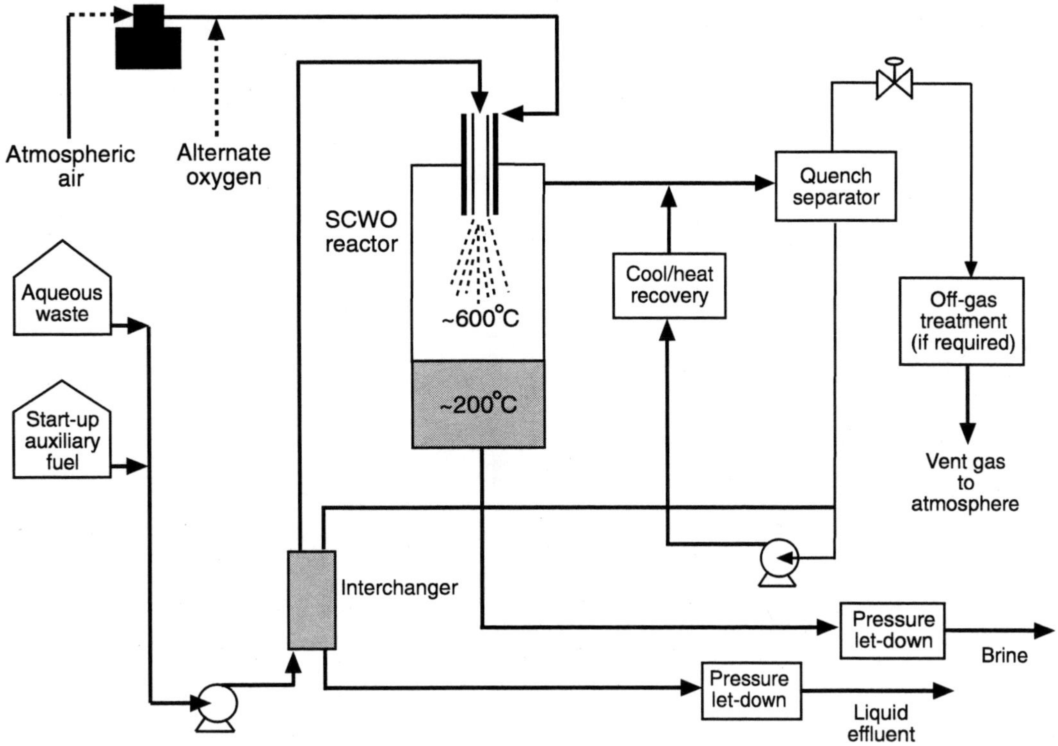

FIGURE F-1 Typical flow sheet for supercritical water oxidation. Source: NRC, 1998.

TECHNICAL CHALLENGES

Although SCWO has been under development for more than 20 years, two major problem areas have prevented its commercial application until recently (Modell, 1985):

- rapid and excessive corrosion of reactor materials as a result of the severe reaction conditions and the formation of acid compounds when heteroatoms are present in the reactor feed
- plugging of reactors by the deposition of inorganic salts formed during SCWO in the presence of dissolved inorganic compounds and heteroatoms

These two problems are generally exacerbated when the feed stream contains acid- and salt-forming heteroatoms, such as Cl, F, P, and S, which are present in the hydrolysates of agents and energetics.

Other problem areas in the application of SCWO are that process kinetics and the effects of process parameters on process design and efficiency are not completely understood. For example, Li and Egiebor (1994) reported that the feed stream preheating rate has a significant effect on the destruction efficiency for organics. The more rapid the preheating, the higher the oxidation efficiency. This was attributed to increased pyrolysis and polymerization of the organic contaminants during slow preheating to form highly refractory compounds resistant to SCWO. Furthermore, Webley and Tester (1991) reported that global kinetic models failed to predict the reaction rates during the SCWO of methane and carbon dioxide. Thus, the fundamental physics of SCWO are not completely understood.

KEEPING UP WITH ONGOING SCWO TESTING

SCWO has been selected by the Army for the treatment VX hydrolysate at the Newport, Indiana, site. The design of that SCWO system is well under way, but significant testing still remains to be done. The Army also plans to use SCWO for the destruction of smokes and dyes at the Pine Bluff Arsenal, Arkansas; the Navy is currently testing two different SCWO designs for onboard treatment of shipboard wastes. The Army should continue to monitor these tests closely and factor the results into its decision to implement SCWO for follow-on ACWA programs.

FINDINGS AND RECOMMENDATIONS FROM THE 1998 NRC REPORT

The following paragraph and the subsequent findings and recommendations are taken directly from the NRC report *Using Supercritical Water Oxidation to Treat Hydrolysate from VX Neutralization* (NRC, 1998). They are reproduced here because the committee considers them applicable to the SCWO technologies evaluated in this study.

Excerpt

Chemical neutralization of VX nerve agent results in the production of a liquid hydrolysate stream that has greatly reduced toxicity compared to the original nerve agent but requires further treatment to meet the requirements of the Chemical Weapons Convention and to be suitable for disposal. After considering several approaches, the U.S. Army has selected SCWO (supercritical water oxidation) as the primary process for treating the hydrolysate from VX neutralization prior to ultimate disposition. The integration of SCWO into the complete process for the destruction of VX stored at Newport, Indiana, also requires an evaporator system after SCWO treatment to allow water to be recycled back into the neutralization process. The evaporation system also produces a dry solid waste stream consisting of salts produced during the neutralization and SCWO treatment steps. Excess condensed water from the evaporator is expected to be of relatively high purity and suitable for discharge. The technology selected for the evaporation process step is mature with considerable full-scale design and operations experience. In contrast, treatment of the hydrolysate will be a new application for SCWO. Thus, the findings and recommendations presented here focus on the use of SCWO for the treatment of VX hydrolysate.

Findings

Finding 1. Limited pilot-scale testing has demonstrated the ability of SCWO to achieve high destruction efficiencies for the organic constituents of VX hydrolysate. Effluent from SCWO treatment of VX hydrolysate has been shown to have negligible acute toxicity in intravenous testing in mice, gavage testing in rats, and dermal testing in rabbits. The separation of salts in the effluents from SCWO through an evaporator system should produce relatively pure water suitable for discharge and solid salts suitable for disposal. Treatment requirements for VX hydrolysate are less stringent than they are for VX because the hydrolysate has low toxicity relative to the agent. However, criteria for process destruction efficiency and final disposal standards have not been established.

Finding 2. Using SCWO to treat VX hydrolysate is significantly different and more complex than previous applications. SCWO systems on a pilot scale have been used to treat several other types of wastes, but SCWO is in commercial operation at only one site. There has been only limited pilot-scale or operational-scale experience with wastes that are similar to VX hydrolysate in being highly corrosive and salt-laden. Operation with VX hydrolysate or appropriate surrogates at design conditions, equipment configuration, or approximate scale for full-scale operations has not been demonstrated. A vertical cylindrical reactor is the only reactor configuration that has been successfully demonstrated to date at pilot scale for the treatment of VX hydrolysate and similar waste streams. Additional development and pilot-scale testing of SCWO technology will be necessary to ensure sustained, reliable operation of a full-scale integrated treatment system. Sufficient time appears to be available in the Army's implementation schedule for the Army to carry out development and testing for using SCWO at the Newport site, provided they are carried out expeditiously.

Finding 3. Pilot-scale operation of SCWO in a vertical cylindrical reactor at the temperature and pressure necessary for the effective destruction of hydrolysate constituents has been limited to one eight-hour and two less than two-hour tests. During pilot-scale testing with hydrolysate, the following factors were identified that could create difficulties in sustaining system performance:

- Large quantities of insoluble salts were produced, which must be effectively managed within, and downstream of, the SCWO reactor.
- Unexpected fluctuations were observed in temperature, pressure, and salt expulsion from the SCWO reactor.

- High levels of corrosion and erosion of materials of construction were observed in the reactor liner and pressure let-down valves.
- The sustained performance and reliability of the pressure let-down system was not demonstrated.

Although at this point in development the Stockpile Committee cannot be certain, it believes that a SCWO system for the treatment of VX hydrolysate with sufficient sustained performance can be achieved with additional development and testing.

Finding 4. Limited bench-scale and pilot-scale tests have demonstrated operating regimes under which SCWO can effectively destroy carbon-phosphorus bonds and oxidize the organic constituents present in VX hydrolysate. The demonstrated conditions for high levels of destruction (> 99 percent) include temperatures between 640°C (1184°F) and 730°C (1346°F) and pressures between 231 and 258 atm (3395 to 3792 psi). At temperatures and pressures below this regime, effluent from SCWO processing may contain significant concentrations of residual organic species that are difficult to destroy, including constituents with carbon-phosphorus bonds.

A basis for the reliable scale-up and operation of SCWO technology for the treatment of VX hydrolysate has not yet been demonstrated. Fundamental knowledge about the following processes within the SCWO reactor is still not available:

- the number and characteristics of the physical phases, including large quantities of entrained and adhered solids and potentially liquid, gas, and supercritical fluid phases
- fluid dynamics and mixing processes complicated by relatively high loadings of insoluble salts
- heterogeneous and homogeneous reaction mechanisms and kinetics
- salt nucleation, particle growth, agglomeration and adhesion mechanisms, and kinetics

Because the understanding of fundamental processes is limited and the process operational data and experience are sparse, empirical design and engineering judgment will be required for the selection of a prudent scale for development prior to full-scale demonstration. This is common engineering practice.

Finding 5. Alkaline VX hydrolysate and its destruction products under SCWO reaction conditions create an extremely corrosive and erosive environment that requires the careful selection of materials of construction. Although preliminary data indicate that certain noble metals, such as platinum and gold, may have acceptable properties, the data currently available are insufficient for the selection of materials of construction. The Army has initiated further testing of materials of construction.

Finding 6. Process monitoring and control strategies for the management of salts within the SCWO reactor and the destruction of the organic constituents of the hydrolysate have not been demonstrated.

Recommendations

Recommendation 1. A pilot-scale SCWO process facility with the critical characteristics of the full-scale design should be constructed and operated to further define operating characteristics and demonstrate sustained continuous operation of the process. Objectives for process development and demonstration should include:

- operation with either hydrolysate or a suitable surrogate to demonstrate reliable operation for periods similar to full-scale design operating cycles
- the development and validation of process monitoring and control strategies for salt management and the destruction of organic constituents
- the definition of stable operating regimes, including the temperature, pressure, and the use of the oxidant (liquid oxygen or compressed air) selected for full-scale operation
- the definition of a basis for process scale-up, operation, and maintenance of a full-scale system
- the development and demonstration of a reliable pressure let-down system

Because the understanding of the fundamental process mechanisms and operating characteristics is limited, the committee recommends that the pilot-scale system be within an order of magnitude of the total mass and heating throughput of a full-scale design unit. Based on testing and reactor scale-ups to date, a vertical cylindrical reactor configuration is recommended as the system that will probably require the least amount of additional development. Other reactor configurations may perform at required levels but would require significant additional development.

Recommendation 2. Testing of materials of construction should be carried out as necessary to finalize the selection of materials for critical components, including the SCWO reactor and the pressure let-down system. Additional pilot-scale testing indicated in Recommendation 1 should include fabrication with the materials of construction selected from testing smaller samples and evaluation of corrosion and erosion rates for critical components.

Recommendation 3. Flexibility and redundancy of critical components should be incorporated into the design of the full-scale system to allow for uncertainties about the basis for scale-up and operation. Trade-offs should be evaluated to establish an appropriate balance between two 100 percent capacity SCWO reactors or a greater number of smaller reactors. The analysis should consider performance uncertainties associated with process scale-up and complexity, as well as the reliability of operating several reactors in parallel.

Recommendation 4. The Army should make provisions for targeted research and development to resolve problems identified during pilot-scale testing and the full-scale implementation of SCWO technology.

Recommendation 5. Requirements for process destruction efficiencies and final disposal standards for all effluent streams from SCWO treatment should be clearly defined to ensure that the final design meets regulatory standards.

REFERENCES

GA (General Atomics). 1997. Assessment of Technologies for Assembled Chemical Weapon Demilitarization. Proposal submitted in response to U.S. Army Solicitation No. DAAM01-97-R-0031, September 15, 1997. San Diego, Calif.: General Atomics.

Li, L., and N.O. Egiebor. 1994. Feedstream preheating effect on supercritical water oxidation of dissolved organics. Energy and Fuels 8: 1126–1130.

Modell, M. 1985. Processing Methods for the Oxidation of Organics in Supercritical Water. U.S. Patent 4,543,190 (September 24, 1985).

NRC (National Research Council). 1998. Using Supercritical Water Oxidation to Treat Hydrolysate from VX Neutralization. Committee on Review and Evaluation of the Army Chemical Stockpile Disposal Program, Board on Army Science and Technology. Washington, D.C.: National Academy Press.

Webley, P.A., and J. Tester. 1991. Oxidation kinetics of carbon monoxide and methane in supercritical water. Energy and Fuels 5: 411–416.

APPENDIX G

Fluid-Jet Cutting of Ordnance and High-Pressure Clean-Out of Energetic Materials

Fluid-jet cutting is a topic of interest to many groups. The fluid used is usually water, and high-pressure water jets have been used by industry to cut through many materials (e.g., metal and plastic) for more than 25 years (Summers, 1997). As part of the ACWA program, Teledyne-Commodore and Parsons-AlliedSignal have proposed using fluid-jet cutting to (1) shear fuzes; (2) precisely section munitions; and (3) gain access to energetic materials by washing out "demilitarized" chunks, slivers, or sections of energetic materials from warheads and rocket motors. Parsons-AlliedSignal proposes using water; Teledyne-Commodore proposes ammonia.

The use of high-pressure water or ammonia to cut explosive-loaded ordnance and/or to wash out energetic materials from ordnance casings is a proven technology. When shearing fuzes or sectioning munitions, the fluid jet often contains an abrasive, such as garnet, and the fluid pressure is normally about 2,722 atm (40,000 psi) to cut through the metal casing. When removing explosives or propellants from inside warheads or rocket motors, the fluid usually does not contain abrasives, and the pressure is normally much lower, about 680 atm (10,000 psi). Critical issues include (1) identification of hazards associated with the specific task; (2) design of the fluid-jet cutting system; (3) determination of processing parameters; (4) containment and segregation of residual metal, downloaded energetic materials, and other refuse; and (5) the development of a preventive maintenance schedule.

DESIGN PARAMETERS AND HAZARDS IDENTIFICATION

Some of the design parameters for fluid-jet cutting of ordnance and some of the hazards associated with its operation are well known. Target responses to the impact of high-velocity, nonabrasive water jets have been analyzed (Kang et al., 1993); the mechanisms and parameters of abrasive water-jet (AWJ) cutting have been examined (P.L. Miller, 1992a); some AWJ explosive safety tests have been evaluated (P.L. Miller, 1992b); and the effects of ultra high-pressure water jets on high explosives have been determined (P.L. Miller, 1992c).

Some design issues associated with fluid-jet cutting systems have been reviewed by a team from Lawrence Livermore National Laboratory and the Pantex Plant in Amarillo, Texas (Kang et al., 1993), who were interested in identifying the physics governing the most efficient mass-removal process when a material is subjected to water-jet impact. Theoretical and experimental investigations were performed on the effects of a nonabrasive water jet impinging on a solid surface. At jet velocities below 1,500 m/s, the maximum impact pressure can be calculated from the pressure across a one-dimensional water-hammer compression wave (Cook et al., 1962). The test data have been correlated to the following expression (Heymann, 1968):

$$\Delta P = P_c - P_u = \rho_u C_u V_j \,[1 + 2\,(V_j/C_u)]$$

where P denotes the pressure, r the liquid density, C the speed of sound, and V_j the jet velocity. The subscripts *u* and *c* signify conditions in the undisturbed liquid region and the compressed liquid region, respectively. Impact and machining experiments were conducted on various materials with water-jet reservoir pressures up to 2,720 atm (40,000 psi). Test results indicated that a maximum mass-removal rate takes place when the distance between the target piece and the nozzle exit (stand-off distance) is several hundred nozzle diameters. At this long stand-off, the jet disintegrates into a series of ligaments and droplets impinging on the surface.

Typical nozzle diameters are about 0.35 mm (0.014 in) to 1.5 mm (0.060 in), depending on the material to be cut and the cutting technique. The techniques can either be optimized for removing material or for cutting and pulverizing material simultaneously for reclamation and reuse. If multiple port nozzles are used, the sum of the port diameters should not exceed a maximum design criterion, such as 1.5 mm (0.060 in). The length-to-diameter ratio of nozzles used for nonabrasive fluid-jet cutting depends on the fluid pressure and cutting attributes but is usually about 50:1.

The parameters that affect safety and risk include water pressure, nozzle design, abrasive concentration and particle size, and cutting procedure. Several hundred thousand explosive-loaded projectiles have been safely cut using fluid-jet technology. Metals ranging from aluminum 6061 to titanium Ti-6Al-4V have been cut using water-jet streams laced with garnet grit. The optimum particle size of garnet grit depends on the metal being cut. The softer the metal, the larger the grit size. A grit size of about 150 microns has been found to be very close to optimal for cutting through steel casings.

Two cutting techniques have also been investigated, cutting laterally across the projectile (like a saw) and cutting rotationally (similar to a lathe). The rotational method is faster. An average time for AWJ cutting through 4.2 inch, Composition B-loaded mortars was 33 seconds using the rotational method and 57 seconds using the lateral method (P.L. Miller, 1992a). Ammonia-jet cutting appears to be even faster. Abrasive ammonia-jet cutting has been reported to be about 25 percent faster than AWJ cutting (Teledyne-Commodore, 1998).

Both water-jet and AWJ cutting of booster and main-charge explosives have been demonstrated to be safe and practical. Even though the impact values from water-jet velocities exceed the threshold impact limits for explosives, the energetic initiation mechanism for water-jet impingement/cutting is different than for "conventional" solid-solid impacts and is well understood. When both projectile diameters and impedance mismatches are taken into account, the results agree with published models for impact velocities (P.L. Miller, 1992b). A substantial safety margin for using nonabrasive water-jet cutting of main-charge explosives has also been documented by the Naval Surface Warfare Center (NSWC) Crane Division (Liddard and Roslund, 1993; Worsey et al., 1990). A graph of the impact initiation probabilities for several high explosives is provided in Figure G-1 as a log-log plot of velocity versus fluid-jet diameter. The lines represent the velocities at which initiation would be expected to occur 50 percent of the time (V50).

Jet impact is the most likely cause of initiation of energetic materials during fluid-jet cutting. The maximum pressure for continuous water flow, with existing equipment, is believed to be about 10,200 atm (150,000 psi). The water jet generated by this driving force has a sonic velocity of about 1,475 m/s (4,900 ft/s), which represents an upper bound for water because it is also the pressure at which water freezes at 25°C (77°F).

In testing by Alliant Techsystems, pentaerythritol tetranitrate (PETN) and trinitrotoluene (TNT) were

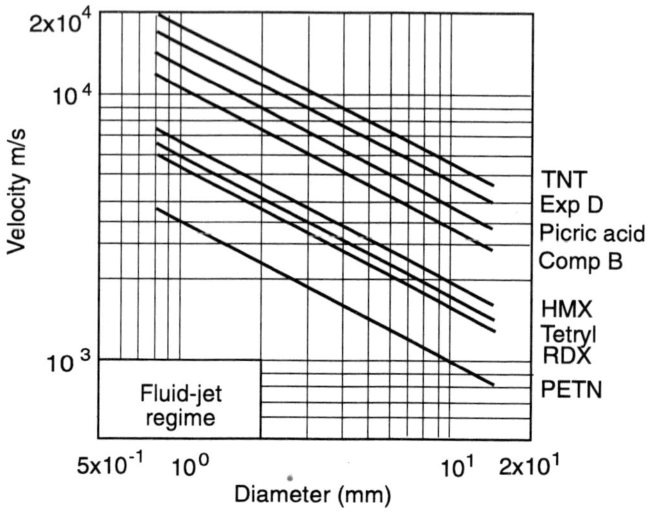

FIGURE G-1 Water-jet velocity at which explosives will initiate 50 percent of the time as a function of the fluid-jet diameter.

selected to represent the sensitivity range of explosives of interest in demilitarization. PETN is an impact-sensitive booster explosive component; TNT is a relatively impact-insensitive explosive used as a main charge. Fifty water-jet impact tests were performed on pressed PETN samples and cast TNT samples. Neither PETN nor TNT was initiated by the impact of water jets at this pressure (P.L. Miller, 1992c).

SURVEY OF PRACTICE AND PRODUCTION

The use of fluid-jet cutting to gain access to munitions for demilitarization and/or resource, reclamation, and reuse has been demonstrated and/or used by the Department of Defense (DOD) and many contractors. The user community can be separated into uses of nonabrasive jets and users of abrasive jets.

A limited survey of nonabrasive water-jet cutting of energetic materials was recently published (Estabrook, 1994). Nonabrasive water jets have been used to download explosives from warheads by the NSWC Indian Head Division at Yorktown, Virginia (Lowell, 1986), and the NSWC Crane Division in Crane, Indiana (Summers et al., 1988; Burch, 1998), as well as at the Western Area Demilitarization Facility in Hawthorne, Nevada (Day and Zimmermann, 1994). Nonabrasive water jets have also been routinely used in the propellant industry to download composite rocket motors to reclaim cases. This technology has been demonstrated by Thiokol, Aerojet GenCorp, and many others. Nonabrasive ammonia jets have been used by the Army in Huntsville, Alabama, to download rocket motors from cases as a first step in a novel recovery process (Melvin, 1992; Morgan, 1994). Carbon dioxide pellets have been entrained in a pressurized pneumatic jet at velocities of 20 m/s to 300 m/s to demonstrate an "environmentally friendly" means of downloading (or "blasting out") explosives from projectiles by the Army at Picatinny Arsenal in New Jersey (Hwang, 1995). The Air Force has even investigated using high-pressure liquid nitrogen as a cryogenic jet to remove propellant from large rocket motors (Coppola, 1995). AWJ has been recently reviewed in the literature (Summers, 1997). In the United States, the AWJ technique combines abrasive with the high-velocity jet stream after it leaves the initial acceleration nozzle. The resulting mixture is then refocused through a second nozzle. AWJ has been used

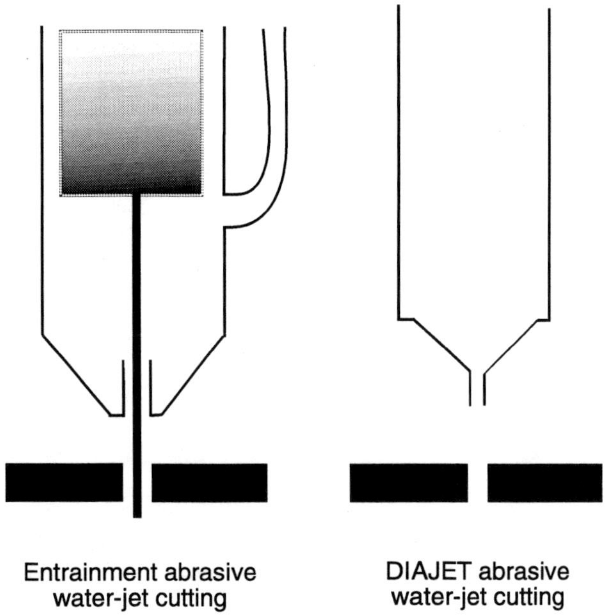

FIGURE G-2 Comparison of the AWJ and ASJ (DIAJET) abrasive-jet cutting techniques.

by Alliant Techsystems to section thousands of projectiles without incident. In the United Kingdom, an alternative method of abrasive slurry jetting (ASJ), marketed under the name "DIAJET," is being developed (D. Miller, 1995). In the ASJ method, the abrasive is fed into the water-jet stream before it is accelerated through a nozzle. These two techniques are compared in Figure G-2. The ASJ technique is potentially more efficient than AWJ because of higher cutting rates at reduced pressures and reduced operating costs. For example, the AWJ operating pressure is typically in the range of 2,382 to 3,743 atm (35,000 to 55,000 psi) for sectioning ordnance; the ASJ operating pressure is between 238 and 680 atm (3,500 and 10,000 psi). ASJ is being implemented by the Defense Evaluation and Research Agency (DERA) in the United Kingdom to demilitarize obsolete ordnance and "render safe" unexploded ordnance, such as abandoned mines. ASJ is also being considered for implementation in the United States as a more efficient method for demilitarizing ordnance (Fossey, 1998).

INCIDENTS ASSOCIATED WITH DEMILITARIZATION OF ORDNANCE

A few incidents have been associated with the demilitarization of ordnance using jet-cutting techniques.

Five recent, diverse incidents are discussed here. The first incident, which occurred in 1996 at the Alliant Techsystems Proving Grounds (Blixrud, 1996), happened during the mechanical defuzing of obsolete 8-inch Navy projectiles immediately prior to demilitarization by jet cutting. The cause of the incident was pinching picrates in the threads while unscrewing a corroded fuze. The lesson learned was that all ordnance must be carefully inspected before processing to identify and remove ordnance that require special treatment from routine operations. This incident might not have occurred if jet cutting had been used to remove the fuze (rather than unscrewing it mechanically).

A second incident involved the fluid-jet nozzle becoming detached from the lance during the high-pressure water wash-out of explosive from a munition warhead. When the nozzle impacted the explosive, it initiated and deflagrated. The lesson learned was that metal parts must be routinely inspected for signs of fatigue.

The third and fourth incidents occurred at Alliant Techsystems during system trials and safety demonstrations. Both incidents involved abrasive water-jet cutting of 20 mm ammunition that contained lead azide, which was press-loaded at high density and deliberately chosen to determine whether initiation would occur. During abrasive water-jet cutting, the lead azide initiated and detonated. After this happened the first time, some design parameters were altered and the tests were repeated. The results were the same the second time.

A fifth incident, which occurred at DERA West Freugh in Scotland in 1995 (Moore, 1999), involved handling ordnance that had been demilitarized by abrasive-jet cutting. About three days after explosive-loaded ordnance had been sectioned, workers performing normal procedures were injured when the ordnance unexpectedly initiated. Evidently, the cause of the initiation was embedded particles from the abrasive-jet cutting operation that had sensitized the explosive to impact, especially when the explosive surface was dehydrated. The lessons learned are that sectioned ordnance should be kept wet while applying impact, shear, or other forces to it and that spent abrasive should not be allowed to "dry out" while it may still be contaminated.

PREVENTATIVE MAINTENANCE

The high-pressure pumps used for jet cutting require extensive preventative maintenance and are, thus, responsible for most of the down time for jet-cutting systems. Wear of the nozzle is a primary concern for safety and performance. Nozzle wear is worst with the AWJ method, followed by the ASJ method, and least with nonabrasive techniques. For AWJ, the nozzle usually clogs at least once during start-up. The more often the system is shut down and restarted, the worse the wear. However, for AWJ, the average service life of a nozzle is about 1,000 hours of operation.

Any closed-loop fluid-jet cutting system has "dead spots" in which sediment, energetic material, metallic particles, and spent abrasive can accumulate. The number of dead spots should be minimized when the system is designed and built. Filter housings, traps, and other solids-capture features in closed-loop systems should have a preventative maintenance plan based on actual processing data or experience to ensure that undesired combinations of trace sediment, energetic materials, metallic particles, and spent abrasives do not dry out sufficiently to undergo exothermic reduction-oxidation reactions. If ammonia jets are used, precautions must be taken to minimize the residence time of materials awaiting further processing to prevent exothermic ammonolysis.

REFERENCES

Blixrud, C. 1996. Investigation Report, 8" Demilitalization. Memorandum from Craig Blixrud, Alliant Techsystems, to M. Dolby and B. Ford, Alliant Techsystems, May 21, 1996.

Burch, D. 1998. Recovery of RDX and HMX from Military Munitions. Presented at the JANAF PDCS/S&EPS Joint Meeting, NASA Johnson Space Center, Houston, Texas, April 21–24, 1998.

Cook, M.A., R.T. Keyes, and W.O. Ursenbach. 1962. Measurements of detonation pressure. Journal of Applied Physics 33: 3413–3421.

Coppola, E.N. 1995. Cryogenic Propellant Removal System (PRS). Pp. 324–332 in Proceedings of the 3rd Global Demilitarization Symposium. Arlington, Va.: Joint Ordnance Commanders Group Demilitarization and Disposal Subgroup.

Day and Zimmerman. 1994. Western Area Demilitarization Facility. Pp. 300–332 in Proceedings of the 2nd Demilitarization Symposium. Arlington, Va.: Joint Ordnance Commanders Group Demilitarization and Disposal Subgroup.

Estabrook, L.C. 1994. Water-Jet Cutting of Energetic Materials. Pp. 200–209 in Proceedings of the Life Cycles of Energetic

Materials 1994 Conference. Paper LA-UR-95-1090. Los Alamos, N.M.: Los Alamos National Laboratory.

Fossey, R.D. 1998. The use of an Abrasive Waterjet System at 700 bar for the Cutting of Military Munitions as Part of a Demilitarization Program. Pp. 453–466 in Proceedings of the 14th International Conference on Jetting Technology. Brugge, Belgium: British Hydrodynamics Research Group.

Heymann, F.J. 1968. On the shock wave velocity and impact pressure in high-speed liquid-solid impact. Journal of Basic Engineering 90: 400–402

Hwang, M. 1995. Carbon Dioxide Blast Demilitarization of Explosive-Loaded Projectiles. Pp. 471–484 in Proceedings of the 3rd Global Demiliterization Symposium. Arlington, Va.: Joint Ordnance Commanders Group Demilitarization and Disposal Subgroup.

Kang, S-W., T. Reitter, G. Carlson, J. Crutchmer, D. Garrett, P. Kramer, and B. Do. 1993. Target Responses to the Impact of High-Velocity, Nonabrasive Water Jets. Pp. 71–86 in Proceedings of the 7th American Water Jet Conference, Vol.1. Seattle, Wash.: Water Jet Technology Association.

Liddard, T.P., and L.S. Roslund. 1993. Projectile/Fragment Impact Sensitivity of Explosives. NSWC T89-184. Crane, Ind.: Naval Surface Warfare Center Crane Division.

Lowell, M.D. 1986. PBX Downloading by High Pressure Water Jet. Presented at the Load, Assemble, and Packaging Section Meeting of the American Defense Preparedness Association (ADPA), Long Beach, California, March 12–13, 1986.

Melvin, W.S. 1992. Critical Fluid Demilitarization and Ingredient Reclamation Technology: Critical Technology Report. Redstone Arsenal, Ala.: U.S. Army Missile Command.

Miller, D. 1995. Abrasive Water Jet Technology for Demilitarization in Europe. Pp. 445–455 in Proceedings of the 3rd Global Demilitarization Symposium. Arlington, Va.: Joint Ordnance Commanders Group Demilitarization and Disposal Subgroup.

Miller, P.L. 1992a. The Mechanisms and Parameters of Abrasive Waterjet (AWJ) Cutting of High-Explosive Projectiles. Presented at the 25th Department of Defense Explosive Safety Board Seminar, Anaheim, California, August 18–21, 1992.

Miller, P.L. 1992b. Abrasive Waterjets (AWJ) Explosive Safety Tests. Presented at the 25th Department of Defense Explosive Safety Board Seminar, Anaheim, California, August 18–21, 1992.

Miller, P.L. 1992c. The Effects of Ultrahigh-Pressure Waterjet Impact on High-Explosives. Presented at the 25th Department of Defense Explosive Safety Board Seminar, Anaheim, California, August 18–21, 1992.

Moore, I. 1999. Telephone conversation between Ian Moore, DERA Shoeburyness, United Kingdom, and Kirk Newman, Naval Surface Warfare Center, Indian Head Division, Yorktown, Virginia, March 3, 1999.

Morgan, M.E. 1994. Critical Fluid Demilitarization and Ingredient Reclamation Technology. Pp. 365–376 in Proceedings of the 3rd Global Demilitarization Symposium. Arlington, Va.: Joint Ordnance Commanders Group Demilitarization and Disposal Subgroup.

Summers, D.A. 1997. Literature Review on the State-of-the-Art in Abrasive Waterjet Cutting. Crane, Ind.: Naval Surface Warfare Center Crane Division.

Summers, D.A., W.Z. Wu, P.N. Worsey, and L.E. Craig. 1988. PBX Demilitarization. NWSC/CR/RDTR-427. Crane, Ind.: Naval Surface Warfare Center Crane Division.

Teledyne-Commodore. 1998. Demonstration of Simulated Chemical Agent Rocket Demilitarization Using Solvated Electron Technology and Ammonia Fluid Jet Technology. Huntsville, Ala.: Teledyne-Commodore.

Worsey, P.N., D.A. Summers, and L.E. Craig. 1990. Waterjet Impact Sensitivity Test Manual. NWSC/CR/RDTR-541. Crane, Ind.: Naval Surface Warfare Center Crane Division.

APPENDIX H

Insights from State Regulators

This appendix provides comments and information provided by state regulators on issues crucial to obtaining environmental permits for any of the technology packages. The committee conducted informal telephone discussions with state regulators in Colorado, Kentucky, and Utah with the understanding that regulators were stating their personal opinions and could not render formal judgments. The following observations emerged from these discussions.

First, the state of Kentucky has a law on the books prohibiting the Kentucky Department of Environmental Protection from *examining* a permit for a chemical-agent treatment process until the process has been satisfactorily demonstrated (Kentucky 1988 Ky. Acts ch. 86, sec.1).

Second, regulatory bodies are concerned about the volume of activated carbon generated by the air-pollution control equipment and how to dispose of it. According to a regulator in Utah, the Tooele facility has replaced large quantities of activated carbon and is now storing the spent material. Facilities that do not have incinerators to treat at least some of the building ventilation air will produce even more carbon. Spent, activated carbon is very dusty and hard to burn, tends to become hot in storage (about 150°F), and may, with time, release adsorbed agent. If the carbon is ultimately disposed of off site, the person opening the drum for inspection (disposal facilities are required to inspect waste prior to accepting it) may be exposed to agent.

Third, there are no hazardous waste landfills in Kentucky, and the disposal of hazardous wastes from any process must take this into account. The "derived-from" rule may also pose a problem. Because Kentucky classifies agent as a hazardous waste, all of the wastes from any of these processes will be hazardous wastes, by definition.

Fourth, regulators from Kentucky and Utah noted that processes proposing to burn a gas produced by the treatment process will have to address several issues. A utility that agree to accept it as a fuel, for example, must provide firm contracts and commitments to do so. If the gas is burned on site, it will be examined by the regulatory authorities for contaminants. If the gas meets fuel specification comparable to EPA's Maximum Available Control Technology Rule (EPA, 1998a), it will pose few problems. If the gas does not meet these specifications, the boiler will have to be permitted like a boiler/industrial furnace or an incinerator for burning hazardous waste.

Fifth, the following comment was included in Utah's evaluation of the ACWA demonstration permit applications and appears to be applicable to all of the full-scale processes. "No process that treats agent should be vented without control into the work space if people are present to operate equipment or to sample. No process that treats agent should be vented without carbon control into the environment." Coupled with the earlier concern about activated carbon, this statement implies that processes that produce large quantities of activated carbon waste may encounter permitting delays.

APPENDIX I

Biographical Sketches of the Committee Members

Robert A. Beaudet, *chair,* received his Ph.D. in physical chemistry from Harvard University. He joined the faculty of the University of Southern California in 1962 as an assistant professor and is now a full professor in the Chemistry Department. He has extensive knowledge of chemical-agent monitoring and detection technologies and has served on several Department of Defense committees on chemical-warfare agents.

Richard J. Ayen received his Ph.D. in chemical engineering from the University of Illinois. Dr. Ayen is currently an independent consultant and was formerly the director of technology for Waste Management, Inc. He has extensive experience in the evaluation and development of new technologies for the treatment of hazardous, radioactive, industrial, and municipal waste.

Joan B. Berkowitz graduated from the University of Illinois with a Ph.D. in physical chemistry. Dr. Berkowitz is currently managing director of Farkas Berkowitz and Company. She has extensive experience in the area of environmental and hazardous-waste management, a comprehensive knowledge of available technologies for the cleanup of contaminated soils and groundwater, and strong background in physical and electrochemistry.

Nosa O. Egiebor graduated from Queens University in Kingston, Ontario, with a Ph.D. in mineral process and reaction engineering. Dr. Egiebor currently holds the Department of Energy Samuel P. Massie Chair of Excellence in Environmental Engineering at Tuskegee University. His areas of expertise span a broad range of topics in environmental engineering; his specific expertise is in the biotreatment of hazardous wastes and supercritical-fluid technology.

Willard C. Gekler graduated from the Colorado School of Mines with a degree in petroleum-refining engineering and pursued additional graduate studies in nuclear engineering at the University of California in Los Angeles. Mr. Gekler is currently vice president, chief engineer at PLG, Inc. His extensive experience includes design and safety analysis of hazardous-materials handling, storage, and waste-treatment systems. He specializes in hazard evaluation, quantitative risk analysis, reliability assessment, and database development for risk and reliability.

Hank C. Jenkins-Smith received his Ph.D. in political science from the University of Rochester. Dr. Jenkins-Smith is currently a professor in the Department of Political Science at the University of New Mexico, where he is also the director of the Institute for Public Policy. His areas of expertise include statistical analysis, measurement of public opinion, politics of risk perception, environmental policy, and public policy.

John L. Margrave, a member of the National Academy of Sciences (NAS), graduated from the University of Kansas with a B.S. in engineering physics and a Ph.D. in physical chemistry. Dr. Margrave is currently the chief scientific officer at the Houston Advanced Research Center and the E. D. Butcher Professor of Chemistry at Rice University. His expertise is in high-temperature chemistry, materials science, and environmental chemistry.

Walter G. May, a member of the National Academy of Engineering (NAE), graduated with a D.Sc. degree in chemical engineering from the Massachusetts Institute of Technology. He was the senior science advisor for Exxon Research and Engineering Company from 1976 until 1983 and professor of chemical engineering at the University of Illinois from 1983 until his retirement in 1991. His expertise is in process design, thermodynamics, chemical-reactor design, separation processes, industrial chemistry and stoichiometry, and chemical-weapons disposal.

Kirk E. Newman received a B.S. in chemistry from the College of William and Mary and an M.S. in chemical engineering from the University of Virginia. Mr. Newman is currently technology group leader for the Yorktown Detachment of the Naval Surface Warfare Center, Indian Head Division. He has extensive experience in the development, processing, and characterization of energetic materials used in military applications.

Jimmie C. Oxley received her Ph.D. in chemistry from the University of British Columbia in Vancouver and is currently an associate professor of chemistry at the University of Rhode Island. Her expertise is in thermal decomposition of energetic materials, explosives chemistry, and explosives safety

William R. Rhyne received a B.S. in nuclear engineering from the University of Tennessee and an M.S. and D.Sc. in nuclear engineering from the University of Virginia. Dr. Rhyne is cofounder and director of H&R Technical Associates, Inc. He has extensive experience in risk and safety analysis associated with the processing and transport of hazardous nuclear materials and chemicals.

Stanley I. Sandler, a member of the NAE, graduated from the University of Minnesota with a Ph.D. in chemical engineering. Currently, he is the Henry Belin du Pont Professor and director of the Center for Molecular and Engineering Thermodynamics at the University of Delaware. His extensive research interests include applied thermodynamics and phase equilibrium, environmental engineering, and separations and purification.

William Randall Seeker received his Ph.D. in engineering (nuclear and chemical) from Kansas State University. He is the senior vice president and member of the Board of Directors of Energy and Environmental Research Corporation. Dr. Seeker has extensive experience in the use of thermal-treatment technologies and environmental-control systems for managing hazardous waste.

Leo Weitzman received his Ph.D. in chemical engineering from Purdue University. He is a consultant with 26 years of experience in the development, design, permitting, and operation of equipment and facilities for treating hazardous wastes and remediation debris. Dr. Weitzman has extensive experience in the disposal of hazardous waste and contaminated materials by thermal treatment, chemical reaction, solvent extraction, biological treatment, and stabilization.